Urban Air Mobility (UAM)

Urban Air Mobility (UAM)

Editors

Ivana Semanjski
Antonio Pratelli
Massimiliano Pieraccini
Silvio Semanjski
Massimiliano Petri
Sidharta Gautama

Basel • Beijing • Wuhan • Barcelona • Belgrade • Novi Sad • Cluj • Manchester

Editors

Ivana Semanjski
Department of Industrial
Systems Engineering and
Product Design
FlandersMake@UGent –
corelab ISyE
Ghent University
Ghent
Belgium

Antonio Pratelli
Department of Civil and
Industrial Engineering
University of Pisa
Pisa
Italy

Massimiliano Pieraccini
Department of Information
Engineering
University of Florence
Florence
Italy

Silvio Semanjski
SEAL Aeronautica
Barcelona
Spain

Massimiliano Petri
Department of Civil and
Industrial Engineering
University of Pisa
Pisa
Italy

Sidharta Gautama
Department of Industrial
Systems Engineering and
Product Design
FlandersMake@UGent –
corelab ISyE
Ghent University
Ghent
Belgium

Editorial Office
MDPI
St. Alban-Anlage 66
4052 Basel, Switzerland

This is a reprint of articles from the Special Issue published online in the open access journal *Drones* (ISSN 2504-446X) (available at: www.mdpi.com/journal/drones/special_issues/uam).

For citation purposes, cite each article independently as indicated on the article page online and as indicated below:

Lastname, A.A.; Lastname, B.B. Article Title. *Journal Name* **Year**, *Volume Number*, Page Range.

ISBN 978-3-0365-9163-6 (Hbk)
ISBN 978-3-0365-9162-9 (PDF)
doi.org/10.3390/books978-3-0365-9162-9

Contents

Elham Fakhraian, Ivana Semanjski, Silvio Semanjski and El-Houssaine Aghezzaf
Towards Safe and Efficient Unmanned Aircraft System Operations: Literature Review of Digital
Twins' Applications and European Union Regulatory Compliance
Reprinted from: *Drones* **2023**, *7*, 478, doi:10.3390/drones7070478 . . . . **1**

**Emanuele Luigi de Angelis, Fabrizio Giulietti, Gianluca Rossetti, Matteo Turci and Chiara
Albertazzi**
Toward Smart Air Mobility: Control System Design and Experimental Validation for an
Unmanned Light Helicopter
Reprinted from: *Drones* **2023**, *7*, 288, doi:10.3390/drones7050288 . . . . **32**

**Jackson Costa, Rubens Matos, Jean Araujo, Jueying Li, Eunmi Choi and Tuan Anh Nguyen
et al.**
Software Aging in Container Orchestration Systems: An Assessment of Effects on Kubernetes
Reprinted from: *Drones* **2023**, *7*, 35, doi:10.3390/drones7010035 . . . . **57**

Abdulghani Mohamed, Matthew Marino, Simon Watkins, Justin Jaworski and Anya Jones
Gusts Encountered by Flying Vehicles in Proximity to Buildings
Reprinted from: *Drones* **2022**, *7*, 22, doi:10.3390/drones7010022 . . . . **79**

Flavia Causa, Armando Franzone and Giancarmine Fasano
Strategic and Tactical Path Planning for Urban Air Mobility: Overview and Application to
Real-World Use Cases
Reprinted from: *Drones* **2022**, *7*, 11, doi:10.3390/drones7010011 . . . . **105**

Ronghua Xu, Sixiao Wei, Yu Chen, Genshe Chen and Khanh Pham
LightMAN: A Lightweight Microchained Fabric for Assurance- and Resilience-Oriented Urban
Air Mobility Networks
Reprinted from: *Drones* **2022**, *6*, 421, doi:10.3390/drones6120421 . . . . **123**

Quan Shao, Jiaming Li, Ruoheng Li, Jiangao Zhang and Xiaobo Gao
Study of Urban Logistics Drone Path Planning Model Incorporating Service Benefit and Risk
Cost
Reprinted from: *Drones* **2022**, *6*, 418, doi:10.3390/drones6120418 . . . . **140**

Matteo Brunelli, Chiara Caterina Ditta and Maria Nadia Postorino
A Framework to Develop Urban Aerial Networks by Using a Digital Twin Approach
Reprinted from: *Drones* **2022**, *6*, 387, doi:10.3390/drones6120387 . . . . **167**

**Luca Bigazzi, Lapo Miccinesi, Enrico Boni, Michele Basso, Tommaso Consumi and
Massimiliano Pieraccini**
Fast Obstacle Detection System for UAS Based on Complementary Use of Radar and
Stereoscopic Camera
Reprinted from: *Drones* **2022**, *6*, 361, doi:10.3390/drones6110361 . . . . **184**

Karolin Schweiger and Lukas Preis
Urban Air Mobility: Systematic Review of Scientific Publications and Regulations for Vertiport
Design and Operations
Reprinted from: *Drones* **2022**, *6*, 179, doi:10.3390/drones6070179 . . . . **200**

Casper Bak Pedersen, Kasper Rosenkrands, Inkyung Sung and Peter Nielsen
Systemic Performance Analysis on Zoning for Unmanned Aerial Vehicle-Based Service Delivery
Reprinted from: *Drones* **2022**, *6*, 157, doi:10.3390/drones6070157 . . . . **255**

Review

Towards Safe and Efficient Unmanned Aircraft System Operations: Literature Review of Digital Twins' Applications and European Union Regulatory Compliance

Elham Fakhraian [1,2,*], **Ivana Semanjski** [1,2], **Silvio Semanjski** [3] and **El-Houssaine Aghezzaf** [1,2]

1   Industrial Systems Engineering and Product Design, Ghent University, 9052 Ghent, Belgium; ivana.semanjski@ugent.be (I.S.); elhoussaine.aghezzaf@ugent.be (E.-H.A.)
2   Industrial Systems Engineering (ISyE), Flanders Make, 9052 Ghent, Belgium
3   SEAL Aeronautica S.L., 08860 Barcelona, Spain; silvio.semanjski@sealaero.com
*   Correspondence: elham.fakhraian@ugent.be

Abstract: Unmanned aerial system/unmanned aircraft system (UAS) operations have increased exponentially in recent years. With the creation of new air mobility concepts, industries use cutting-edge technology to create unmanned aerial vehicles (UAVs) for various applications. Due to the popularity and use of advanced technology in this relatively new and rapidly evolving context, a regulatory framework to ensure safe operations is essential. To reflect the several ongoing initiatives and new developments in the domain of European Union (EU) regulatory frameworks at various levels, the increasing needs, developments in, and potential uses of UAVs, particularly in the context of research and innovation, a systematic overview is carried out in this paper. We review the development of UAV regulation in the European Union. The issue of how to implement this new and evolving regulation in UAS operations is also tackled. The digital twin (DT)'s ability to design, build, and analyze procedures makes it one potential way to assist the certification process. DTs are time- and cost-efficient tools to assist the certification process, since they enable engineers to inspect, analyze, and integrate designs as well as express concerns immediately; however, it is fair to state that DT implementation in UASs for certification and regulation is not discussed in-depth in the literature. This paper underlines the significance of UAS DTs in the certification process to provide a solid foundation for future studies.

Keywords: digital twin (DT); unmanned aerial system/unmanned aircraft system (UAS); unmanned aerial vehicle (UAV); drone; urban air mobility (UAM); advanced air mobility (AAM); European Union (EU) regulation; regulatory framework

Citation: Fakhraian, E.; Semanjski, I.; Semanjski, S.; Aghezzaf, E.-H. Towards Safe and Efficient Unmanned Aircraft System Operations: Literature Review of Digital Twins' Applications and European Union Regulatory Compliance. *Drones* **2023**, *7*, 478. https://doi.org/10.3390/drones7070478

Academic Editor: Pablo Rodríguez-Gonzálvez

Received: 7 June 2023
Revised: 10 July 2023
Accepted: 18 July 2023
Published: 20 July 2023

1. Introduction

In recent years, new innovative technologies, such as unmanned aerial vehicle (UAVs) and vertical take-off and landing (VTOL) aircraft, have led to the creation of new air mobility concepts [1]. UAVs operate in various sectors: agriculture, inspection, media, and entertainment. UAVs' operational and technological capabilities have evolved. They are expected to gain greater freedom of use and enter the area of commercial flights in the near future. Currently, most UAV civil operations are conducted in low-level uncontrolled or segregated controlled airspace due to safety concerns [2]. Operations in high-risk environments set higher requirements to overcome related risks: collisions with civil aircraft, injuries, and accidents due to UAV operation errors. The prevailing measures in UAS management necessitate the thorough consideration and addressing of concerns pertaining to scalability, compliance, cybersecurity, privacy, limitations in real-time monitoring, and the intricate regulatory landscape, which often entail significant investments of time and resources. One of the possible solutions is to leverage digital twin (DT) technology to map the physical space during UAV operation into the virtual space to assess the risk related to the operation beforehand. The utilization of DT has become an engaging subject today [3]. A DT is a virtual replica of real-world entities or processes. DTs develop models to simulate future scenarios and employ historical as well as real-time data to illustrate the past and present [4]. DTs can gain new and unexpectedly detailed insights into how machines and operations work in addition to how to improve them using sensors, cost-efficient and more secure data storage, powerful computers to analyze data, and artificial intelligence [5]. DT allows engineers to check, analyze, and integrate designs as well as express concerns immediately [6]. For example, DT helps anticipate when a machine may fail based on data analysis, which allows the boosting of productivity through preventive maintenance [7]. DTs' application is mainly grouped into the manufacturing [8], aviation, automotive, education and research [9], and healthcare and medicine fields [10]. DT technology is expected to change the "rules of the game" in aviation manufacturing in the future [11]. The aviation community is fostering an aspiration to offer air mobility as an alternative for everyday transportation needs, commonly known as urban air mobility (UAM) and advanced air mobility (AAM) [12]. AAM encompasses a broad concept that enables individuals to access on-demand air mobility, cargo and package delivery, healthcare applications, and emergency services through an interconnected multimodal transportation network [13]. Achieving this system necessitates the seamless integration of air traffic management systems, ground control systems, and communication networks to facilitate effective communication between AAM vehicles and ground systems to ensure safe and efficient operations. As a result, the aviation industry is actively working towards developing an innovative aerospace framework that promotes shared aerospace practices, ensuring the safety, sustainability, and efficiency of air traffic operations [14]. A wide range of literature has been published to explore operational strategies and expectations in the context of AAM [15–28]. Currently, NASA, in collaboration with the FAA, other federal partner agencies, industry, and academia, is actively engaged in research and development efforts to establish the infrastructure, information architecture, concepts of operation, operations management tools, software functions, and other functional components of AAM [29]. Nevertheless, several challenges have the potential to affect the growth of AAM. These challenges include autonomous flight capabilities, the availability of necessary infrastructure for take-off and landing, integration into existing airspace as well as other transportation modes, and competition with shared automated vehicles [30].

UAM, a subset of AAM, is anticipated to yield substantial economic benefits while posing notable developmental challenges. UAM necessitates the development of sophisticated urban-capable vehicles and the establishment of an airspace system capable of efficiently managing high-density operations [12]. According to the European Union Aviation Safety Agency (EASA), UAM is defined as "a new safe, secure and more sustainable air transportation system for passengers and cargo in urban environments, enabled by new technologies and integrated into multimodal transportation systems. The transportation is performed by electric aircraft taking off and landing vertically, remotely piloted or with a pilot on board" [31]. The EASA further predicts that, by 2030, approximately 340 million people residing in EU cities will experience UAM [31]. The concept of urban aerial transportation is not novel, as historical examples of UAM services date back to the 1940s [32]. A notable instance of these historical examples is New York Airways, which operated commercial helicopter-based passenger transport services from 1953 to 1979. However, due to a series of fatal accidents and crashes, New York Airways ultimately ceased operations and filed for bankruptcy. Although this particular chapter of urban aerial mobility concluded abruptly, modern-day congested metropolises have witnessed the resurgence of diverse helicopter transport services [33]. Similar to other transportation systems, UAM necessitates the establishment of infrastructure encompassing the physical ground infrastructure for vehicles as well as the implementation of digital technology and telecommunications for effective traffic management. An essential element for the successful introduction of UAM is the development of appropriate regulations, including the definition of certification standards and policies that govern UAM operations. Addressing these regulatory aspects is crucial to ensure the safe and efficient integration of UAM into existing transportation frameworks [34].

A wide range of literature has been published to answer the research question of how to safely integrate unmanned aircraft systems (UASs) into UAM and AAM within the context of regulation. Studies have addressed key concerns about privacy, the operation of civilian drone regulations, and the social as well as ethical implications of this integration. Winkler et al. [35] highlighted the concerns and needs for privacy and the operation of civilian drone regulations. Clarke investigated the impacts of civilian drone regulation on behavioral privacy [36] and public safety [37]. Thomasen [38] evaluated the impact of robots (including drones) and their regulation on public spaces. In this paper, the authors also examined the technology's impacts on women's privacy and related regulations [39]. Merkert et al. [40] used a theoretical road pricing framework to analyze drone operators' willingness to pay for low-altitude airspace management (LAAM). West et al. [41] reviewed the public's opinions on drone policy. Li and Kim [42] studied the dynamics of local drone policy adoption in California. Nelson and Gorichanaz [43] investigated the emergence of drones and evolving regulation in 20 cities in Southern California. However, in the available literature and official documentation, there was no agreed and consolidated definition of UAM in Europe until recent years, when the EASA introduced the UAM concept as "The safe, secure and sustainable air mobility of passengers and cargo enabled by new generation technologies integrated into a multimodal transportation system conducted in to, within or out of urban environments" [1]. The EASA is also establishing a regulatory framework addressing the safety, security, and environmental aspects of UASs to ensure their acceptance and adoption by European citizens. Some elements of this regulatory framework have already been established; for example, Regulation (EU) 2019/947, Regulation (EU) 2019/945, Regulation (EU) 2021/664, Regulation (EU) 2021/665, and Regulation (EU) 2021/666 [1].

In parallel with the establishment of regulatory frameworks, the potential of [5,8–11] DT utilization in the aviation industry has been explored and documented in numerous pieces of the scientific literature [44–49]. DTs can be used in any stage of the aircraft life cycle [50–60], such as design, manufacturing, operations, and maintenance. DTs can also be implemented on components as well as systems [61–70] that provide a comprehensive view of an aircraft and its individual parts. It allows for monitoring and analysis at different levels, enabling engineers to assess the performance and health of specific components as well as understand the overall behavior and interactions within the system. Various research efforts have been conducted to use DT in UASs [3,71–90], addressing challenges and opportunities of UASs within this dynamic and evolving field. However, despite the significant discussion of DTs in the general aviation literature, especially in relation to manufacturing and maintenance, more effort and attention need to be devoted to the application of DTs in UASs [71].

Overall, the aviation industry is subjected to an international framework, yet it requires additional efforts to establish a similar framework for UAS operations [91]. Considering the strong ongoing developments in this domain, the approach to UAS certification does not evolve with the same dynamic [6], and the UAS European Union (EU) regulatory framework was fragmented before 2020, mainly considered in quite local and regional contexts. However, some significant steps have recently been made in this aspect, particularly since 2020, and the EU legal framework of UASs is undergoing changes to provide uniform regulation. One of the aims of this paper is to also bring these developments to the closer attention of the research community in order to support strongly evolving research efforts, as this aspect has so far been generally understated across the scientific literature. Understanding the appropriate operational category presented by the EASA for UASs helps to gain more insights into the requirements of authorizations and certification. However, when developing a product that requires regulatory certification, this is only one half of the matter. The separation between design and analysis activity is one of the critical gaps in the certification process. DTs facilitate engineering and manufacturing teams to design and build products better and faster. It also helps them to check, analyze, and integrate designs as well as express concerns instantly [6]. This paper provides a comprehensive overview of the developed UAS regulation in the European Union provided by the EASA and examines the potential of DTs to assist the certification process. This paper aims to make a bridge between DTs, UASs, and the EU regulatory framework to present a reliable basis for future studies. The structure of the paper is as follows: Section 2 is structured into three subsections. The first subsection provides an overview of the research methodology. The second subsection introduces the current and upcoming European regulatory framework for UASs. The third subsection illustrates the concept and applications of DTs. Section 3 provides a valuable resource by analyzing the existing relevant literature and highlighting important trends as well as developments. Section 4 presents the links and potential to use DTs to assist the certification by drones' EU regulatory framework. forming an outlook for future studies and applications. Section 5 presents the conclusion, where some key lessons learned based on the existing body of literature are presented.

2. Materials and Methods

One of the essential steps toward determining the potential of DTs in the certification process is specifying the related regulation in the context of operational robustness and airworthiness. Airworthiness concerns the safety standards in all construction aspects: structural strength, safeguard provisions, design requirements relating to aerodynamics, performance, and electrical as well as hydraulic systems [92]. Robustness refers to the characteristic of mitigation measures resulting from combining the improvements in safety provided by mitigation measures and the levels of assurance as well as integrity in attaining the desired safety enhancement [93]. In general, international and national regulations are focused on safety. However, small drones avoid many of these requirements, as they pose fewer risks [91]. UAV operations are a relatively new concept and have significant potential in combination with new technologies, resulting in new applications (with their required regulations). DTs are also a relatively new concept accepted in various industries and have great potential for UAV operations. A DT is a description of a component, product, or system providing a series of interconnected relevant digital models containing engineering data, operation data, and behavior descriptions obtained from simulations. It can be modified as a real-world system can be developed through its life cycle. A DT is used to develop solutions that are applicable to actual systems in addition to describing the behavior. It can be applied to testing and simulation, enabling users to observe how new behaviors are exhibited and find answers to their problems [94].

In the legal context, it is essential to acknowledge and understand the distinct terminology used when referring to drones, as they may carry different legal implications. The term "drone" was first used in 1935 and is nowadays quite accepted by both the media and the general public [95]. Alongside "drone", the most frequently used terms are "unmanned aerial vehicle" (UAV) and "unmanned aircraft system/unmanned aerial system" (UAS). The terms "drone" and "unmanned aerial vehicle" (UAV) stand out as referring only to a flying platform (the airplane and its payload). The phrase "unmanned aerial system" (UAS) is the most well known term for an entire system (a flying platform and ground station). "Unmanned aircraft system" (UAS) is widely used by the Federal Aviation Administration (FAA), European Aviation Safety Agency (EASA), and International Civil Aviation Organization (ICAO). Hence, it is better to utilize the term "unmanned aircraft systems" when referring to UASs in this study. It is essential to utilize the correct terminology in order to deliver the concepts in the debate properly [95].

Official documents and legislations mainly use the terms "UAV" and "UAS". While professional drone users are familiar with these terms and use them, the terms "UAV" and "UAS" are less familiar to the public, especially when abbreviated [95]. People might therefore have few or no associations with these terms, so the term "drone" is occasionally used in conjunction with these terms for simpler demonstration in documents. In this work, we make an effort to use the terminology accurately, considering the references to prevent misconception.

This section is divided into two subsections: The first subsection introduces the existing and upcoming European regulatory framework for UASs. The second subsection illustrates different DTs' methodologies.

2.1. Research Methodology

To answer the research question of how DT can assist the certification process, we provide a detailed and comprehensive analysis of the state of the art through the following source databases: Google Scholar, Scopus, Springer, Science Direct, and the European Union Aviation Safety Agency. We instigated a data search by combining the keywords "Unmanned Aerial Vehicle", "Unmanned Aircraft System", "Unmanned Aerial System", "UAV", "UAS", and "drone" in combination with "digital twins", "DT", "certification", "regulation", "European Union (EU) regulation", "regulatory framework", "Urban Air Mobility", "UAM", "Advanced Air Mobility", and "AAM". In the literature search, we identified relevant articles according to the title and context of the study. A total of 121 references, which were best-aligned with the scope and objectives of our research, were selected, of which 20 articles were directly relevant to the scope of DT applications for UASs. The results sections of the selected references were analyzed to gain valuable insights in this domain.

The first step to answering the identified research questions is to investigate the existing and upcoming European regulatory framework for UASs and to understand the concept, methodologies, and applications of DTs.

2.2. European Union Regulatory Framework

Until 2020, the Member States regulated civil drones with an operating mass of less than 150 Kg, and the EASA handled civil drones with an operating mass of over 150 Kg. The fragmentation in the extent, content, and level of national detail led to unreached conditions for the joint recognition of operational authorization between the EU Member States [91]. Fortunately, the EASA is providing uniform regulation for the EU legal framework of UASs since 2020 [96]. Figure 1 presents an overview of European Union regulatory framework progress over time.

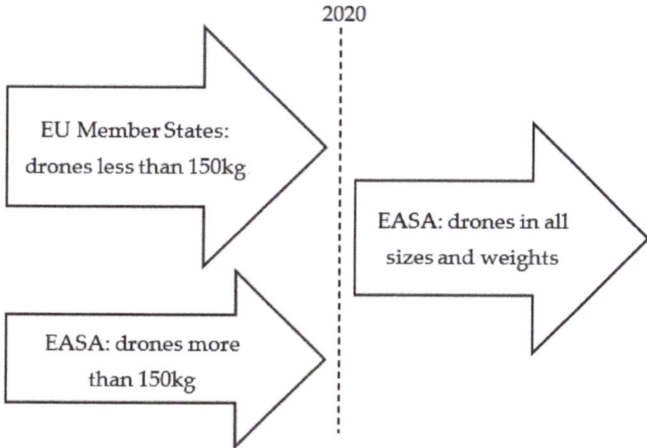

Figure 1. Chronological evolution of the European Union regulatory framework.

2.2.1. Access Rules for Unmanned Aircraft Systems (Regulations (EU) 2019/947 and 2019/945)

The operational framework for civil drones in the Europian Union (EU) is Regulations 2019/947 and 2019/945. These regulations conduct a risk-based approach, considering the weight, specifications, and intended operation of civil drones [97]. Regulation 2019/947 was expected to be implemented on 1 July 2020; however, due to the COVID-19 crisis, it was delayed to 31 December 2020 [96].

Civil Drone Operation Categories in the European Union Regulatory Framework

Regulation 2019/947 presents three risk-based categories for civil drone operations, shown in Figure 2: the open, specific, and certified categories [97]. The definition of each category is as follows:

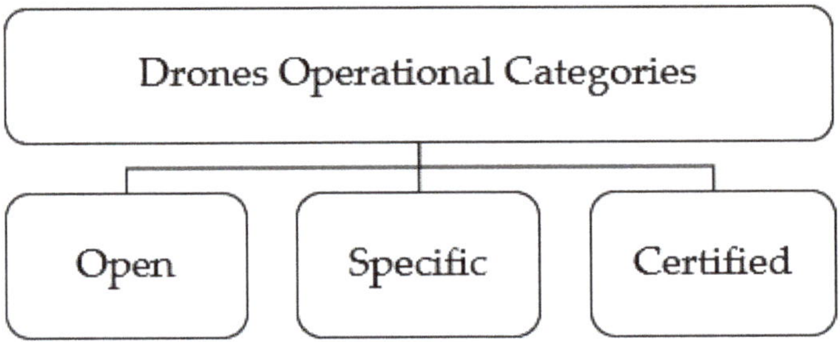

Figure 2. Categorization of UAS operations under EU regulation.

1. The open category (low-risk): Drones in low-risk operations (e.g., leisure drone activities and low-risk commercial activities) are in the open category. This category is specified by three subcategories: A1, flying over people but not over assemblies of people; A2, flying close to people; and A3, flying far from people. Each subcategory has requirements based on UAS's weight (the operational weight is less than 25 Kg) [98].

2. The specific category (medium-risk): Operations that carry more risks and are not in the scope of the open category's operations are in the specific category. In this category, operational authorization (issued by the competent authority of registration) is required based on the risk assessment outcome conducted under Article 11 of Regulation (EU) 2019/947, unless the operation is a standard scenario (STS): a predefined operation described in the appendix of EU Regulation 2019/947 [99].

3. The certified category (high-risk): UAS high-risk operations and future drones onboard passenger flights (e.g., air taxis) are in the certified category. These UASs must always be certified, the UAS operator will need air operator approval issued by the competent authority, and the remote pilot must hold a pilot license. In the future, drone automation will reach fully autonomous UAS operations. The safety approach of these flights will be very similar to manned aviation. Almost all aviation regulations will need to be amended, and the EASA decided to conduct this major task in multiple phases [100].

Overall, drone operations with any of the below conductions are certainly in the certified category:

- A UAS with a dimension of 3 m or more flying over assemblies of people (operation of a less than 3 m UAS flying over assemblies of people may be in the specific category unless the risk assessment outcome indicates that is in the certified category).
- Transport of people.
- Transport of dangerous goods (the payload is not in a crash-protected container) [93].

Operational Risk Assessment for Drones in Specific Category

UAS operational risk assessment is divided into three categories: standard scenarios (STSs), predefined risk assessment (PDRA), and specific operation risk assessment (SORA) [93]. The definition of each category is as follows:

1. Standard scenario (STS): Due to the lower risks in UAS operations in STSs listed in Table 1, a declaration may be submitted.

Table 1. List of standard scenarios (STSs) [93.].

STS#	Edition/ Date	UAS Characteristics	BVLOS/VLOS [2]	Overflown Area	Maximum Range from Remote Pilot	Maximum Height	Airspace
STS-01	June 2020	Bearing a C5 class marking (maximum characteristic dimensions of up to 3 m and MTOM [1] of up to 25 kg)	VLOS	Controlled ground area that might be located in a populated area	VLOS	120 m	Controlled or uncontrolled, with a low risk of encounter with manned aircraft
STS-02	June 2020	Bearing a C6 class marking (maximum characteristic dimensions of up to 3 m and MTOM of up to 25 kg)	BVLOS	Controlled ground area that is entirely located in a sparsely populated area	2 km with an AO [3] 1 km, if no AO	120 m	Controlled or uncontrolled, with a low risk of encounter with manned aircraft

[1] Maximum take-off mass. [2] Beyond visual line of sight/visual line of sight. [3] Airspace observer.

2. Predefined risk assessment (PDRA): PDRA is considered the most common operation in Europe, and instead of conducting a full risk assessment, an authorization request may be submitted based on the PDRAs listed in Table 2. PDRAs are described in a generic way to provide flexibility, while STSs are detailed. The two types of PDRAs are PDRAs derived from STSs (a UAS operator conducts similar operations without the UAS class label mandated in STSs) and generic PDRAs. A PDRA with the letter "G" is a generic PDRA, and those with an "S" are PDRAs derived from STSs [93].

Table 2. List of predefined risk assessments (PDRAs) [93].

PDRA#	Edition/ Date	UAS Characteristics	BVLOS/VLOS	Overflown Area	Maximum Range from Remote Pilot	Maximum Height	Airspace	AMC# [1] Article 11
PDRA-S01	1.0/July 2020	Maximum characteristic dimension of up to 3 m and MTOM of up to 25 kg	VLOS	Controlled ground area that might be located in a populated area	VLOS	120 m	Controlled or uncontrolled, with a low risk of an encounter with manned aircraft	AMC4
PDRA-S02	1.0/July 2020	Maximum characteristic dimension of up to 3 m and MTOM of up to 25 kg	BVLOS	Controlled ground area that is entirely located in a sparsely populated area	2 km with an AO, 1 km if no AO	120 m	Controlled or uncontrolled, with a low risk of an encounter with manned aircraft	AMC5
PDRA-G01	1.1/July 2020	Maximum characteristic dimension of up to 3 m and typical kinetic energy of up to 34 kJ	BVLOS	Sparsely populated area	If no AO, up to 1 km	150 m (operational volume)	Uncontrolled, with a low risk of an encounter with manned aircraft	AMC2

Table 2. *Cont.*

PDRA#	Edition/ Date	UAS Characteristics	BVLOS/VLOS	Overflown Area	Maximum Range from Remote Pilot	Maximum Height	Airspace	AMC#[1] Article 11
PDRA-G02	1.0/July 2020	Maximum characteristic dimension of up to 3 m and typical kinetic energy of up to 34 kJ	BVLOS	Sparsely populated area	N/a	As established for the reserved airspace	As reserved for the operation	AMC3

[1] Acceptable means of compliance.

3. Specific operation risk assessment (SORA): SORA evaluates the UAS operation risks, considering any class, size, and type of operation [93]. Figure 3 demonstrates the SORA methodology.

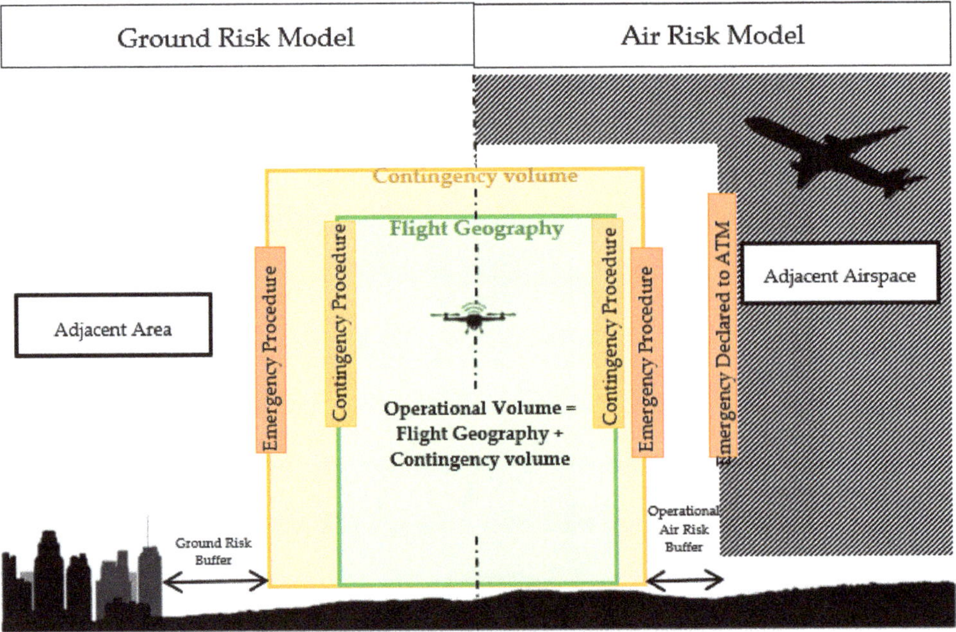

Figure 3. Graphical representation of SORA [93].

SORA defines risk as "the combination of the frequency (probability) of an occurrence and its associated level of severity". Risk mitigations and operational safety objectives (OSOs) can be demonstrated at different robustness levels presented by SORA: low, medium, and high. SORA focuses on the assessment of air and ground risks. Figure 4 presents the required workflow to conduct SORA. Ten steps are required to conduct SORA, and some of these steps may be repeated in different environments [22]. It is important to verify the operational feasibility before starting SORA. The operation must not be categorized as the open category or certified category, must not be covered by an STS or a PDRA, and not be subjected to a specific NO-GO from the competent authority [93].

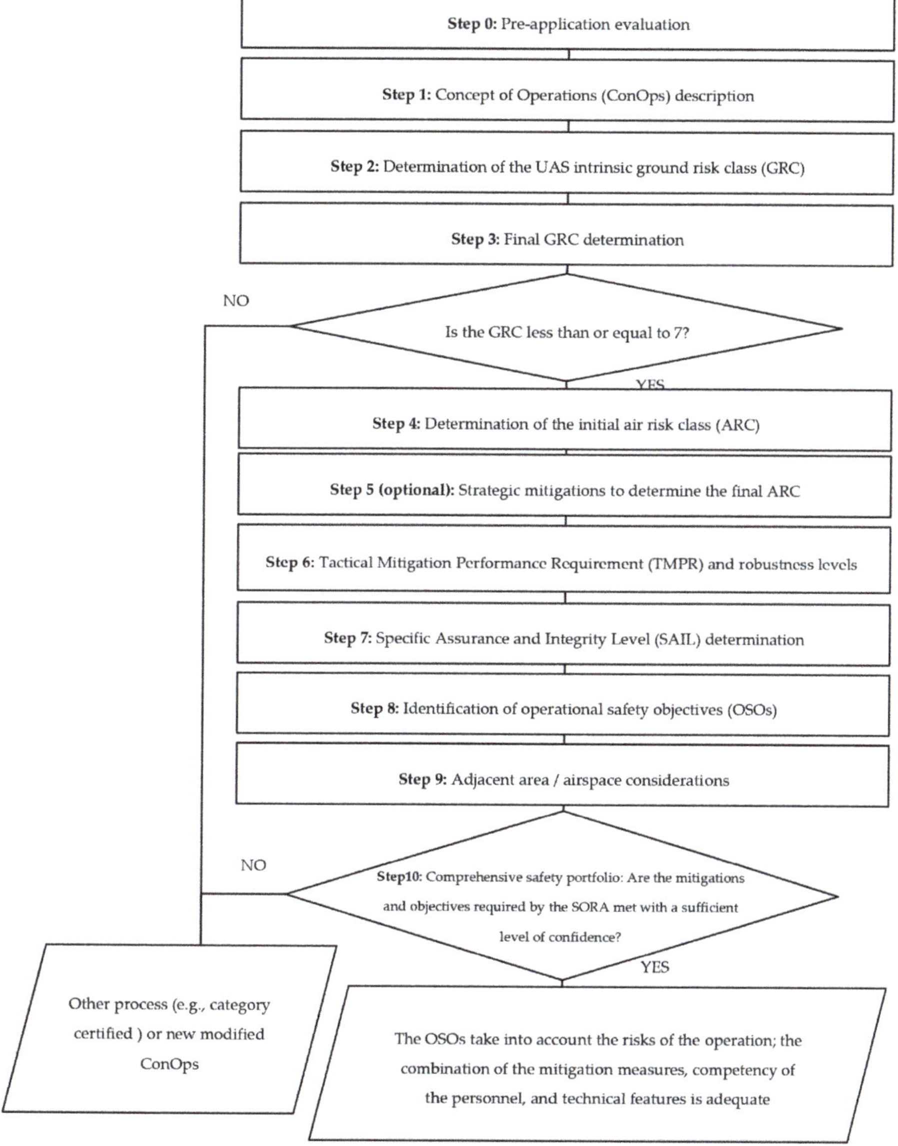

Figure 4. Workflow for conducting SORA: ten steps and iterations [93].

To ensure safety in UAS operations, especially in populated areas, the design verification of drones by the EASA is needed depending on the risk level of operations [101]:

- In high-risk operations (i.e., SAIL V and VI according to SORA), the EASA will issue a type certificate according to Part 21 (Regulation (EU) 748/2012). Easy Access Rules for Airworthiness and Environmental Certification (Regulation (EU) No. 748/2012) contains the applicable rules for the airworthiness and environmental certification of aircraft and related products, parts, and appliances, as well as for the certification of design and production organizations [102].

- In medium-risk operations (i.e., SAIL III and IV according to SORA), a design verification report will be applied [101].

2.2.2. Commission Implementing Regulation (EU) in U-Space (Regulations (EU) 2021/664, 2021/665, and 2021/666)

U-space is a set of services and procedures to ensure safe and efficient airspace accessibility for a large number of UAS operations, with the purpose of achieving automated UAS management and integration. The European Commission adopted and published a regulatory framework for U-space in April 2021. This regulatory package is going to implement three regulations as of January 2023 [103]:

1. Regulation (EU) 2021/664 regulates the technical and operational requirements for the U-space system [104].
2. Regulation (EU) 2021/665 amends Regulation (EU) 2017/373 to establish requirements for air traffic management and air navigation service providers in the U-space designated in controlled airspace [105].
3. Regulation (EU) 2021/666 modifies Regulation (EU) 923/2012 to establish the rules for the presence and requirements for manned aviation operating in U-space airspace [106].

2.2.3. EASA Artificial Intelligence Roadmap (Autonomous and Automatic UASs)

Autonomous and automatic UASs are reaching a level of complexity and development such that they are expected to conduct safe operations in urban air mobility (UAM). Automatic UAVs operate on predetermined routes, and remote pilots intervene in the case of unforeseen events. In autonomous UAVs, artificial intelligence (AI) must conduct a safe flight (without a pilot's intervention) and cope with unforeseen conditions as well as unpredictable emergencies. Automatic UAV operations are allowed in all categories. Autonomous UAVs only operate in the specific category and certified category (where the Regulation includes more flexible tools to verify requirements and the level of robustness); they are not allowed in the open category [107].

One of the key research questions is how these operations can safely be used in UAM [108]. In 2020, the EASA published a human-centric approach for the safe use of AI in aviation, entitled "EASA AI roadmap". Figure 5 presents the trustworthy AI building blocks: AI trustworthiness analysis, learning assurance, AI explainability, and AI safety risk mitigation [109]. The EASA AI roadmap's deliverables timeline foresees the first approvals of AI in 2025 [110].

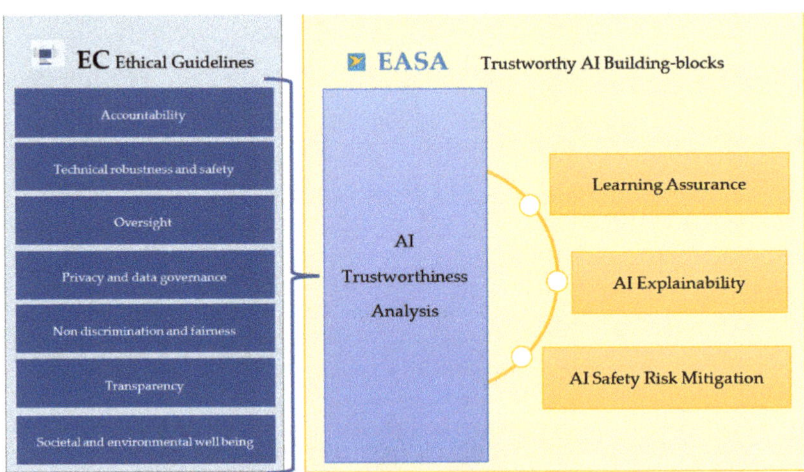

Figure 5. EASA trustworthy AI building blocks [109].

2.3. Digital Twins

The digital twin concept was first used in the manufacturing literature in 2010 as "a digital representation of an asset (e.g., physical objects, processes, devices) containing the model of its data, its functionalities and communication interfaces" [111], providing the elements and dynamics of asset operation throughout its life cycle [112]. Various DT definitions exist in the current literature depending on the domains and industries [113]. A list of DT definitions based on domains is as follows:

1. Aerospace industry: "A Digital Twin is an integrated multiphysics, multiscale, probabilistic simulation of an as-built vehicle or system that uses the best available physical models, sensor updates, fleet history, etc., to mirror the life of its corresponding flying twin. The Digital Twin is ultra-realistic and may consider one or more important and interdependent vehicle systems, including airframe, propulsion and energy storage, life support, avionics, thermal protection, etc." [114].
2. Manufacturing industry: "The Digital Twin is a set of virtual information constructs that fully describes a potential or actual physical manufactured product from the micro atomic level to the macro geometrical level. At its optimum, any information that could be obtained from inspecting a physical manufactured product can be obtained from its Digital Twin" [112].
3. Construction industry: "Digital twin construction (DTC) is a new mode for managing production in construction that leverages the data streaming from a variety of site monitoring technologies and artificially intelligent functions to provide accurate status information and to proactively analyze and optimize ongoing design, planning, and production" [115].
4. Service infrastructure: "a dynamic virtual representation of a physical object or system across its lifecycle, using real-time data to enable understanding, learning and reasoning" [116].
5. Healthcare: "A digital twin is a digital representation of a physical asset reproducing its data model, its behavior and its communication with other physical assets. Digital twins act as a digital replica for the physical object or process they represent, providing nearly real-time monitoring and evaluation without being in close proximity" [111].

DTs in various industries have approximately the same features and application purposes. The main components for generating DT models are physical elements/assets, linked data, and virtual models [113]. DTs can be categorized as follows:

1. Static DT: A static DT is developed (with the design information in a digital format) before the manufacturing process [117].
2. Dynamic DT: With the help of real-time sensors mounted on a product, a dynamic digital is obtained. These sensors allow us to access real-time information. The data obtained from the physical machine by the sensors are transferred to a virtual machine. The virtual machine uses trained simulation- and data-driven models on the received data to present the needed information about the physical machine [118]. With the help of artificial intelligence and data analytics, the DT gains the potential to reach autonomous decision making [113].

Static DT is the simplest way of implementing DT, and dynamic DT is the most complex one. As the level of details and information increases, the complexity and cost of DTs increase. Figure 6 presents the relationship between DTs and business value. Code green is simple design data, code yellow is the design and manufacturing data, and red is the dynamic DT that also includes operational field data [117].

Figure 7 illustrates the DT complexities (three main complexity levels) and time horizon approximations (three main life cycle stages of a physical system with the related DT applications) [119].

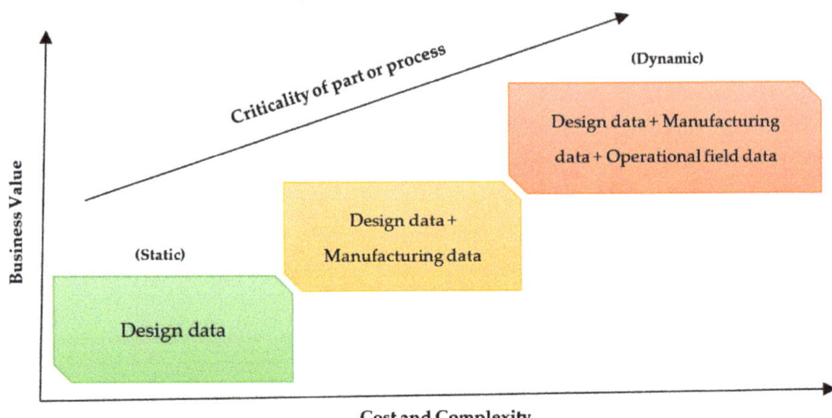

Figure 6. Types of digital twins and business value [117].

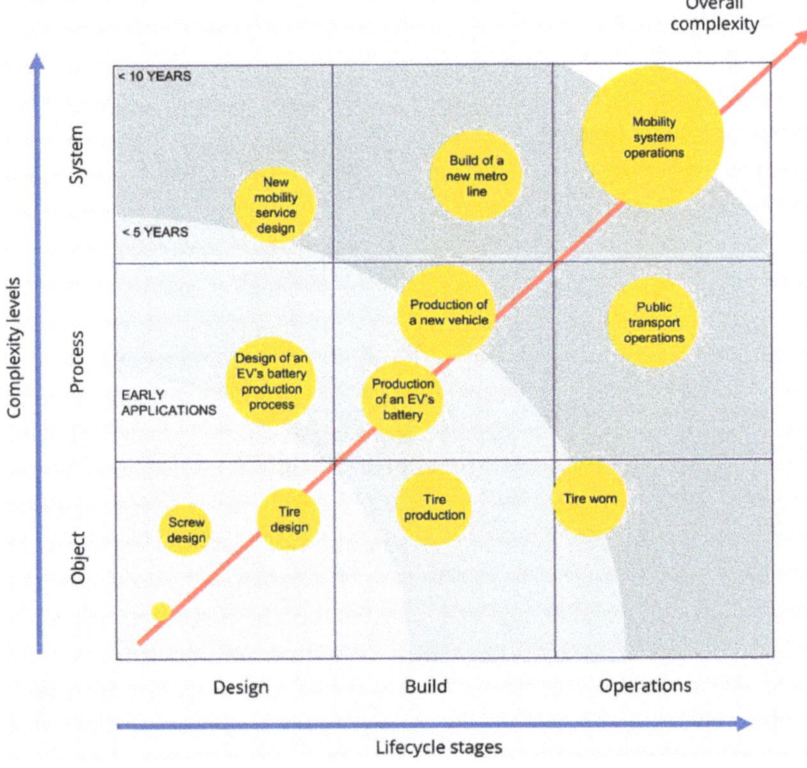

Figure 7. Digital twin complexities and time horizon approximations [119].

Before developing and implementing the DT, various research questions must be answered. Semeraro et al. [120] presented Table 3 to summarize the key research questions of DTs answered by the literature so far.

Table 3. List of digital twin research questions [120].

Research Question	Answers
"What is a Digital Twin?" Definition	"A set of adaptive models that emulate the behaviour of a physical system in a virtual system getting real time data to update itself along its life cycle. The digital twin replicates the physical system to predict failures and opportunities for changing, to prescribe real time actions for optimizing and/or mitigating unexpected events observing and evaluating the operating profile system"
"Where is appropriate to use a Digital Twin?" Contexts and use cases	1. Healthcare Improving operational efficiency of healthcare operations 2. Maritime and Shipping Design customization 3.Manufacturing Product development and predictive manufacturing 4. City Management Modeling and simulation of smart cities 5. Aerospace Predictive analytics to foresee future aircraft problems
"Who is doing Digital Twins?" Platforms	GE Predix; SIEMENS PLM; Microsoft Azure; IBM Watson; PTC Thing Worx; Aveva; Twin Thread; DNV-GL; Dassault 3D Experience; Sight Machine; and Oracle Cloud
"When and Why has a Digital Twin to be developed?" Life cycle and functions	1. In the design phase The digital twin is used to help designers to configure and validate product development quicker, accurately interpreting market demands and the customer preferences 2. In the production phase The digital twin shows great potential in real-time process control and optimization, as well as accurate prediction 3. In the service phase The digital twin can monitor the health of a product and perform diagnoses as well as prognoses
"How to design and implement a Digital Twin?" Architecture and components	The physical layer involves various subsystems and sensory devices that collect data and working parameters The network layer connects the physical to the virtual, sharing data and information The computing layer consists of virtual models emulating the corresponding physical entities

It is important to distinguish between the concepts "digital twin", "digital shadow", and "digital model". Figure 8 highlights the differences in these concepts by focusing on the data transfer among physical and virtual twins [119].

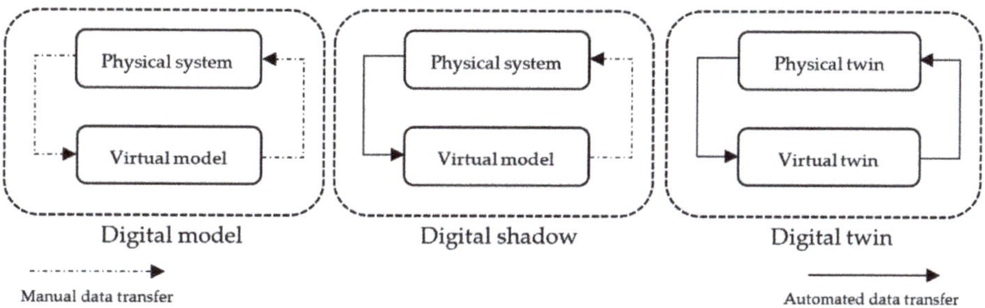

Figure 8. Data transfer comparison between physical systems and virtual models in a digital twin, digital shadow, and digital model [119].

Figure 9 demonstrates the important risks and challenges when developing DTs [117]. Modeling a digital copy of a physical system to perform real-time validation and optimization is a complex task as it involves sensors, multifunctional models, multisource data, services, etc. A DT requires an accurate model of reality and a large amount of data. It can potentially be used in life cycle assessments; however, the development of standards-based interoperability is important and challenging for evaluating DT applications along the entire life cycle. A few contributions also focused on DT applications for improving sustainability performance [120].

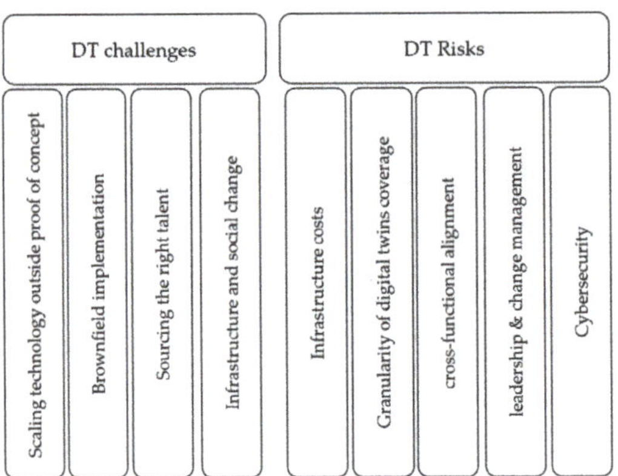

Figure 9. Digital twins' risks and challenges [117].

To comprehensively understand the state of knowledge on the application of DTs in UASs, as well as their benefits and challenges, a synthesis of the literature that integrates various subtopics is crucial. The implementation of DTs has been widely explored in aviation-related scientific literature. For example, the EU-funded project Secure Urban Air Mobility for European Citizens (AURORA) is planning to develop and integrate safety-critical technologies to support autonomous UAS flights in urban environments. Figure 10 presents examples of DT applications in this project [121].

(a)

(b)

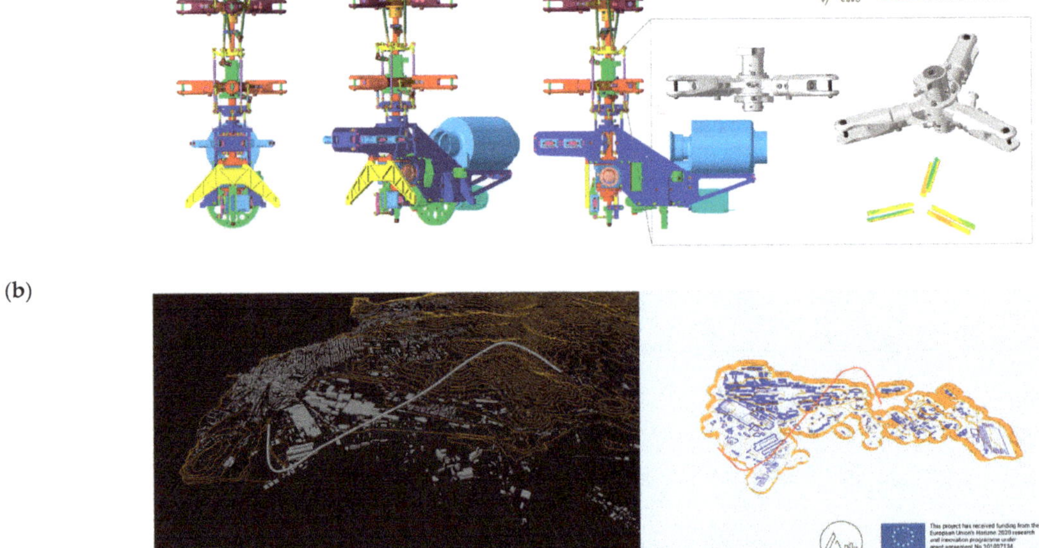

Figure 10. The EU-funded AURORA project: (**a**) digital representation of the rotor (digital twin for manufacturing and digital certification) and (**b**) digital representation of the flight path (digital twin for solutions testing) [121].

The German Aerospace Center (DLR) has also established an internal project to identify techniques, technologies, and processes for DTs [44]. Liu et al. [45] reviewed the overall framework for creating a DT in combination with the industrial Internet of things (IIoT) to enhance the autonomy of aerospace platforms. Liao et al. [46] presented the findings of research conducted at the National Research Council of Canada (NRC), which included a review and evaluation of DT concepts and digital threads, particularly the airframe digital twin (ADT) framework used by the United States Air Force (USAF), as well as a feasibility and adaptability study of the ADT for use with Royal Canadian Air Force (RCAF) aircraft. Aydemir et al. [47] reviewed the available approaches, technologies, and challenges of DTs for aircraft applications. Mendi et al. [48] evaluated DT applications and their advantages in military aviation. Ibrion et al. [49] presented DTs' risks and challenges in the marine industry by learning from the aviation industry. DTs can be effectively utilized in any stage of the aircraft life cycle, encompassing the design, manufacturing, operation, and maintenance phases. DTs enable engineers to create virtual prototypes and simulate various scenarios, allowing for the efficient optimization of aerodynamic performance, structural integrity, and overall aircraft functionality in the design stage. DTs can facilitate real-time monitoring and quality control, ensuring that components are produced to precise specifications and tolerances during manufacturing. DTs, based on their level of complexity, have the potential for real-time data collection and analysis, offering insights into the operation phase, including aircraft performance, fuel efficiency, and operational safety. DTs can also support predictive maintenance by continuously monitoring the health of aircraft systems and components, as well as detecting potential issues before they lead to failures or disruptions. Leveraging DTs throughout the aircraft life cycle can enhance decision making, improve safety, reduce costs, and ultimately maximize the overall performance and lifespan of the aircraft. For instance, Tuegel et al. [50] proposed the airframe DT structural modeling concept to design and maintain airframes (which has the potential to improve US Air Force aircraft management over the life cycle) by creating a tail number computational model and structural management plans for each aircraft. Seshadri et al. [51] suggested employing DTs to manage the structural health of damaged aircraft using guided wave responses. A genetic algorithm (GA) optimization evaluates the cumulative signal responses at preselected sensor locations to estimate the size, position, and orientation of the damage. Mandolla et al. [52] implemented a DT for additive manufacturing in the aerospace industry by utilizing blockchain solutions. This work highlights how businesses utilizing the blockchain can create secure and connected manufacturing infrastructure and provides a conceptual solution to securing and organizing the data generated by an end-to-end additive manufacturing process in the aerospace industry. Zhang et al. [53] established a digital-thread-based modeling digital twin (DTDT) framework for an aircraft assembly system, enhancing the controllability and traceability of the manufacturing process and product quality through improved data management. Tyncherov et al. [54] proposed DT modeling of aircraft operational life cycle by presenting aircraft systems' DTs with operational and maintenance environments as a cloud of data considering machine learning (ML) methods to improve prediction and planning accuracy. Tuegel et al. [55] reengineered the aircraft structural life prediction process to high-performance digital computing, presenting a conceptual model of DTs for predicting aircraft structure life and assuring its structural integrity. Ríos et al. [56] discussed an aircraft avatar implication through an industrialization-focused perspective while reviewing the various topics involved in an aircraft's digital counterpart development (i.e., product identification, product life cycle, and product information). Strelets et al. [57] created a DT in a uniform information environment of the product life cycle, which, as the virtual copy of a product, is convenient to use at all stages of the life cycle. Liang et al. [58] presented a real-time displacement detection DT in aircraft assembly. Zhang et al. [59] proposed an effective simulation and optimization containing heuristic algorithms and applied them to a DT-based aircraft part production workshop. Singh et al. [60] presented an information management (IM) framework for DTs in aircraft manufacturing, with a case study for aircraft structure damage tolerance, demonstrating the different phases of IM (from identification to retrieval and retention).

The existing body of aviation-related scientific literature extensively explores the potential of DTs and highlights their versatile applications, including their effectiveness not only in system-level implementation but also at the individual-component level. Employing DTs at these different levels can unlock new insights and ultimately advance the state of knowledge in the field. For example, Lei et al. [61] modeled a DT for tooth surface grinding, considering the low-risk transmission performance of non-orthogonal aviation spiral bevel gears. Zakrajsek et al. [62] developed a DT for a specific aircraft tire at touchdown to improve tire touchdown wear prediction. Xu et al. [63] suggested DT optimization with several DT modules for a system to virtually simulate as well as optimize the parameters, performance, and manufacturing. The DT modules make corrections during the optimization using real-time feedback data from manufacturing measurements and performance testing. Borgo et al. [64] presented a DT of a ground steering system and systematically analyzed the effect of uncertainties and sensor faults with estimation algorithms (least squares estimation and soft computing approach) under several scenarios. Hu et al. [65] developed a DT decision-making approach to generate reconfigurable fixturing schemes optimization for the trimming operation of aircraft skins. Peng et al. [66] provided an online fault diagnosis system for the TFE-731 turbofan engine and used model-based and data-driven approaches to create DTs of the engine parameters. Li et al. [67] used the concept of dynamic Bayesian networks (DBNs) to develop a health monitoring model for aircraft. An example of the proposed method is also illustrated on an aircraft wing's fatigue crack growth [68]. Kosova et al. [69] developed a DT and used ML for a health-monitoring system (limited to aircraft hydraulic systems) to diagnose system failures in the early stages using 20 failure scenarios. Laukotka et al. [70] implemented DTs for civil aviation, aircraft, and aircraft cabins, based on modular product family design and model-based systems engineering.

Various research efforts have been diligently conducted to explore and harness the potential of DTs in UASs. The application of DTs in UASs has emerged and prompted researchers to utilize the benefits of this technology, aiming to enhance design, operation and mission planning, and maintenance practices, leading to more reliable, efficient, and capable UASs. However, after reviewing DTs throughout the entire life cycle of the aviation system, Xiong et al. [71] concluded that while aviation DTs are frequently utilized in manufacture and maintenance, more effort and attention are required for UAV DT applications. Lv et al. [72] also reviewed AI applications in DTs in aerospace, intelligent manufacturing, unmanned vehicles, and smart city transportation. Salinger et al. [73] presented a hardware testbed for a self-aware UAV to advance dynamic data-driven application system (DDDAS) development. Self-awareness refers to a vehicle's ability to collect information about itself and utilize that knowledge to complete missions through dynamic decision making on board. Kapteyn et al. [74] combined reduced-order models with Bayesian estimation to create a data-driven DT for a 12 ft wingspan UAV to enable the aircraft to adjust its mission plan in the event of structural damage or deterioration. The authors further advanced the methodology using interpretable ML [75]. Alaez et al. [76] modeled a DT of a VTOL UAV using the Gazebo robotics simulator, compared the UAV's take-off, hovering, and landing operation with and without a wind physics model, and tested it in different wind speeds and directions. Yang et al. [77] proposed a DT for a multirotor UAV with a simulation system, a physical UAV, and a service center for advanced capability training as well as algorithm verification. The authors also demonstrated a DT simulation platform for verification that further simulates and tracks the life cycle of a multirotor UAV [78]. Lv et al. [3] analyzed the effects and limitations of UAVs in 5G/B5G wireless communication and developed a UAV DT 5G communication channel model using deep learning (DL) to further reduce UAV limitations. Moorthy et al. [79] designed a UAV network simulator focusing on high-fidelity UAV flight control by using two simulators they developed in prior years: UBSim (a Python-based event-driven simulator) and UB-ANC (a simulation framework used to design, implement, and test various UAV networking applications). Wu et al. [80] addressed the security concerns that arise when a drone system is attacked and investigated the computational intelligence of drone information systems and DTs of drone networks based on DL. Shen et al. [81] proposed a DT with deep reinforcement learning (DRL) (in which a DT of a multi-UAV system is built into a central server to train a DRL model) to solve the flocking motion problem of multi-UAV systems. Lv et al. [82] developed a UAV DT to provide medical resources quickly and accurately to analyze the feasibility of UAV DTs during COVID-19 prevention and used DL algorithms to construct a UAV DT information forecasting model. Fraser et al. [83] used DT and data-driven approaches to investigate the general susceptibilities of UAVs against contemporary cyber threats. Kapteyn et al. [84] suggested a probabilistic graphical model representing the DT and its physical asset for a UAV using experimental data to calibrate the DT. The UAV encounters an in-flight damage event and the DT is updated using sensor data. Riordan et al. [85] presented a DT to evaluate UAS-mounted LiDAR ability to detect small-object air collision risks, considering the Hamburg port with its aerial hazards (e.g., birds, drones, helicopters, and low-flying aircraft). Iqbal et al. [86] presented a DT with a runtime trust assessment for an autonomous food delivery drone system to evaluate the trusted execution of intelligent agents (autonomous drones or other vehicles). Grigoropoulos et al. [87] employed DTs and simulations to support offline validation and runtime checking in a platform as a service (PaaS) system for drone applications. Lee et al. [88] proposed a DT with a model-based system engineering methodology for a UAS capable of route selection in a military case study, where the route optimization module suggests an optimal path based on inputs such as potential damage. Lei et al. [89] created a DT to define the physical entity of a UAV swarm and track its life cycle. The UAV swarm's behaviors are investigated using an ML-based decision model. Wang et al. [90] combined DTs and convolutional neural networks (CNNs) for a UAV autonomous network to explore the airspace structure and safety performance of the UAV system. The presented literature emphasizes the significance of exploring and utilizing DTs in UASs. These case studies highlight the significance of DTs in addressing various challenges and opportunities of UASs associated with topics such as driving technological advancements, decision-making processes, and operational efficiency within this dynamic and evolving field. Digital twin technology has the potential to address some of these challenges and complement existing measures in UAV management. By modeling a digital copy of UASs and their operational environment, DTs can provide real-time monitoring, analyses, and optimization of UAS operations. This can enhance situational awareness, enable predictive maintenance, improve traffic management, and support decision-making processes. DTs can also facilitate data integration and interoperability across different systems, enabling a more comprehensive and coordinated approach to UAV management. However, it is important to note that DTs are not a standalone solution but should be integrated into a holistic framework that considers regulatory, technical, and operational aspects. Overall, the unique role of using a DT to facilitate UAV certification and regulation lies in the ability to model a digital copy of a physical system for real-time validation and optimization. However, this task is inherently complex and presents several challenges, as depicted in Figure 9, which offers an overview of the risks and challenges associated with the overall DT process. One challenge is the requirement for an accurate model of reality, which necessitates a deep understanding of the physical system and its operational characteristics. Additionally, as demonstrated in Figure 8, the creation of DTs necessitates the transfer of data between a physical system and a virtual model. Depending on the complexity level of a DT, this process involves handling a large amount of data from various sources, including sensors and sometimes even multifunctional models. Ensuring the accuracy and reliability of these data is crucial for the effectiveness of a DT. Furthermore, integrating a DT into UASs to assist the certification process requires careful consideration of legal and regulatory requirements. These challenges highlight the need for careful planning, robust data management, and close collaboration between experts in UAV certification and DT technology to successfully utilize DT in the context of UAV certification.

3. Results

UAVs are becoming popular. Autonomous (artificial intelligence applications) and automatic UAVs are expected to conduct safe operations, and they will enter UAM to transport goods and individuals in the near future. A wide range of literature is published to answer the research questions of "how to adapt UAV applications to regulations" and "how to adapt DT applications to UAV". However, it is fair to state that there is not much literature considering the use of DT applications in UAVs for certification and regulation. This lack of literature is inevitable in the early stages of new, emerging concepts. In order to fill this gap, we conducted a literature review considering a total of 121 references. Table 4 provides a comprehensive collection of references along with the keywords that are closely aligned with our research concepts. They serve as concise descriptors that capture the essence of the paper's content and help identify its key focus areas. The inclusion of these relevant keywords allows for a focused exploration and clear navigation of the existing literature, facilitating the identification of common themes, connections, and relationships across the literature. By including associated keywords in the table, we aimed to provide additional information and context about the content of each reference. We have systematically identified and classified the references into key focus areas: DTs, general aviation, UAVs/UASs, UAM/AAM, and regulation. By organizing the references under these categories, the table allows for a clear understanding of the primary themes and topics covered in each reference, enhancing the clarity and structure of our research with a more organized exploration. While the references consider multiple topics and overlap across the key focus areas, we have made an effort to present the primary purpose of each paper and provide associated keywords to highlight key themes and connections that contribute to a more comprehensive understanding of our research concepts and emphasize the various aspects explored in the literature.

Table 4. Compilation of references and their associated keywords relevant to our research concepts.

Reference Number	Year	Type	Key Focus	Related Keywords
[1]	2022	Regulatory document	Regulation	EASA regulations, operation of air taxis in cities
[2]	2021	Journal article	UASs/UAVs	Airspace organization and management, air traffic control, air traffic management, air traffic service provision, unmanned aircraft system, UAS traffic management
[3]	2021	Journal article	DT, UASs/UAVs	Unmanned aerial vehicles, deep learning, digital twins
[4]	2022	Other	DTs	Digital twins
[5]	2023	Other	DTs	Digital twins
[6]	2022	Other	General aviation	Aerospace certification, digital twins
[7]	2020	Other	DTs	Digital twins
[8]	2019	Journal article	DTs	Artificial intelligence, digital twins, human–computer interaction, machine learning
[9]	2018	Conference proceeding	DTs	Digital twins, learning theories, situational awareness
[10]	2021	Journal article	DTs	Digital twins, manufacturing system design, smart manufacturing
[11]	2020	Conference proceeding	DTs	Digital twin concept, digital twin application
[12]	2021	Journal article	UASs/UAVs	eVTOL, rotorcraft, design, advanced air mobility, urban air mobility

Table 4. *Cont.*

Reference Number	Year	Type	Key Focus	Related Keywords
[13]	2022	Journal article	UAM/AAM	Advanced air mobility, urban air mobility, emergency response, air ambulance, electric vertical take-off and landing, VTOL, eVTOL
[14]	2023	Journal article	UAM/AAM	Advanced air mobility, connected eVTOL, operations, infrastructure, communications, sustainability
[15]	2021	Conference proceeding	UAM/AAM	Surveillance, traffic control, aircraft navigation, safety, air traffic control, active appearance model
[16]	2020	Conference proceeding	UAM/AAM	Urban air mobility, aircraft performance, flight trajectory, autonomous systems, flight control, flight operation, detect and avoid
[17]	2022	Conference proceeding	UAM/AAM	Urban air mobility, aerial photography, conventional takeoff and landing, airspace management, short take-off and landing, federal aviation regulation, commercial aircraft
[18]	2021	Conference proceeding	UAM/AAM	Urban air mobility, autonomous systems, human automation interaction, ground control station, air transportation, national aeronautics and space administration, small unmanned aircraft systems
[19]	2021	Conference proceeding	UAM/AAM	Safety management, urban air mobility, airspace management, unmanned aircraft systems, supersonic aircraft, national airspace system, flight operations quality assurance, aeronautical information service
[20]	2022	Conference proceeding	UAM/AAM	Urban air mobility, aeronautics, special-use airspace, federal aviation administration, heliports, aviation, take-off and landing
[21]	2022	Conference proceeding	UAM/AAM	Flight testing, aviation, urban air mobility, propeller blades, true airspeed, flight path angle, vertical take-off and landing
[22]	2021	Conference proceeding	UAM/AAM	Urban air mobility, airspace class, air transportation, vertical take-off and landing, rotorcrafts, airspace system, helicopters, fixed-wing aircraft
[23]	2023	Conference proceeding	UAM/AAM	Urban air mobility, landing lights, flight testing, flight management system, flight control system, flight vehicle
[24]	2023	Conference proceeding	UAM/AAM	Urban air mobility, image registration, Federal Aviation Administration, vision-based navigation, heliports, instrument landing system
[25]	2021	Conference proceeding	UAM/AAM	Urban air mobility, airspace, software architecture, aeronautics, Federal Aviation Administration, aviation, unmanned aerial vehicle, aerospace industry
[26]	2022	Conference proceeding	UAM/AAM	Air mobility, Federal Aviation Administration, guidance system, sensor fusion, landing lights
[27]	2023	Conference proceeding	UAM/AAM	Air mobility, optical sensor, aviation, radar measurement, detect and avoid, take-off and landing
[28]	2022	Conference proceeding	UAM/AAM	Airspace, urban air mobility, near-mid-air collision, target level of safety, air traffic controller, helicopters, air traffic management, flight planning

Table 4. *Cont.*

Reference Number	Year	Type	Key Focus	Related Keywords
[29]	2021	Journal article	UAM/AAM	Advanced air mobility, cost–benefit analysis, ARIMA forecasting, electric vertical take-off and landing aircraft, small unmanned aircraft system, green transportation
[30]	2021	Journal article	UAM/AAM	Advanced air mobility, urban air mobility, on-demand air mobility, air taxi, vertical take-off and landing
[31]	2023	Other	UAM/AAM	Urban air mobility
[32]	2021	Journal article	UAM/AAM	Urban air mobility, air taxi, electric vehicle, autonomous vehicle, ride hailing, carsharing
[33]	2020	Book		On-demand mobility, transport modeling, urban air mobility, vertical take-off, landing
[34]	2020	Journal article	UAM/AAM	Urban air mobility, vehicle concepts, policy, transport simulation, infrastructure
[35]	2018	Journal article	Regulation	Drones, aircraft, atmospheric modeling, guidelines, FAA, government policies
[36]	2014	Journal article	Regulation	Remotely piloted aircraft (RPA), UAV
[37]	2014	Journal article	Regulation	Co-regulation, self-regulation, aviation safety, drone, RPA, UAV
[38]	2020	Journal article	Regulation	Drone, regulation
[39]	2016	Journal article	Regulation	Privacy regulation, drone privacy
[40]	2021	Journal article	UASs/UAVs	WTP for drone flying, road pricing for drone airspace
[41]	2019	Journal article	Regulation	Drone, regulation
[42]	2022	Journal article	Regulation	Drone regulation, local policy adoption
[43]	2019	Journal article	Regulation	Drones, regulation, policy
[44]	2020	Other	General aviation	Digital twin, data management
[45]	2018	Conference proceeding	DTs	Digital twin
[46]	2020	Journal article	General aviation	Airframe digital twin, digital thread, individual aircraft tracking
[47]	2020	Conference proceeding	General aviation	Commercial aircraft, machine learning, airspace, artificial intelligence, neural networks, aircraft production, aviation
[48]	2022	Journal article	DTs	Digital twins, military aircraft, aircraft propulsion
[49]	2019	Conference proceeding	DTs	Digital twins, aviation industry
[50]	2012	Conference proceeding	DTs	Aircraft structures, high-performance computing structural modeling, air forces, flight dynamics
[51]	2017	Conference proceeding	General aviation	Aircraft structures, genetic algorithm, structural damage
[52]	2019	Journal article	DTs	Digital technology, digital twin, aircraft industry
[53]	2022	Journal article	General aviation	Digital twin, digital thread, aircraft assembly
[54]	2020	Conference proceeding	General aviation	Aircraft maintenance, aircraft life cycle, digital twin
[55]	2011	Journal article	General aviation	Aircraft structural life prediction, digital twin
[56]	2015	Conference proceeding	DTs	Product avatar, digital twin, digital counterpart, aircraft avatar
[57]	2020	Conference proceeding	DTs	Product life cycle, digital twin, aircraft

Table 4. *Cont.*

Reference Number	Year	Type	Key Focus	Related Keywords
[58]	2020	Journal article	General aviation	Aircraft manufacture, digital twin
[59]	2022	Journal article	DTs	Digital twin shop floor, large-scale problem optimization, simulation
[60]	2021	Conference proceeding	General aviation	Digital twin, aircraft manufacturing
[61]	2022	Journal article	General aviation	Non-orthogonal aviation spiral bevel gears, free-form tooth surface grinding, digital twin modeling
[62]	2017	Conference proceeding	General aviation	Flight data, flight operation, flywheels, structural health monitoring
[63]	2021	Journal article	General aviation	Optimization, digital twin, virtual modules
[64]	2020	Conference proceeding	General aviation	Digital twin, virtual sensing, aircraft ground-steering system
[65]	2022	Journal article	General aviation	Aircraft skin, digital twin, layout optimization
[66]	2022	Journal article	DTs	Autoregressive moving average (ARMA) model, turbofan engine modeling
[67]	2017	Conference proceeding	DTs	Aircraft wings, stochastic crack growth models, surrogate model, mathematical models
[68]	2017	Journal article	General aviation	Aircraft wings, stochastic crack growth models, fatigue cracking, airframes
[69]	2022	Journal article	General aviation	Digital twin, aircraft hydraulics, ensemble learning
[70]	2021	Conference proceeding	General aviation	Digital twin, aviation, aircraft cabins
[71]	2022	Journal article	General aviation	Digital twin, aviation industry
[72]	2021	Journal article	DTs	Digital twin, artificial intelligence, autonomous driving
[73]	2020	Conference proceeding	DTs	Digital twin, self-aware unmanned vehicle
[74]	2022	Journal article	DTs, UASs/UAV	Digital twin, model updating, unmanned aerial vehicle
[75]	2020	Conference proceeding	DTs, UASs/UAVs	Machine learning, unmanned aerial vehicle, recurrent neural network
[76]	2022	Journal article	DTs, UASs/UAVs	VTOL, UAV, digital twin, aerodynamic coefficients, gazebo, wind model
[77]	2021	Conference proceeding	DTs, UASs/UAVs	Digital twin, UAV, virtual and real interaction
[78]	2020	Conference proceeding	DTs, UASs/UAVs	UAV, digital twin, simulation
[79]	2022	Conference proceeding	DTs, UASs/UAVs	Unmanned aerial vehicle (UAV), multifidelity simulation
[80]	2022	Conference proceeding	DTs, UASs/UAVs	Unmanned aerial vehicle, deep learning, digital twins
[81]	2022	Journal article	DTs, UASs/UAVs	Deep reinforcement learning (DRL), digital twin (DT), multi-UAV systems
[82]	2021	Journal article	DTs, UASs/UAVs	Unmanned aerial vehicles, digital twins, deep learning
[83]	2021	Conference proceeding	DTs, UASs/UAVs	Digital twin, machine learning, UAV, UAS, cybersecurity
[84]	2021	Journal article	DTs, UASs/UAVs	Digital twin
[85]	2021	Conference proceeding	DTs, UASs/UAVs	Unmanned aerial systems, detect and avoid, data-driven simulation
[86]	2022	Conference proceeding	DTs, UASs/UAVs	Modeling, autonomous drones, digital twin
[87]	2020	Conference proceeding	DTs, UASs/UAVs	Drones, simulation environment, digital twin
[88]	2021	Journal article	DTs, UASs/UAVs	Digital twin, model-based systems engineering

Table 4. *Cont.*

Reference Number	Year	Type	Key Focus	Related Keywords
[89]	2021	Journal article	DTs, UASs/UAVs	data models, unmanned aerial vehicles, integrated circuit modeling, digital twin, computational modeling, machine learning algorithms, real-time systems
[90]	2022	Journal article	DTs, UASs/UAVs	unmanned aerial vehicles, safety, aircraft, aircraft navigation, security, monitoring
[91]	2018	Book	Regulation, UASs/UAVs	European policies, civil drones, safety, security
[92]	2012	Book	General aviation	Aircraft structures
[94]	2018	Conference proceeding	DTs	Digital twin, simulation, cyber-physical system
[95]	2016	Book	Regulation, UASs/UAVs	Drone laws, RPAS, UAS, UAV, commercial drones, autonomous aviation
[96]	2023	Other	Regulation	EASA Provisions, EU Regulations 2019/947 and 2019/945
[97]	2023	Regulatory document	Regulation	Civil drones, unmanned aircraft
[98]	2023	Regulatory document	Regulation	Open category of civil drones
[99]	2023	Regulatory document	Regulation	Specific category of civil drones
[100]	2023	Regulatory document	Regulation	Certified category of civil drones
[93]	2022	Regulatory document	Regulation	Rules For Unmanned Aircraft Systems, Regulation (EU) 2019/947, Regulation (EU) 2019/945
[101]	2021	Regulatory document	Regulation	EASA guidelines, The Design Verification of Specific Category Drones
[102]	2023	Regulatory document	Regulation	Rules for Airworthiness and Environmental Certification, Regulation (EU) No 748/2012
[103]	2021	Other	Regulation	EU regulatory for U-space
[104]	2021	Regulatory document	Regulation	Regulation (EU) 2021/664
[105]	2021	Regulatory document	Regulation	Regulation (EU) 2021/665
[106]	2021	Regulatory document	Regulation	Regulation (EU) 2021/666
[107]	2023	Other	Regulation	Autonomous drones, automatic drones
[108]	2022	Conference proceeding	UAS/UAV	Adversarial machine learning, aviation, urban air mobility, pilot, convolutional neural network, unmanned aircraft system, cyber–physical system
[109]	2020	Regulatory document	Regulation	EASA AI roadmap, AI in aviation
[110]	2022	Conference proceeding	Regulation	Reinforcement learning, aviation, European Aviation Safety Agency, artificial intelligence, neural networks, urban air mobility, unmanned aircraft system, air traffic management, continuing airworthiness
[111]	2020	Journal article	DTs	Digital twin
[112]	2017	Other	DTs	Digital twin
[113]	2021	Conference proceeding	DTs	Digital twin technologies
[114]	2012	Conference proceeding	General aviation	Digital twin, air forces, NASA Goddard Space Flight Center
[115]	2020	Journal article	DTs	Digital twin
[116]	2018	Journal article	DTs	Digital twin
[117]	2022	Other	DTs	Digital twin

Table 4. *Cont.*

Reference Number	Year	Type	Key Focus	Related Keywords
[118]	2022	Book	DTs	Digital twin, digital manufacturing, digital technologies in manufacturing, digital image processing
[119]	2023	Book	DTs, UASs/UAVs	Digital twin, smart urban mobility, UAV
[120]	2021	Journal article	DTs	Digital twin, cyber–physical systems
[121]	2023	Other	DTs, UASs/UAVs	Intelligent urban air mobility, digital twin, autonomous flight

A time frame of two decades was chosen for conducting the literature review, since the term "digital twin" was first introduced in 2010 [111]. However, in this section, as we discussed in the research methodology in Section 2.1, we only analyzed articles within the scope of DT applications for UASs. DT applications in UASs are relatively new, resulting in the majority of the relevant literature having been published within recent years. Although research on DTs in UAS applications has recently gained momentum, there remains a substantial amount of work to be undertaken toward the further exploration and understanding of the potential value and significance that DTs can bring to the field of UAS applications. Figure 11 provides a word cloud visualization that depicts the frequency of selected keywords (DT, UAV, AI, drone, UAS, certification, regulation, and VTOL) within publications related to the applications of DTs in UASs. These specific keywords were carefully selected during the research process, and the word cloud offers a concise representation of the pathway to the literature review scope.

Figure 11. Keyword analysis word cloud for DT applications in publication on UASs.

While the complexity of UASs is fast evolving, only 40% of the publications briefly mentioned certification and regulation when using DTs in UASs, and not many scientific literature efforts focus on the use of DT applications in UASs for certification and regulation, as shown in Figure 12. DTs facilitate designing, building, and analyzing procedures. DTs are very good and relatively time- as well as cost-efficient tools to assist the certification process, since they help engineers check, analyze, and integrate designs as well as express concerns instantly.

Autonomous (with the help of AI and without a pilot's intervention) UAVs are expected to conduct safe operations and cope with unforeseen conditions. As presented in Figure 13, half of the publications considering the use of DT applications in UASs mentioned autonomous flight operations, and 38% of these publications also discussed the use of AI, which leads to the key research question of how these operations can be safely conducted. In UASs' EU operational scope, the EASA published the "EASA AI roadmap" as a human-centric approach to the safe use of AI in aviation.

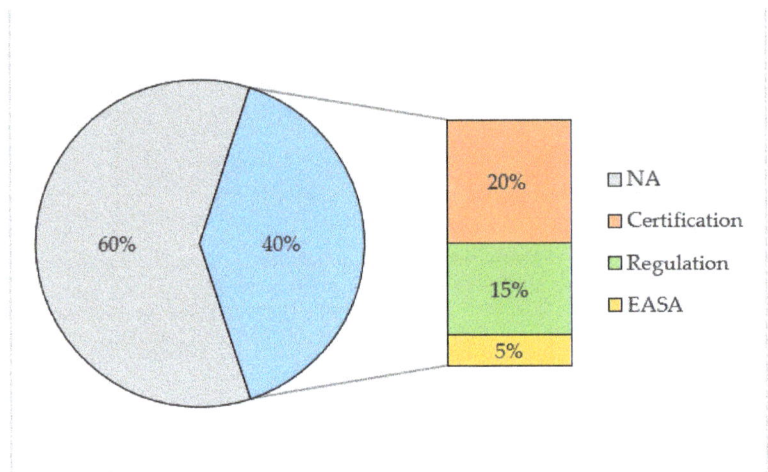

Figure 12. Keyword analysis focused on certification and regulatory frameworks for DT applications in publication on UASs.

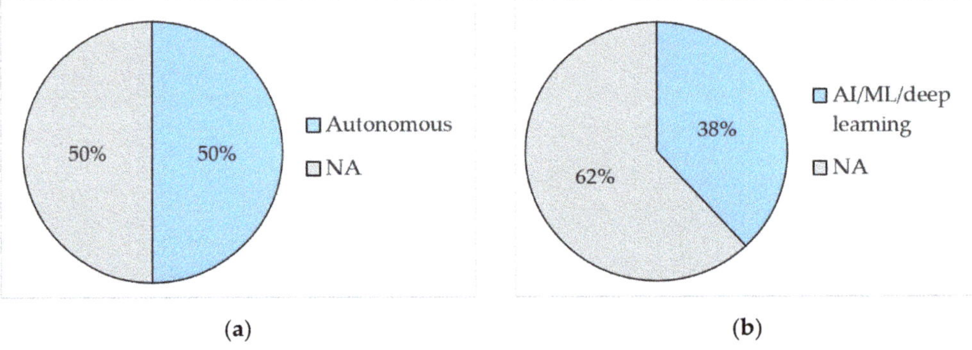

(**a**)   (**b**)

Figure 13. Keyword analysis focused on (**a**) autonomous flights and (**b**) the use of AI/ML/deep learning approaches for DT applications in publications on UASs. (**a**) Autonomous flights mentioned in publications. (**b**) AI/ML/deep learning approaches mentioned in publications.

4. Discussion

Flying cars and aerial transportation systems are some of the distinctive features of the future cities described in science fiction films and books. This is one of the basic concepts accepted by society when imagining the future, and with today's technological advancements we wonder if conducting safe automatic and autonomous flights for metropolitan areas is a few steps away in the near future. The establishment of a socially acceptable regulatory framework is necessary to transform this vision into a reality.

The regulatory framework for UASs in the European Union was fragmented before 2020, as shown in Figure 1, with each EU Member State being responsible for drones with a maximum take-off mass (MTOM) of less than 150 kg while the EASA was in charge of drones with an MTOM exceeding this weight. The transition to new regulations began in 2020, and the EASA is now responsible for drones of all weights and size. Nonetheless, this regulatory framework is still in its early stages, and further developments are expected.

Implementing this evolving regulatory framework presents a significant challenge in UAS operations. DTs can potentially offer a solution by facilitating the design, construction, and analysis processes. They are time- and cost-efficient tools to assist the certification process, since they help engineers check, analyze, and integrate designs as well as express concerns instantly. However, only a limited number of publications (40%) briefly mentioned certification and regulation when discussing the application of DTs in UASs, as shown in Figure 12. Therefore, efforts need to be carried out to emphasize the importance of DTs in assisting the certification process within UAS operations.

In Figure 13, it is notable that autonomous flight operations were mentioned in 50% of the papers examining the use of DTs in UASs, and AI applications were discussed in 38% of the publications. Autonomous and automatic UASs are expected to conduct safe operations in UAM. A quick comparison of autonomous and automatic flights can show that there is no human safety net present during an autonomous flight in the event of unforeseen circumstances. Therefore, precise regulations must be created in the context of AI to ensure autonomous flights are safely conducted. In 2020, the EASA also published the first guidance, the EASA AI roadmap, for the safe use of artificial intelligence in aviation. However, we are still a long way from the dream of having science fiction flying transport systems coming true, as the timeline outlined in the EASA AI roadmap document predicts the first approvals of AI in 2025.

The research subject of how to adapt UASs with UAM and regulation is studied in a wide range of the literature from across the world. However, the implementation of DTs in UASs for assisting the certification process and considering regulation, especially within the context of the EU regulatory framework, remains relatively unexplored. The concept of drone regulation, particularly in relation to EU legislation and the integration of UAM for cargo and passenger transport, is still relatively new. The development of regulations, as well as applying these regulations to UAS operational categories, requires the consideration of numerous criteria and parameters to ensure a robust level of safety and seamless flight operations. Moreover, due to safety concerns and ongoing regulation development, UAV autonomous flights are not currently being carried out in most European countries. The lack of literature and documents is inevitable in the early stages of a new concept's development. Consequently, one of the challenges lies in staying informed about evolving regulations and keeping track of the developments and changes that emerge in this field.

Overall, this paper highlights the necessity of further research on and the exploration of DT applications in UASs, particularly concerning certification and regulation. It is essential to recognize that DTs cannot function as standalone solutions but should be seamlessly integrated into a comprehensive framework that takes into account regulatory, technical, and operational aspects. The distinct advantage of employing DTs to facilitate UAS certification and regulation lies in their ability to create a digital replica of a physical system, enabling real-time validation and optimization. Nonetheless, this task is inherently complex and presents several challenges, such as the necessity of an accurate model of reality and handling a large amount of data from various sources. Furthermore, utilizing DTs to assist the certification process requires careful consideration of legal and regulatory requirements. It is crucial to address these challenges and associated complexities to pave the way for the successful implementation of UASs in UAM.

5. Conclusions

The popularity of UAV operations has increased, and new air mobility concepts have emerged over the past years. It is essential to develop regulations in this new technological context that effectively address the challenges and opportunities presented by UASs. There are various levels of ongoing activities and recent advances in UAS regulatory frameworks, especially in the domain of European Union (EU) regulations. Due to the growing demands, advancements, and possible applications of UASs, particularly in research and innovation, there is a need for a systematic overview. To bridge this gap, we present a comprehensive overview of the developed UAS regulations in the European Union and explore the concept of DTs as well as their potential applications in the UAS domain. We aimed to conduct a systematic review to provide a structured methodology that synthesizes multiple studies to offer a comprehensive and unbiased assessment of DTs' applications in UASs with EU regulatory compliance. Despite limited scholarly focus on the implementation of DTs in UASs considering certification and regulation, we analyzed the existing literature to identify and emphasize the important trends and developments. The overall challenges and the importance of UAS DTs are highlighted to provide a robust foundation for future studies on UAS DTs and their compliance with the EU regulatory framework.

Author Contributions: Conceptualization, I.S.; methodology, E.F.; writing—original draft preparation, E.F.; writing—review and editing, S.S., I.S., E.-H.A. and E.F.; supervision, I.S. and E.-H.A.; project administration, I.S. All authors have read and agreed to the published version of the manuscript.

Funding: This research was funded by the European Union's Horizon 2020 research and innovation program (H2020-MG-3-6-2020 Research and Innovation Action "Towards sustainable urban air mobility"), grant number 101007134.

Data Availability Statement: Not applicable.

Conflicts of Interest: The authors declare no conflict of interest.

References

1. European Union Aviation Safety Agency (EASA). 'EASA Publishes World's First Rules for Operation of Air Taxis in Cities'. Available online: https://www.easa.europa.eu/newsroom-and-events/press-releases/easa-publishes-worlds-first-rules-operation-air-taxis-cities (accessed on 30 June 2022).
2. Liu, Z.; Cai, K.; Zhu, Y. Civil unmanned aircraft system operation in national airspace: A survey from Air Navigation Service Provider perspective. *Chin. J. Aeronaut.* **2021**, *34*, 200–224. [CrossRef]
3. Lv, Z.; Chen, D.; Feng, H.; Lou, R.; Wang, H. Beyond 5G for digital twins of UAVs. *Comput. Netw.* **2021**, *197*, 108366. [CrossRef]
4. 'European Digital Twin of the Ocean (European DTO)'. Available online: https://research-and-innovation.ec.europa.eu/funding/funding-opportunities/funding-programmes-and-open-calls/horizon-europe/eu-missions-horizon-europe/restore-our-ocean-and-waters/european-digital-twin-ocean-european-dto_en (accessed on 21 April 2022).
5. The NLR Digital Twins. Available online: https://www.nlr.org/capabilities/digital-twins/ (accessed on 31 January 2023).
6. Eliminating Aerospace Certification Gaps Utilizing the Electrical Systems Digital Twin. Available online: https://www.plm.automation.siemens.com/global/en/webinar/aerospace-certification/97877 (accessed on 28 January 2022).
7. ELISE-Digital Twin. Available online: https://joinup.ec.europa.eu/collection/elise-european-location-interoperability-solutions-e-government/glossary/term/digital-twin#:~:text=Digital%20twins%20create%20a%20virtual,(2020)%2C%20European%20Data%20Strategy (accessed on 1 January 2022).
8. Barricelli, B.R.; Casiraghi, E.; Fogli, D. A Survey on Digital Twin: Definitions, Characteristics, Applications, and Design Implications. *IEEE Access* **2019**, *7*, 167653–167671. [CrossRef]
9. David, J.; Lobov, A.; Lanz, M. Learning Experiences Involving Digital Twins. In Proceedings of the IECON 2018—44th Annual Conference of the IEEE Industrial Electronics Society, Washington, DC, USA, 21–23 October 2018.
10. Leng, J.; Wang, D.; Shen, W.; Li, X.; Liu, Q.; Chen, X. Digital twins-based smart manufacturing system design in Industry 4.0: A review. *J. Manuf. Syst.* **2021**, *60*, 119–137. [CrossRef]
11. Wu, J.; Yang, Y.; Cheng, X.U.N.; Zuo, H.; Cheng, Z. The Development of Digital Twin Technology Review. In Proceedings of the 2020 Chinese Automation Congress, CAC 2020, Shanghai, China, 6–8 November 2020; Institute of Electrical and Electronics Engineers Inc.: Piscataway, NJ, USA, 2020; pp. 4901–4906. [CrossRef]
12. Johnson, W.; Silva, C. NASA concept vehicles and the engineering of advanced air mobility aircraft. *Aeronaut. J.* **2022**, *126*, 59–91. [CrossRef]
13. Goyal, R.; Cohen, A. Advanced Air Mobility: Opportunities and Challenges Deploying eVTOLs for Air Ambulance Service. *Appl. Sci.* **2022**, *12*, 1183. [CrossRef]
14. Al-Rubaye, S.; Tsourdos, A.; Namuduri, K. Advanced Air Mobility Operation and Infrastructure for Sustainable Connected eVTOL Vehicle. *Drones* **2023**, *7*, 319. [CrossRef]
15. de Oliveira, Í.R.; Neto, E.C.P.; Matsumoto, T.T.; Yu, H. Decentralized air traffic management for advanced air mobility. In Proceedings of the Integrated Communications, Navigation and Surveillance Conference, ICNS, Dulles, VA, USA, 20–22 April 2021; Institute of Electrical and Electronics Engineers Inc.: Piscataway, NJ, USA, 2021. [CrossRef]
16. Wing, D.J.; Chancey, E.T.; Politowicz, M.S.; Ballin, M.G. Achieving resilient in-flight performance for advanced air mobility through simplified vehicle operations. In Proceedings of the AIAA AVIATION 2020 FORUM, Reno, NV, USA, 15–19 June 2020; American Institute of Aeronautics and Astronautics Inc.: Reston, VA, USA, 2020. [CrossRef]
17. Garrow, L.A.; German, B.; Schwab, N.T.; Patterson, M.D.; Mendonca, N.; Gawdiak, Y.O.; Murphy, J.R. A Proposed Taxonomy for Advanced Air Mobility. In Proceedings of the AIAA AVIATION 2022 Forum, Reston, VA, USA, 29 June 2022. [CrossRef]

18. Chancey, E.T.; Politowicz, M.S.; Le Vie, L. Enabling Advanced Air Mobility Operations through Appropriate Trust in Human-Autonomy Teaming: Foundational Research Approaches and Applications. In Proceedings of the AIAA Scitech 2021 Forum, Reston, VA, USA, 19–21 January 2021. [CrossRef]

19. Ellis, K.K.; Krois, P.; Koelling, J.; Prinzel, L.J.; Davies, M.; Mah, R. A Concept of Operations (ConOps) of an In-time Aviation Safety Management System (IASMS) for Advanced Air Mobility (AAM). In Proceedings of the AIAA Scitech 2021 Forum, Reston, VA, USA, 19–21 January 2021. [CrossRef]

20. Mendonca, N.; Murphy, J.; Patterson, M.D.; Alexander, R.; Juarex, G.; Harper, C. Advanced Air Mobility Vertiport Considerations: A List and Overview. In Proceedings of the AIAA AVIATION 2022 Forum, Reston, VA, USA, 29 June 2022; American Institute of Aeronautics and Astronautics: Reston, VA, USA, 2022. [CrossRef]

21. Pascioni, K.A.; Watts, M.E.; Houston, M.; Lind, A.; Stephenson, J.H.; Bain, J. Acoustic Flight Test of the Joby Aviation Advanced Air Mobility Prototype Vehicle. In Proceedings of the 28th AIAA/CEAS Aeroacoustics 2022 Conference, Reston, VA, USA, 14–17 June 2022; American Institute of Aeronautics and Astronautics: Reston, VA, USA, 2022. [CrossRef]

22. Alvarez, L.E.; Jones, J.C.; Bryan, A.; Weinert, A.J. Demand and Capacity Modeling for Advanced Air Mobility. In Proceedings of the AIAA AVIATION 2021 FORUM, Reston, VA, USA, 2–6 August 2021; American Institute of Aeronautics and Astronautics: Reston, VA, USA, 2021. [CrossRef]

23. Kannan, K.; Baculi, J.E.; Lombaerts, T.; Kawamura, E.; Gorospe, G.E.; Holforty, W.; Ippolito, C.A.; Stepanyan, V.; Dolph, C.; Brown, N. A Simulation Architecture for Air Traffic Over Urban Environments Supporting Autonomy Research in Advanced Air Mobility. In Proceedings of the AIAA SCITECH 2023 Forum, Reston, VA, USA, 25 January 2023; American Institute of Aeronautics and Astronautics: Reston, VA, USA, 2023. [CrossRef]

24. Kawamura, E.; Dolph, C.; Kannan, K.; Lombaerts, T.; Ippolito, C.A. Simulated Vision-based Approach and Landing System for Advanced Air Mobility. In Proceedings of the AIAA SCITECH 2023 Forum, Reston, VA, USA, 25 January 2023; American Institute of Aeronautics and Astronautics: Reston, VA, USA, 2023. [CrossRef]

25. Van Dalsem, W.; Shetye, S.; Das, A.N.; Krishnakumar, K.S.; Lozito, S.; Freeman, K.; Swank, A.; Shannon, P.; Tomljenovic, L. A Data & Reasoning Fabric to Enable Advanced Air Mobility. In Proceedings of the AIAA Scitech 2021 Forum, Reston, VA, USA, 19–21 January 2021; American Institute of Aeronautics and Astronautics: Reston, VA, USA, 2021. [CrossRef]

26. Kawamura, E.; Kannan, K.; Lombaerts, T.; Ippolito, C.A. Vision-Based Precision Approach and Landing for Advanced Air Mobility. In Proceedings of the AIAA SCITECH 2022 Forum, Reston, VA, USA, 3–7 January 2022; American Institute of Aeronautics and Astronautics: Reston, VA, USA, 2022. [CrossRef]

27. Lombaerts, T.; Kannan, K.; Kawamura, E.; Dolph, C.; Stepanyan, V.; Gorospe, G.E.; Ippolito, C.A. Distributed Ground Sensor Fusion Based Object Tracking for Autonomous Advanced Air Mobility Operations. In Proceedings of the AIAA SCITECH 2023 Forum, Reston, VA, USA, 25 January 2023; American Institute of Aeronautics and Astronautics: Reston, VA, USA, 2023. [CrossRef]

28. Chen, S.; Wei, P.; Evans, A.D.; Egorov, M. Estimating Airspace Resource Capacity for Advanced Air Mobility Operations. In Proceedings of the AIAA AVIATION 2022 Forum, Reston, VA, USA, 29 June 2022; American Institute of Aeronautics and Astronautics: Reston, VA, USA, 2022. [CrossRef]

29. Dulia, E.F.; Sabuj, M.S.; Shihab, S.A.M. Benefits of advanced air mobility for society and environment: A case study of Ohio. *Appl. Sci.* **2022**, *12*, 207. [CrossRef]

30. Goyal, R.; Reiche, C.; Fernando, C.; Cohen, A. Advanced air mobility: Demand analysis and market potential of the airport shuttle and air taxi markets. *Sustainability* **2021**, *13*, 7421. [CrossRef]

31. European Union Aviation Safety Agency (EASA). 'Urban Air Mobility'. Available online: https://www.easa.europa.eu/en/what-is-uam (accessed on 5 July 2023).

32. Garrow, L.A.; German, B.J.; Leonard, C.E. Urban air mobility: A comprehensive review and comparative analysis with autonomous and electric ground transportation for informing future research. *Transp Res. Part C Emerg Technol.* **2021**, *132*, 103377. [CrossRef]

33. Rothfeld, R.; Straubinger, A.; Fu, M.; Al Haddad, C.; Antoniou, C. Chapter13—Urban air mobility. In *Demand for Emerging Transportation Systems*; Elsevier: Amsterdam, The Netherlands, 2020; pp. 267–284. [CrossRef]

34. Straubinger, A.; Rothfeld, R.; Shamiyeh, M.; Büchter, K.D.; Kaiser, J.; Plötner, K.O. An overview of current research and developments in urban air mobility—Setting the scene for UAM introduction. *J. Air Transp. Manag.* **2020**, *87*, 101852. [CrossRef]

35. Winkler, S.; Zeadally, S.; Evans, K. Privacy and civilian drone use: The need for further regulation. *IEEE Secur. Priv.* **2018**, *16*, 72–80. [CrossRef]

36. Clarke, R. The regulation of civilian drones' impacts on behavioural privacy. *Comput. Law Secur. Rev.* **2014**, *30*, 286–305. [CrossRef]

37. Clarke, R.; Moses, L.B. The regulation of civilian drones' impacts on public safety. *Comput. Law Secur. Rev.* **2014**, *30*, 263–285. [CrossRef]

38. Thomasen, K. Robots, Regulation, and the Changing Nature of Public Space. *Ott. Law Rev.* **2020**, *51*, 275–312.

39. Thomasen, K. Beyond Airspace Safety: A Feminist Perspective on Drone Privacy Regulation. *Can. J. Law Technol.* **2016**, *16*. [CrossRef]

40. Merkert, R.; Beck, M.J.; Bushell, J. Will It Fly? Adoption of the road pricing framework to manage drone use of airspace. *Transp. Res. Part A Policy Pract.* **2021**, *150*, 156–170. [CrossRef]

41. West, J.P.; Klofstad, C.A.; Uscinski, J.E.; Connolly, J.M. Citizen support for domestic drone use and regulation. *Am. Politics Res.* **2019**, *47*, 119–151. [CrossRef]
42. Li, X.; Kim, J.H. Managing disruptive technologies: Exploring the patterns of local drone policy adoption in California. *Cities* **2022**, *126*, 103736. [CrossRef]
43. Nelson, J.; Gorichanaz, T. Trust as an ethical value in emerging technology governance: The case of drone regulation. *Technol. Soc.* **2019**, *59*, 101131. [CrossRef]
44. Meyer, H.; Zimdahl, J.; Kamtsiuris, A.; Meissner, R.; Raddatz, F.; Haufe, S.; Bäßler, M. *Development of a Digital Twin for Aviation Research*; DGLR: Bonn, Germany, 2020; pp. 1–8. [CrossRef]
45. Liu, Z.; Meyendorf, N.; Mrad, N. The role of data fusion in predictive maintenance using digital twin. In *AIP Conference Proceedings*; American Institute of Physics Inc.: College Park, MD, USA, 2018. [CrossRef]
46. Liao, M.; Renaud, G.; Bombardier, Y. Airframe digital twin technology adaptability assessment and technology demonstration. *Eng. Fract. Mech.* **2020**, *22*, 106793. [CrossRef]
47. Aydemir, H.; Zengin, U.; Durak, U.; Hartmann, S. The digital twin paradigm for aircraft—Review and outlook. In Proceedings of the AIAA Scitech 2020 Forum, Orlando, FL, USA, 6–10 January 2020; American Institute of Aeronautics and Astronautics Inc.: Reston, VA, USA, 2020; pp. 1–12. [CrossRef]
48. Mendi, A.F.; Erol, T.; Dogan, D. Digital Twin in the Military Field. *IEEE Internet Comput.* **2022**, *26*, 33–40. [CrossRef]
49. Ibrion, M.; Paltrinieri, N.; Nejad, A.R. On Risk of Digital Twin Implementation in Marine Industry. In *Journal of Physics: Conference Series*; Institute of Physics Publishing: Bristol, UK, 2019. [CrossRef]
50. Tuegel, E.J. The airframe digital twin: Some challenges to realization. In *Collection of Technical Papers—AIAA/ASME/ASCE/AHS/ASC Structures, Structural Dynamics and Materials Conference*; American Institute of Aeronautics and Astronautics Inc.: Reston, VA, USA, 2012. [CrossRef]
51. Seshadri, B.R.; Krishnamurthy, T. Structural health management of damaged aircraft structures using the digital twin concept. In Proceedings of the 25th AIAA/AHS Adaptive Structures Conference, Grapevine, TX, USA, 9–13 January 2017; American Institute of Aeronautics and Astronautics Inc.: Reston, VA, USA, 2017. [CrossRef]
52. Mandolla, C.; Petruzzelli, A.M.; Percoco, G.; Urbinati, A. Building a digital twin for additive manufacturing through the exploitation of blockchain: A case analysis of the aircraft industry. *Comput. Ind.* **2019**, *109*, 134–152. [CrossRef]
53. Zhang, Q.; Zheng, S.; Yu, C.; Wang, Q.; Ke, Y. Digital thread-based modeling of digital twin framework for the aircraft assembly system. *J. Manuf. Syst.* **2022**, *65*, 406–420. [CrossRef]
54. Tyncherov, T.; Rozkova, L. Aircraft Lifecycle Digital Twin for Defects Prediction Accuracy Improvement. In Proceedings of the 19th International Conference on Reliability and Statistics in Transportation and Communication, Riga, Latvia, 16–19 October 2020; Springer International Publishing: Berlin/Heidelberg, Germany, 2020; pp. 54–63. [CrossRef]
55. Tuegel, E.J.; Ingraffea, A.R.; Eason, T.G.; Spottswood, S.M. Reengineering Aircraft Structural Life Prediction Using a Digital Twin. *Int. J. Aerosp. Eng.* **2011**, *2011*, 154798. [CrossRef]
56. Ríos, J.; Hernández, J.C.; Oliva, M.; Mas, F. Product avatar as digital counterpart of a physical individual product: Literature review and implications in an aircraft. In *Transdisciplinary Lifecycle Analysis of Systems*; IOS Press: Amsterdam, The Netherlands, 2015; pp. 657–666. [CrossRef]
57. Strelets, D.Y.; Serebryansky, S.A.; Shkurin, M.V. Concept of Creation of a Digital Twin in the Uniform Information Environment of Product Life Cycle. In Proceedings of the 13th International Conference Management of Large-Scale System Development (MLSD), Moscow, Russia, 28–30 September 2020; Institute of Electrical and Electronics Engineers Inc.: Piscataway, NJ, USA, 2020. [CrossRef]
58. Liang, B.; Liu, W.; Liu, K.; Zhou, M.; Zhang, Y.; Jia, Z. A displacement field perception method for component digital twin in aircraft assembly. *Sensors* **2020**, *20*, 5161. [CrossRef]
59. Zhang, Z.; Guan, Z.; Gong, Y.; Luo, D.; Yue, L. Improved multi-fidelity simulation-based optimisation: Application in a digital twin shop floor. *Int. J. Prod. Res.* **2022**, *60*, 1016–1035. [CrossRef]
60. Singh, S.; Shehab, E.; Higgins, N.; Fowler, K.; Erkoyuncu, J.A.; Gadd, P. Towards Information Management Framework for Digital Twin in Aircraft Manufacturing. *Procedia CIRP* **2021**, *96*, 163–168. [CrossRef]
61. Lei, D.; Rong, K.; Song, B.; Ding, H.; Tang, J. Digital twin modeling for tooth surface grinding considering low-risk transmission performance of non-orthogonal aviation spiral bevel gears. *ISA Trans.* **2022**, *128*, 646–663. [CrossRef]
62. Zakrajsek, A.J.; Mall, S. The development and use of a digital twin model for tire touchdown health monitoring. In Proceedings of the 58th AIAA/ASCE/AHS/ASC Structures, Structural Dynamics, and Materials Conference, Grapevine, TX, USA, 9–13 January 2017; American Institute of Aeronautics and Astronautics Inc.: Reston, VA, USA, 2017. [CrossRef]
63. Xu, Z.; Ji, F.; Ding, S.; Zhao, Y.; Zhou, Y.; Zhang, Q.; Du, F. Digital twin-driven optimization of gas exchange system of 2-stroke heavy fuel aircraft engine. *J. Manuf. Syst.* **2021**, *58*, 132–145. [CrossRef]
64. Borgo, M.D.; Elliott, S.J.; Tehrani, M.G.; Stothers, I.M. Virtual Sensing of Wheel Position in Ground-Steering Systems for Aircraft Using Digital Twins. In *Model Validation and Uncertainty Quantification, Volume 3, Proceedings of the 38th IMAC, A Conference and Exposition on Structural Dynamics 2020, Houston, TX, USA, 10–13 February 2020*; Springer International Publishing: Berlin/Heidelberg, Germany, 2020; pp. 107–118. [CrossRef]
65. Hu, F. Digital Twin-Driven Reconfigurable Fixturing Optimization for Trimming Operation of Aircraft Skins. *Aerospace* **2022**, *9*, 154. [CrossRef]

66. Peng, C.C.; Chen, Y.H. Digital Twins-Based Online Monitoring of TFE-731 Turbofan Engine Using Fast Orthogonal Search. *IEEE Syst. J.* **2022**, *16*, 3060–3071. [CrossRef]
67. Li, C.; Mahadeven, S.; Ling, Y.; Wang, L.; Choze, S. A dynamic Bayesian network approach for digital twin. In Proceedings of the 19th AIAA Non-Deterministic Approaches Conference, Grapevine, TX, USA, 9–13 January 2017; American Institute of Aeronautics and Astronautics Inc.: Reston, VA, USA, 2017. [CrossRef]
68. Li, C.; MahaDeVan, S.; Ling, Y.; Choze, S.; Wang, L. Dynamic Bayesian network for aircraft wing health monitoring digital twin. *AIAA J.* **2017**, *55*, 930–941. [CrossRef]
69. Kosova, F.; Unver, H.O. A digital twin framework for aircraft hydraulic systems failure detection using machine learning techniques. *Proc. Inst. Mech. Eng. C J. Mech. Eng. Sci.* **2022**, *237*, 1563–1580. [CrossRef]
70. Laukotka, F.; Hanna, M.; Krause, D. Digital twins of product families in aviation based on an MBSE-assisted approach. *Procedia CIRP* **2021**, *100*, 684–689. [CrossRef]
71. Xiong, M.; Wang, H. Digital twin applications in aviation industry: A review. *Int. J. Adv. Manuf. Technol.* **2022**, *121*, 5677–5692. [CrossRef]
72. Lv, Z.; Xie, S. Artificial intelligence in the digital twins: State of the art, challenges, and future research topics. *Digit. Twin* **2022**, *1*, 12. [CrossRef]
73. Salinger, S.J.; Kapteyn, M.G.; Kays, C.; Pretorius, J.V.R.; Willcox, K.E. A Hardware Testbed for Dynamic Data-Driven Aerospace Digital Twins. In Proceedings of the Dynamic Data Driven Applications Systems: Third International Conference, DDDAS 2020, Boston, MA, USA, 2–4 October 2020; pp. 37–45. [CrossRef]
74. Kapteyn, M.G.; Knezevic, D.J.; Huynh, D.B.P.; Tran, M.; Willcox, K.E. Data-driven physics-based digital twins via a library of component-based reduced-order models. *Int. J. Numer. Methods Eng.* **2022**, *123*, 2986–3003. [CrossRef]
75. Kapteyn, M.G.; Knezevic, D.J.; Willcox, K.E. Toward predictive digital twins via component-based reduced-order models and interpretable machine learning. In Proceedings of the AIAA Scitech 2020 Forum, Orlando, FL, USA, 6–10 January 2020; American Institute of Aeronautics and Astronautics Inc.: Reston, VA, USA, 2020. [CrossRef]
76. Aláez, D.; Olaz, X.; Prieto, M.; Villadangos, J.; Astrain, J.J. VTOL UAV digital twin for take-off, hovering and landing in different wind conditions. *Simul. Model Pract. Theory* **2022**, *123*, 102703. [CrossRef]
77. Yang, Y.; Meng, W.; Li, H.; Lu, R.; Fu, M. A Digital Twin Platform for Multi-Rotor UAV. In Proceedings of the Chinese Control Conference, CCC, Shanghai, China, 26–28 July 2021; IEEE Computer Society: Piscataway, NJ, USA, 2021; pp. 7909–7913. [CrossRef]
78. Yang, Y.; Meng, W.; Zhu, S. A Digital Twin Simulation Platform for Multi-rotor UAV. In Proceedings of the 2020 7th International Conference on Information, Cybernetics, and Computational Social Systems, ICCSS 2020, Guangzhou, China, 13–15 November 2020; Institute of Electrical and Electronics Engineers Inc.: Piscataway, NJ, USA, 2020; pp. 591–596. [CrossRef]
79. Moorthy, S.K.; Harindranath, A.; McManus, M.; Guan, Z.; Mastronarde, N.; Bentley, E.S.; Medley, M. A Middleware for Digital Twin-Enabled Flying Network Simulations Using UBSim and UB-ANC. In Proceedings of the 18th Annual International Conference on Distributed Computing in Sensor Systems (DCOSS), Los Angeles, CA, USA, 30 May–1 June 2022; Institute of Electrical and Electronics Engineers Inc.: Piscataway, NJ, USA, 2022; pp. 322–327. [CrossRef]
80. Wu, J.; Guo, J.; Lv, Z. Deep Learning Driven Security in Digital Twins of Drone Network. In Proceedings of the IEEE International Conference on Communications, Seoul, Republic of Korea, 16–20 May 2022; Institute of Electrical and Electronics Engineers Inc.: Piscataway, NJ, USA, 2022; pp. 1–6. [CrossRef]
81. Shen, G.; Lei, L.; Li, Z.; Cai, S.; Zhang, L.; Cao, P.; Liu, X. Deep Reinforcement Learning for Flocking Motion of Multi-UAV Systems: Learn from a Digital Twin. *IEEE Internet Things J.* **2022**, *9*, 11141–11153. [CrossRef]
82. Lv, Z.; Chen, D.; Feng, H.; Zhu, H.; Lv, H. Digital Twins in Unmanned Aerial Vehicles for Rapid Medical Resource Delivery in Epidemics. In *IEEE Transactions on Intelligent Transportation Systems*; Institute of Electrical and Electronics Engineers Inc.: Piscataway, NJ, USA, 2021. [CrossRef]
83. Fraser, B.; Al-Rubaye, S.; Aslam, S.; Tsourdos, A. Enhancing the Security of Unmanned Aerial Systems using Digital-Twin Technology and Intrusion Detection. In Proceedings of the AIAA/IEEE Digital Avionics Systems Conference-Proceedings, San Antonio, TX, USA, 3–7 October 2021; Institute of Electrical and Electronics Engineers Inc.: Piscataway, NJ, USA, 2021. [CrossRef]
84. Kapteyn, M.G.; Pretorius, J.V.R.; Willcox, K.E. A probabilistic graphical model foundation for enabling predictive digital twins at scale. *Nat. Comput. Sci.* **2021**, *1*, 337–347. [CrossRef]
85. Riordan, J.; Manduhu, M.; Black, J.; Dow, A.; Dooly, G.; Matalonga, S. LiDAR Simulation for Performance Evaluation of UAS Detect and Avoid. In Proceedings of the International Conference on Unmanned Aircraft Systems (ICUAS), Athens, Greece, 15–18 June 2021; Institute of Electrical and Electronics Engineers Inc.: Piscataway, NJ, USA, 2021; pp. 1355–1363. [CrossRef]
86. Iqbal, D.; Buhnova, B. Model-based Approach for Building Trust in Autonomous Drones through Digital Twins. In Proceedings of the IEEE International Conference on Systems, Man, and Cybernetics (SMC), Prague, Czech Republic, 9–12 October 2022; IEEE: Piscataway, NJ, USA, 2022; pp. 656–662. [CrossRef]
87. Grigoropoulos, N.; Lalis, S. Simulation and Digital Twin Support for Managed Drone Applications. In Proceedings of the IEEE/ACM 24th International Symposium on Distributed Simulation and Real Time Applications (DS-RT), Prague, Czech Republic, 14–16 September 2020; IEEE: Piscataway, NJ, USA, 2020; pp. 1–8. [CrossRef]
88. Lee, E.B.K.; Van Bossuyt, D.L.; Bickford, J.F. Digital twin-enabled decision support in mission engineering and route planning. *Systems* **2021**, *9*, 82. [CrossRef]

89. Lei, L.; Shen, G.; Zhang, L.; Li, Z. Toward Intelligent Cooperation of UAV Swarms: When Machine Learning Meets Digital Twin. *IEEE Netw.* **2021**, *35*, 386–392. [CrossRef]

90. Wang, W.; Li, X.; Xie, L.; Lv, H.; Lv, Z. Unmanned Aircraft System Airspace Structure and Safety Measures Based on Spatial Digital Twins. *IEEE Trans. Intell. Transp. Syst.* **2022**, *23*, 2809–2818. [CrossRef]

91. de Miguel Molina, M. *Ethics and Civil Drones: European Policies and Proposals for the Industry*; Springer Briefs in Law: Berlin/Heidelberg, Germany, 2018. [CrossRef]

92. Megson, T.H.G. *Aircraft Structures for Engineering Students*, 5th ed.; No. 1.; Butterworth-Heinemann: Oxford, UK, 2012. [CrossRef]

93. European Union Aviation Safety Agency (EASA). 'Easy Access Rules for Unmanned Aircraft Systems (Regulation (EU) 2019/947 and Regulation (EU) 2019/945)'. Available online: https://www.easa.europa.eu/document-library/easy-access-rules/easy-access-rules-unmanned-aircraft-systems-regulation-eu (accessed on 28 September 2022).

94. Boschert, S.; Heinrich, C.; Rosen, R. Next generation digital twin. In *Proceedings of TMCE*; CFP: Palmas de Gran Canaria, Spain, 2018; pp. 209–218.

95. Custers, B. *The Future of Drone Use Opportunities and Threats from Ethical and Legal Perspectives*; TMC Asser Press Springer: Berlin/Heidelberg, Germany, 2016.

96. European Union Aviation Safety Agency (EASA). 'EASA Provisions: The Applicability Dates under EU Regulation 2019/947 and 2019/945'. Available online: https://www.easa.europa.eu/the-agency/faqs/drones-uas (accessed on 1 February 2023).

97. European Union Aviation Safety Agency (EASA). 'Civil Drones (Unmanned Aircraft)'. Available online: https://www.easa.europa.eu/domains/civil-drones (accessed on 1 February 2023).

98. European Union Aviation Safety Agency (EASA). 'Open Category of Civil Drones'. Available online: https://www.easa.europa.eu/domains/civil-drones/drones-regulatory-framework-background/open-category-civil-drones (accessed on 1 February 2023).

99. European Union Aviation Safety Agency (EASA). 'Specific Category of Civil Drones'. Available online: https://www.easa.europa.eu/domains/civil-drones/drones-regulatory-framework-background/specific-category-civil-drones (accessed on 1 February 2023).

100. European Union Aviation Safety Agency (EASA). 'Certified Category of Civil Drones'. Available online: https://www.easa.europa.eu/domains/civil-drones/drones-regulatory-framework-background/certified-category-civil-drones (accessed on 1 February 2023).

101. European Union Aviation Safety Agency (EASA). 'EASA Issues Guidelines for the Design Verification of Drones Operated in the Specific Category'. Available online: https://www.easa.europa.eu/newsroom-and-events/press-releases/easa-issues-guidelines-design-verification-drones-operated (accessed on 8 April 2021).

102. European Union Aviation Safety Agency (EASA). 'Easy Access Rules for Airworthiness and Environmental Certification (Regulation (EU) No 748/2012)'. Available online: https://www.easa.europa.eu/document-library/general-publications/easy-access-rules-initial-airworthiness (accessed on 31 March 2023).

103. Elisabeth Landrichter, The New EU Regulatory Framework for U-Space. European Civil Aviation Conference Bulletin on Unmanned Aircraft Systems. Available online: https://www.ecac-ceac.org/activities/unmanned-aircraft-systems/uas-bulletin/22-uas-bulletin/505-uas-bulletin-2-the-new-eu-regulatory-framework-for-u-space (accessed on 1 December 2021).

104. European Union Aviation Safety Agency (EASA). 'Commission Implementing Regulation (EU) 2021/664'. Available online: https://www.easa.europa.eu/en/document-library/regulations/commission-implementing-regulation-eu-2021664 (accessed on 29 April 2021).

105. European Union Aviation Safety Agency (EASA). 'Commission Implementing Regulation (EU) 2021/665'. Available online: https://www.easa.europa.eu/en/document-library/regulations/commission-implementing-regulation-eu-2021665-0 (accessed on 29 April 2021).

106. European Union Aviation Safety Agency (EASA). 'Commission Implementing Regulation (EU) 2021/666'. Available online: https://www.easa.europa.eu/en/document-library/regulations/commission-implementing-regulation-eu-2021666 (accessed on 29 April 2021).

107. European Union Aviation Safety Agency (EASA). 'The Difference between Autonomous and Automatic Drones'. Available online: https://www.easa.europa.eu/the-agency/faqs/regulations-uas-drone-explained (accessed on 2 February 2023).

108. Torens, C.; Jünger, F.; Schirmer, S.; Schopferer, S.; Maienschein, T.; Dauer, J.C. Machine Learning Verification and Safety for Unmanned Aircraft-A Literature Study. In Proceedings of the AIAA Scitech 2022 Forum, Reston, VA, USA, 3–7 January 2022.

109. European Union Aviation Safety Agency (EASA). 'EASA Artificial Intelligence Roadmap—A Human Centric Approach to AI in Aviation'. Available online: https://www.easa.europa.eu/newsroom-and-events/news/easa-artificial-intelligence-roadmap-10-published (accessed on 7 February 2020).

110. Torens, C.; Durak, U.; Dauer, J.C. Guidelines and Regulatory Framework for Machine Learning in Aviation. In Proceedings of the AIAA Scitech Forum, Reston, VA, USA, 3–7 January 2022.

111. Croatti, A.; Gabellini, M.; Montagna, S.; Ricci, A. On the Integration of Agents and Digital Twins in Healthcare. *J. Med. Syst.* **2020**, *44*, 161. [CrossRef]

112. Grieves, M.; Vickers, J. Digital twin: Mitigating unpredictable, undesirable emergent behavior in complex systems. In *Transdisciplinary Perspectives on Complex Systems—New Findings and Approaches*; Springer International Publishing: Berlin/Heidelberg, Germany, 2017; pp. 85–113. [CrossRef]

113. Feng, H.; Chen, Q.; de Soto, B.G. Application of digital twin technologies in construction: An overview of opportunities and challenges. In Proceedings of the 38th International Symposium on Automation and Robotics in Construction (ISARC), Dubai, United Arab Emirates, 2 November 2021; pp. 979–986. [CrossRef]
114. Glaessgen, E.H.; Stargel, D.S. The Digital Twin Paradigm for Future NASA and US. In Air Force Vehicles. In Proceedings of the 53rd AIAA/ASME/ASCE/AHS/ASC Structures, Structural Dynamics and Materials Conference, Honolulu, HI, USA, 26 April 2012. [CrossRef]
115. Sacks, R.; Brilakis, I.; Pikas, E.; Xie, H.S.; Girolami, M. Construction with digital twin information systems. *Data-Centric Eng.* **2020**, *1*, e14. [CrossRef]
116. Bolton, R.N.; McColl-Kennedy, J.R.; Cheung, L.; Gallan, A.; Orsingher, C.; Witell, L.; Zaki, M. Customer experience challenges: Bringing together digital, physical and social realms. *J. Serv. Manag.* **2018**, *29*, 776–808. [CrossRef]
117. Krishnan, M.S. Digital Twins. Available online: https://www.coursera.org/learn/digital-twins (accessed on 30 November 2022).
118. Pal, S.K.; Mishra, D.; Pal, A.; Dutta, S.; Chakravarty, D.; Pal, S. *Digital Twin—Fundamental Concepts to Applications in Advanced Manufacturing*; The Springer Series in Advanced Manufacturing; Springer: Berlin/Heidelberg, Germany, 2022.
119. Semanjski, I. *Smart Urban Mobility, Transport Planning in the Age of Big Data and Digital Twins*; Elsevier: Amsterdam, The Netherlands, 2023.
120. Semeraro, C.; Lezoche, M.; Panetto, H.; Dassisti, M. Digital twin paradigm: A systematic literature review. *Comput. Ind.* **2021**, *130*, 103469. [CrossRef]
121. AURORA Consortium, 'AURORA H2020 Project'. Available online: https://aurora-uam.eu/ (accessed on 28 February 2023).

Article

Toward Smart Air Mobility: Control System Design and Experimental Validation for an Unmanned Light Helicopter

Emanuele Luigi de Angelis[1],*, Fabrizio Giulietti [1], Gianluca Rossetti [2], Matteo Turci [2] and Chiara Albertazzi [3]

[1] Department of Industrial Engineering, CIRI Aerospace, University of Bologna, 47121 Forlí, Italy; fabrizio.giulietti@unibo.it
[2] Zephyr S.r.l., 47014 Meldola, Italy; gianluca.rossetti@zephyraerospace.com (G.R.)
[3] Curti Costruzioni Meccaniche S.p.A., 48014 Castel Bolognese, Italy
* Correspondence: emanuele.deangelis4@unibo.it; Tel.: +39-0543-374-453

Abstract: Light helicopters are used for a variety of applications, attracting users from private and public market segments because of their agility and convenient storage capabilities. However, most light helicopters on the market today are designed and manufactured with technologies dating back to the 1980s, with safety issues to be addressed by advanced design methods, more powerful engines, and innovative solutions. In this regard, the DISRUPT (Development of an innovative and safe ultralight, two-seater turbine helicopter) project, led by Curti Aerospace Division (Italy) and co-funded by the EU H2020 program, is a state-of-the-art concept for a novel ultralight helicopter equipped with a ballistic parachute. In order to validate the first parachute ejection in a safe scenario, a dronization process was selected as a viable solution to be performed in collaboration with the University of Bologna. In the present paper, the steps followed to transform the helicopter into an unmanned vehicle are detailed according to the model-based design approach, with particular focus on mathematical modeling, control system design, and experimental validation. Obtained results demonstrate the feasibility of using a civil helicopter first as a remotely-piloted vehicle and then as a highly-automated personal transportation system in the framework of smart and sustainable air mobility.

Keywords: urban air mobility; helicopter; parachute; model-based design; control system; flight testing

Citation: de Angelis, E.L.; Giulietti, F.; Rossetti, G.; Turci, M.; Albertazzi, C. Toward Smart Air Mobility: Control System Design and Experimental Validation for an Unmanned Light Helicopter. *Drones* **2023**, *7*, 288. https://doi.org/10.3390/drones7050288

Academic Editor: Pablo Rodríguez-Gonzálvez

Received: 20 March 2023
Revised: 19 April 2023
Accepted: 21 April 2023
Published: 25 April 2023

1. Introduction

The interest in Urban Air Mobility (UAM) had a step increase over the last few years [1]. On the one hand, the slow growth rate of ground infrastructure led to critical traffic congestion in urban areas. On the other hand, the increasing demand for moving people and payloads further and faster drove the attention of the research community and stakeholders toward the exploitation of the vertical dimension [2]. For example, Amazon and Google pioneered the testing of urban parcel delivery by means of multirotor aircraft [3,4]. In such a way, they paved the way for a wide range of studies on highly-automated low-altitude vehicles as an alternative means of transportation, where "the regular Joe" is capable of performing a mission without having the skills of a licensed pilot [5–7]. In this respect, two early attempts that investigated concepts of operation and technologies for a new personal transportation system based on both an aerial platform and a ground infrastructure were, respectively, PPlane (2009–2013) and myCopter (2011–2014), projects funded by the European Commission under the 7th Framework Program (FP7) [8,9].

By taking advantage of consolidated experience in conventional aviation, high reliability of onboard systems, and rapid improvement of electrical propulsion performance, manufacturers and transport stakeholders (such as Airbus, Volocopter, and Uber) investigated concepts for personal air transportation systems. With the aim of playing a lead role in this new raising market, they considered electric platforms with Vertical Take-Off and Landing (VTOL) capabilities as key elements for the next generation of controlled airspace [10,11].

Among all the above-mentioned projects and applications, it is acknowledged that a cost-effective solution to sustainable Urban Air Mobility and Delivery (UAMD) is represented by the use of small/light aircraft, where onboard flight control systems, supported by Air Traffic Management (ATM) technology, will provide safe navigation in dynamic scenarios and weather conditions in the presence of other sky users [12]. Transforming a conventional aircraft (both fixed and rotary-wing) into a Remotely–Piloted Aerial System (RPAS) may represent a successful strategy for different reasons. First of all, available light/ultralight conventional aircraft have already passed through several design, test, and certification steps with the aim of fulfilling reliability, performance, and flying quality requirements [13]. Moreover, reversible control chains can be easily replaced by Electro-Mechanical Actuators (EMA), controlled by dedicated onboard avionics. Starting from this design bias, researchers can thus focus on the design and experimental validation of all other technologies allowing for UAMD (including Guidance, Navigation, and Control (GNC) systems, telemetry, communication, and ATM devices) in addition to ground handling facilities. In this respect, thanks to their compact size and peculiar VTOL configuration, civil ultralight helicopters represent suitable test-beds for performing the transition toward a highly-automated personal transportation system.

By focusing on the very recent past, examples of the transition of conventional helicopters into RPASs can be dated back to 2004, when the Unmanned Little Bird demonstrator, derived by Boeing from a civil MD 530F, made its first autonomous flight (with a safety pilot). In particular, a pre-programmed 20-min armed intelligence, surveillance, and reconnaissance mission was performed around the United States Army's Yuma Proving Ground facility [14]. In 2006, Northrop Grumman introduced the MQ–8 Fire Scout unmanned helicopter family, obtained from Schweizer 333 and Bell 407, designed to provide reconnaissance, situational awareness, aerial fire, and precision targeting support for ground, air, and sea forces [15]. In 2008, an unmanned, highly-automated version of the Kaman K-MAX helicopter took its maiden flight, with the aim of operating in combat scenarios as well as in civilian situations involving chemical, biological, or radiological hazards [16]. Later on, Eurocopter launched a series of flights for a new rotary-wing solution designed to expand the mission capabilities of Eurocopter helicopters [17]. The Optionally–Piloted Vehicle (OPV) program, based on the EC145 helicopter platform (now Airbus Helicopters H145), was revealed during a demonstration flight: after an automatic takeoff, an EC145 flew a circuit via pre-programmed waypoints and performed a mid-route hover to deploy a load from the external sling. The EC145 continued on a return route segment representing a typical observation mission, followed by an automatic landing. Finally, Sikorsky demonstrated its OPV Matrix Technology on a modified S–76B helicopter called the Sikorsky Autonomy Research Aircraft (SARA). Since 2013, the program has made progress with more than 300 h of autonomous flight with the aim of improving decision-aiding for manned operations, while enabling both unmanned and reduced-crew operations [18].

This paper presents the results of a research work performed within DISRUPT (2016–2018), a collaborative project co-funded by the EU within the H2020 program and led by Curti Aerospace Division. Specifically, DISRUPT proposed a new light rotorcraft configuration, the two-seater Curti Zefhir helicopter, that features a turbine engine and an emergency ballistic parachute to respectively enhance flight performance and increase passenger safety (see Figure 1). PBS Velká Bíteš manufactures the turboshaft engine, derated from 160 to 105 kW of maximum continuous power. While ballistic parachutes have been certified on some fixed-wing aircraft, such as Cirrus light airplanes, their installation on helicopters is a challenging proposition due to the overhead presence of rotating blades. Contained in a non-rotating pod above the main rotor, the parachute solution proposed by Curti and Junkers ProFly thus becomes a backup for conditions where autorotation cannot be performed, such as (a) flight control failure or loss of maneuverability, (b) flying over an area where emergency landing cannot be safely performed, or (c) flight conditions that prevent restoring rotor rotation speed [19].

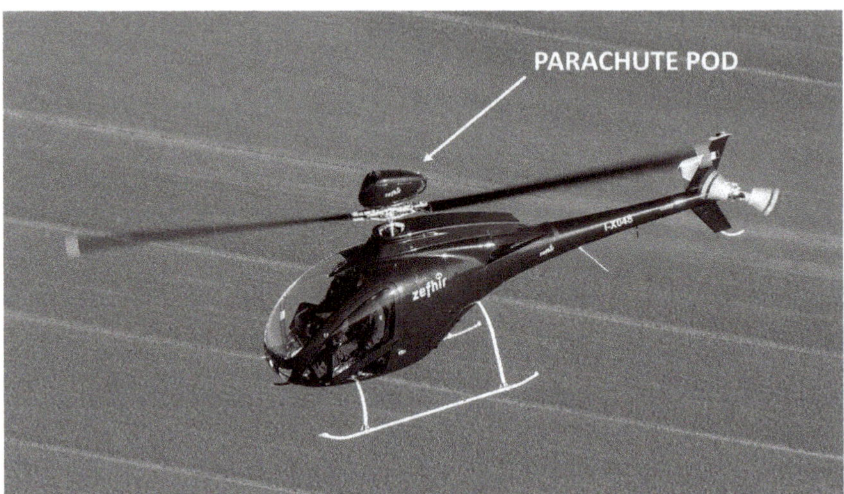

Figure 1. Zefhir helicopter (courtesy of Curti Aerospace Division).

Although the main objectives of DISRUPT were not strictly related to the main topics of UAM, the need for a remotely-piloted configuration arose immediately; since the experimental validation of the parachute system with the full-scale helicopter was one of the main expected results, the transition toward an unmanned configuration became a mandatory activity to perform the ejection test without a human pilot on board. A crucial but challenging step of the process was the design of a stabilization system, intended as a flexible and reliable software/hardware solution allowing the pilot to manage the ejection task while reducing the workload required by control action. Helicopters generally show nonlinear, complex dynamics that might manifest some unstable flight characteristics in limited zones of the flight envelope. In the particular case of a radio-controlled rotorcraft, without the direct perception of linear accelerations and attitude motion, the remote piloting of a helicopter is indeed an extremely hazardous task [20,21]. Hence, an Automatic Flight Control System (AFCS) was designed, tested, and implemented, allowing the pilot to safely control the aircraft in terms of desired attitude.

The main goal of the paper is to present for the first time a detailed description of all the phases allowing the successful transition of a conventional light helicopter into a RPAS while investigating the validity of a rescue system in the framework of future UAM applications. According to the Model-Based Design (MBD) philosophy, (1) mission requirements are listed and (2) system architecture is defined. Furthermore, (3) an accurate 6DOF nonlinear model is implemented in the Matlab/Simulink environment, which includes helicopter subsystems, environmental effects, and sensor and actuator behavior. (4) The mathematical model is validated and refined by using flight data collected during an identification campaign. (5) After the analysis of open loop dynamic modes, (6) an attitude control system allowing the remote pilot to easily control the aircraft is designed, implemented, and validated by means of both (7) Hardware–In–the–Loop (HIL) techniques and (8) flight tests.

The paper is structured as follows. Section 2 addresses the outline of mission requirements and the selection of system components. The entire simulation model, the trim and stability analysis, and the model validation procedure are presented in Sections 3 and 4, respectively. Control system design, implementation, and HIL validation are described in Section 5. Experimental results validating the AFCS performance and reporting the parachute recovery mission are finally summarized in Section 6. A section of concluding remarks ends this paper.

The successful outcome of the ejection test and the interest that has arisen in several journals and broadcast media prove the relevance of the research activity presented in this paper [22,23]. Zefhir is currently the only civil helicopter equipped with a ballistic parachute. Indeed, such a test has never been filmed or documented in the entire history of aerospace technology. However, due to the highly-classified nature of the data involved in the early stages of aircraft development, a detailed description of helicopter features and both numerical and experimental results is omitted in the present framework. The focus of the analysis is thus placed on the description of methodological aspects, with particular attention to both numerical and experimental validations, supported by results available in the literature. Furthermore, the comparison between experimental data and the results of simulations is possibly characterized in terms of relative errors, while the description of the technological setup is circumscribed to functional aspects. The uniqueness of the experiment and the absence of strict performance requirements finally vindicate the limits posed by the novelty of the proposed control approach. In this respect, the necessity to rapidly design a safe single-case ejection test necessarily restricts the degree of experimentation, driving the MBD workflow to focus on long-standing results in the field of PID control. Although the latter does not guarantee optimality, it takes advantage of (1) a reduced number of involved parameters; (2) simple implementation and low computational cost; (3) the possibility to perform dedicated flight tests aiming at characterizing the closed-loop dynamic behavior one axis at a time while evaluating the effects of single gain contribution; and (4) an intuitive sizing procedure, suitable for collaboration with the candidate pilot to pursue a set of prescribed handling qualities. Alternative control techniques, such as robust nonlinear and adaptive control that involve the stabilization of vehicle speed components, are currently under experimental validation by the authors, provided small-scale rotorcraft are adopted as test beds in the direction of safe, scalable, and high-performance air mobility and delivery scenarios [24].

2. Mission Requirements and System Architecture

2.1. Mission Requirements

Mission systems and subsystems are grouped into the ground segment and the flight segment:

- Ground segment or Ground Control Station (GCS): the complete set of ground-based systems used to control and monitor the flight segment. The main components include the human-machine interface, computer, telemetry, and aerials for the control, video, and data link to and from the unmanned vehicle.
- Flight segment: the helicopter is equipped with the necessary avionics to perform a remotely–piloted flight. The main components include sensors, actuators for rotor blade pitch angle control, an onboard computer, and aerials for the control, video, and data links to and from the ground segment.

The final mission is defined by the following phases (Figure 2):

1. Pre-flight checks: the systems involved in the mission are prepared and visually checked. The helicopter is placed on flat terrain at a safety distance from the GCS. The airfield is required to be clear of obstacles while the mission airspace is circumscribed by a radius of 5 km and a height of 500 m with respect to the GCS.
2. Avionics power-on: both the ground and the flight segment subsystems are activated. Telemetry data are received by the GCS, and software/hardware verification checks are performed. The pilot validates the correct actuation of control commands.
3. Engine start: the ignition procedure is started by the pilot's action and the turbine reaches the idle condition.
4. Take-off and climb: the helicopter takes-off and climbs out of ground effect at a controlled rate until reaching 300 m above the airfield.
5. Cruise: the helicopter is stabilized in steady level flight at about 30 kts.

6. Engine shutdown and parachute ejection: the pilot performs the termination procedure, which includes engine shutdown and parachute ejection.
7. Descent: the helicopter descends with a stabilized speed and lands within the prescribed area.

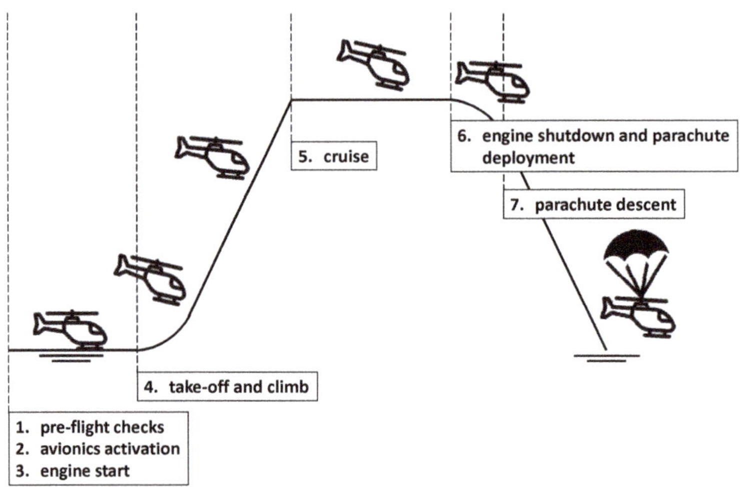

Figure 2. Mission phases definition.

The mission is performed in the visual line of sight. However, telemetry information needs to be available to both the GCS crew and the pilot. Given the intrinsic dynamic instability of the helicopter, fuselage attitude stabilization algorithms are required to assist the pilot throughout the mission profile. Conversely, no closed-loop control is applied to the MR collective pitch. The use of an independent Flight Termination System (FTS) is mandatory to stop the engine in case of emergency.

2.2. System Architecture

The selection of components for both the ground and flight segments is performed on the basis of mission requirements. At the same time, the MBD approach is adopted to define systems and subsystems as the result of an iterative process, where the making of a simulation model represents the core of control system development (see Sections 3 and 5). In what follows, the equipment list is presented, while the unmanned system layout is sketched in Figure 3.

The GCS is made of:

- the control module, where a modified commercial-off-the-shelf Radio Controller (RC) is used as a human-machine interface. Commands from the pilot, represented by stick deflections and switch activation inputs, are generated as Pulse-Width Modulated (PWM) signals and collected via the Pulse-Position Modulation (PPM) protocol. The PPM signal is finally provided to an integrated micro-controller board and output to the communication module according to serial protocol;
- the monitoring module, represented by a rugged laptop, where a graphical user interface is designed to display telemetry data, plan the mission, and send high-level commands via an Ethernet TCP/IP connection to a Real-Time Computer (RTC1) for data acquisition and processing;
- the communication module, which provides an RX/TX radio link to the flight segment. An ethernet switch is used to collect data from the monitoring module, while a ground-based radio modem is connected to a pair of 8 dBi 2.4 GHz directional patch antennas (respectively characterized by right-hand circular and vertical polarization).

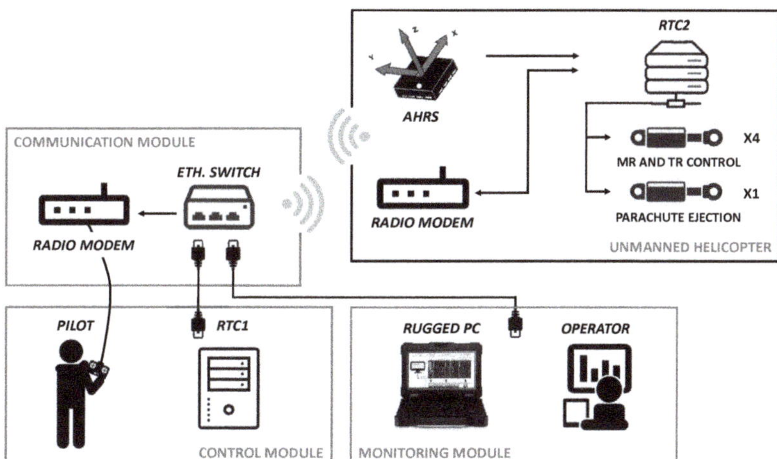

Figure 3. The unmanned system setup.

The helicopter is equipped with:

- a corresponding radio modem. Data are output via serial protocol and converted to a widespread standard industrial bus for communication with the Flight Management System (FMS);
- a Real-Time Computer (RTC2) performing FMS data acquisition and control tasks;
- a combined navigation and Attitude and Heading Reference System (AHRS) to estimate attitude information in a dynamic environment, along with position and velocity. Data are output via serial protocol and converted to a standard industrial bus for communication with the FMS;
- a set of 4 EMAs controls the collective, lateral, and longitudinal blade pitches of the MR and the collective pitch of the TR. An additional EMA is used for parachute deployment actuation. An FTS, based on a separate 868 MHz radio system, allows the Fuel Shut-Off Valve (FSOV) to close for emergency engine shutdown. The EMA and FTS selected for the experiment are devices available in the civil market.

3. System Modeling

Starting from the definition of reference frames, a 6 degrees-of-freedom model is adopted to represent the helicopter, with general expressions for the kinematics and dynamics of a rigid body with a center of gravity CG.

3.1. Reference Frames

Three right-handed orthogonal reference frames are introduced, according to the definitions in [25]:

1. an Earth-fixed North-East-Down frame, $\mathcal{F}_E = \{O_E; x_E, y_E, z_E\}$: the origin, O_E, is arbitrarily fixed to a point on the Earth's surface, x_E aims in the direction of the geodetic North, z_E points downwards along the Earth's ellipsoid normal, and y_E completes a right-handed triad. This frame is assumed to be inertial under the assumption of a flat and non-rotating Earth;
2. a Local Vertical-Local Horizontal frame, $\mathcal{F}_H = \{CG; x_H, y_H, z_H\}$: the origin is located at the vehicle's center of gravity, CG. Under the hypothesis of a flat Earth, $\mathcal{F}_H$ has axes parallel to $\mathcal{F}_E$;
3. a body-fixed frame, $\mathcal{F}_B = \{CG; x_B, y_B, z_B\}$: the x_B–axis is positive out the nose of the rotorcraft in its plane of symmetry, z_B is perpendicular to x_B in the same plane of symmetry, pointing downwards, and y_B completes a right-handed triad;

4. an aircraft reference frame, $\mathcal{F}_A = \{O_A; x_A, y_A, z_A\}$, used to locate CG and all helicopter components: axes are parallel to the body-fixed frame axes, such that $x_A = -x_B$, $y_A = y_B$, and $z_A = -z_B$. The origin is located ahead and below the rotorcraft at some arbitrary point within the plane of symmetry. Stations (ST) are measured positive aft along the longitudinal axis. Buttlines (BL) are lateral distances, positive to the pilot's right, and waterlines (WL) are measured vertically, positive upwards. A sketch of the rotorcraft, including the selected $\mathcal{F}_A$ frame, is reported in Figure 4. The positions of the main components, expressed in $\mathcal{F}_A$, are listed in Tables 1 and 2, together with relevant helicopter data.

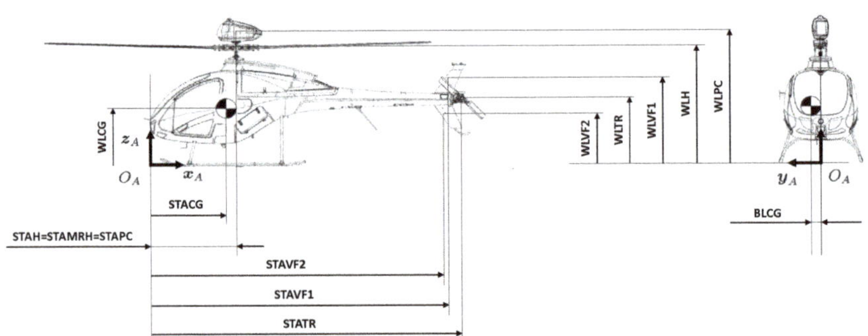

Figure 4. Sketch of Zefhir helicopter (courtesy of Curti Aerospace Division).

Table 1. MR and TR relevant parameters.

Parameter	Symbol	Computer Mnemonic	Value	Units
Main Rotor				
MR radius	R_{MR}	ROTOR	3.8	m
MR chord	c_{MR}	CHORD	0.195	m
MR rotational speed	Ω_{MR}	OMEGA	528.5	rpm
MR Lock number	γ_{MR}	GAMMA	4.25	-
MR hinge offset	ε	EPSLN	0	percent/100
MR flapping spring constant	K_β	AKBETA	0	N m/rad
MR tangent of δ_3	K_1	AKONE	0	-
MR solidity	σ_{MR}	SIGMA	0.0327	-
MR hub stationline	STA_H	STAH	2	m
MR hub buttline	BL_H	BLH	0	m
MR hub waterline	WL_H	WLH	2.4	m
Tail Rotor				
TR radius	R_{TR}	RTR	0.57	m
TR chord	c_{TR}	cTR	0.12	m
TR rotational speed	Ω_{TR}	OMTR	3 061.8	rpm
TR tangent of δ_3	K_{1TR}	FKITR	1	-
TR solidity	σ_{TR}	STR	0.0382	-
MR hub stationline	STA_{TR}	STATR	6.4	m
MR hub buttline	BL_{TR}	BLTR	−0.25	m
MR hub waterline	WL_{TR}	WLTR	1.34	m

Table 2. Fuselage, empennages, and miscellaneous components location.

Parameter	Symbol	Computer Mnemonic	Value	Units
Fuselage (Fus.)				
Fus. aerodynamic ref. point stationline	STA_{RPF}	STARPF	0	m
Fus. aerodynamic ref. point buttline	BL_{RPF}	BLRPF	0	m
Fus. aerodynamic ref. point waterline	WL_{RPF}	STARPF	0	m

Table 2. *Cont.*

Parameter	Symbol	Computer Mnemonic	Value	Units
Horizontal stabilizer (HS)				
HS stationline	STA_{HS}	STAHS	6.199	m
HS buttline	BL_{HS}	BLHS	0.435	m
HS waterline	WL_{HS}	WLHS	1.394	m
Upper vertical fin (VF1)				
VF1 stationline	STA_{VF1}	STAVF1	6.1	m
VF1 buttline	BL_{VF1}	BLVF1	0.052	m
VF1 waterline	WL_{VF1}	WLVF1	1.683	m
Lower vertical fin (VF2)				
VF2 stationline	STA_{VF2}	STAVF2	6.069	m
VF2 buttline	BL_{VF2}	BLVF2	0.048	m
VF2 waterline	WL_{VF2}	WLVF2	0.996	m
Main rotor hub (MRH)				
MRH stationline	STA_{MRH}	STAMRH	2	m
MRH buttline	BL_{MRH}	BLMRH	0	m
MRH waterline	WL_{MRH}	WLMRH	2.4	m
Parachute canopy (PC)				
PC stationline	STA_{PC}	STAPC	2	m
PC buttline	BL_{PC}	BLPC	0	m
PC waterline	WL_{PC}	WLPC	2.468	m

Let $s(\cdot) = \sin(\cdot)$, $c(\cdot) = \cos(\cdot)$. Vector transformation between $\mathcal{F}_H$ and $\mathcal{F}_B$ is provided by the rotation matrix [12]

$$
R(\alpha) = \begin{bmatrix}
c\theta\,c\psi & c\theta\,s\psi & -s\theta \\
s\phi\,s\theta\,c\psi - c\phi\,s\psi & s\phi\,s\theta\,s\psi + c\phi\,c\psi & s\phi\,c\theta \\
c\phi\,s\theta\,c\psi + s\phi\,s\psi & c\phi\,s\theta\,s\psi - s\phi\,c\psi & c\phi\,c\theta
\end{bmatrix}
\tag{1}
$$

obtained by a 3-2-1 Euler rotation sequence where $\alpha = [\phi,\, \theta,\, \psi]^T$ describes the attitude of the rotorcraft in terms of the classical 'roll', 'pitch', and 'yaw' angles, respectively. The following notation is adopted: if w is an arbitrary vector, its components are transformed from $\mathcal{F}_H$ to $\mathcal{F}_B$ through $w_B = R\,w_H$. In what follows, the subscript B will be dropped for simplicity.

3.2. Rigid Body Dynamics

Vehicle dynamics is described by Newton–Euler equations of motion projected in $\mathbb{F}_B$, namely:

$$
\dot{\mathbf{v}} = -\omega \times \mathbf{v} + \mathbf{F}/m
\tag{2}
$$

$$
\dot{\omega} = J^{-1}[-\omega \times (J\,\omega) + \mathbf{M}]
\tag{3}
$$

where $v = [u,\, v,\, w]^T$ is linear velocity, $\omega = [p,\, q,\, r]^T$ is angular velocity,

$$
J = \begin{bmatrix}
J_{xx} & -J_{xy} & -J_{xz} \\
-J_{xy} & J_{yy} & -J_{yz} \\
-J_{xz} & -J_{yz} & J_{zz}
\end{bmatrix}
\tag{4}
$$

is the inertia tensor about CG with respect to $\mathcal{F}_B$, and m is the total mass of the rotorcraft. $\mathbf{F} = [F_x,\, F_y,\, F_z]^T$ and $\mathbf{M} = [M_x,\, M_y,\, M_z]^T$ are the external force and moment vectors, respectively.

The external force acting on the rotorcraft is made of gravity, $F^{(g)}$, and aerodynamic, $F^{(a)}$, contributions. Taking into account Equation (1), gravity force vector expressed in the body frame is

$$F^{(g)} = R(\alpha) \begin{bmatrix} 0 \\ 0 \\ m\,g \end{bmatrix} = m\,g \begin{bmatrix} -\sin\theta \\ \sin\phi\,\cos\theta \\ \cos\phi\,\cos\theta \end{bmatrix} \tag{5}$$

where g is gravitational acceleration, described by means of WGS84 Taylor series model [26].

Rotorcraft attitude kinematics, that relates the generalized velocity $\dot{\alpha}$ and the angular velocity ω is given by [12]:

$$\dot{\alpha} = \begin{bmatrix} 1 & \sin\phi\,\tan\theta & \cos\phi\,\tan\theta \\ 0 & \cos\phi & -\sin\phi \\ 0 & \sin\phi/\cos\theta & \cos\phi/\cos\theta \end{bmatrix} \omega \tag{6}$$

while the position of the helicopter $p_E = [x_E, y_E, z_E]^T$, with components expressed in the inertial frame $\mathcal{F}_E$, is obtained from the equation:

$$\dot{p}_E = R(\alpha)^T v \tag{7}$$

3.3. Aerodynamic Forces and Moments

The characterization of aerodynamic force, $F^{(a)} = [X, Y, Z]^T$, and moment, $M^{(a)} = [L, M, N]^T$, is performed on the basis of the model detailed in [25], whose nomenclature is adopted in the present work. A conventional single MR helicopter with teetering configuration and counterclockwise rotation are considered. Contributions are provided by the main rotor (MR), tail rotor (TR), fuselage (F), horizontal stabilizer (HS), upper and lower vertical fins (VF1 and VF2), main rotor hub (MRH), and parachute canopy (PC). Air parameters are calculated from the International Standard Atmosphere (ISA) model as a function of rotorcraft altitude [27].

3.3.1. MR and TR Modeling

The following assumptions and simplifications are made about the MR model: (a) rotor blades are rigid in bending and torsion; (b) flapping angles are small, and the analysis follows the simple strip theory [28]; (c) the effects of aircraft motion on blade flapping are limited to those related to the angular accelerations $\dot{p}$ and $\dot{q}$, the angular rates p and q, and the normal acceleration component $\dot{w}$; (d) blade flow stall is disregarded; (e) rotor inflow is uniform, and no inflow dynamics is modeled; (f) main rotor blade flapping is approximated by the first harmonic terms with time-varying coefficients, that is

$$\beta(t) = a_0 - a_1 \cos\xi - b_1 \sin\xi \tag{8}$$

where a_0 is treated as a preset constant (coning angle) and ξ is blade azimuth. Coefficients $a_1(t)$ and $b_1(t)$ respectively represent the longitudinal and lateral tilt of the rotor tip–path plane, obtained as solutions to the equations in Appendix C of [25] with null hinge offset ratio, $\epsilon = 0$, flapping spring constant, $K_\beta = 0$, and pitch-flap coupling ratio, $\tan\delta_3 = 0$. Finally, the MR shaft is aligned with z_B.

The tail rotor is modeled according to a teetering configuration without cyclic pitch. Provided that the flapping frequency is typically much higher than that of the MR system, TR tip-path plane dynamics are neglected, no flapping spring constant is considered, and the pitch-flap coupling ratio, δ_{3TR}, is characterized by a non-null value (see Appendix D in [25]).

Contrary to some of the assumptions provided in [25], the blades of both MR and TR are characterized by cambered airfoils with a lift-curve slope $a < 2\pi$ 1/rad and a zero-lift angle of attack $\alpha_0 \neq 0$. The rotor blade profile drag coefficient, C_d, is calculated as

$$C_d = 0.008 + 0.3\left(\frac{6\,C_T}{\sigma\,a}\right)^2 + \Delta C_d \tag{9}$$

where C_T is the rotor thrust coefficient, σ is rotor solidity, and ΔC_d is the extra drag coefficient determined by flow compressibility effects. Let M_{90} be the Mach number evaluated at the tip of the advancing blade, where $\xi = 90$ deg. In order to estimate the extra drag, the approximate model proposed by Prouty and described in [28] is adopted, where

$$\Delta C_d(M_{90}) = \begin{cases} 12.5(M_{90} - M_{dr})^3 & \text{for } M_{90} \geq M_{dr} \\ 0 & \text{otherwise} \end{cases} \tag{10}$$

and $M_{dr} = 0.74$ is the drag-rise Mach number. With respect to the characterization of rotor inflow, a number of non-ideal effects are considered, based on the approach in [28], for the characterization of forces and moments. A constant tip-loss factor $B < 1$ is adopted to account for blade tip losses. Other non-ideal effects, including nonuniform inflow, wake swirl and contraction, and blade interference, are accounted for by an induced power factor k_i, assumed to be a constant. The MR in-ground effect is provided by the model in ref. [29], and the inflow iterative scheme is solved according to Halley's method with a damping coefficient equal to 0.01 [30].

Cockpit/RC control of MR is provided by pilot commands in terms of lateral cyclic δ_a, longitudinal cyclic δ_e, and collective δ_c. All commands are expressed in terms of non-dimensional variables, such that $\delta_a \in [-1, +1]$ (positive direction: right to generate $L > 0$), $\delta_e \in [-1, +1]$ (positive direction: aft to generate $M > 0$), and $\delta_c \in [-1, +1]$ (positive direction: up to generate $Z < 0$). Onboard control of the tail rotor is performed by pedal commands, expressed as $\delta_p \in [-1, +1]$ (positive direction: right pedal forward to generate $N > 0$). The transformation of pilot commands into blade pitch angles is provided by a set of low-order polynomial functions, $A_{1s} = C_1(\delta_a)$, $B_{1s} = C_2(\delta_e)$, $\theta_0 = C_3(\delta_c)$, and $\theta_{0TR} = C_4(\delta_p)$, provided by the manufacturer. A_{1s} and B_{1s}, respectively, represent the lateral and longitudinal cyclic pitch angles measured from the MR hub plane in $\mathcal{F}_B$. Rotor blades are modeled with a linear twist, such that θ_0 is the blade collective pitch ideally extrapolated to the rotor center and θ_{tw} is the total blade twist angle (tip minus root pitch angle). No twist characterizes TR blades, where collective pitch is identified by θ_{0TR}.

An additional degree of freedom is related to the power plant made of free turbines, MR, and TR transmissions. In particular, MR and TR rotational speeds vary according to the current torque requirements and the engine power available. Changes in speed cause the free turbine governor to vary fuel flow to change the available power and maintain the desired angular rate. The engine dynamic model is found in [25]. For the sake of brevity, details are not provided in the present paper. Modeling parameters in terms of maximum available power, engine dynamics, specific fuel consumption, and mechanical transmission efficiency are provided by the manufacturer.

3.3.2. Fuselage, Empennages, and Miscellaneous Components

With respect to fuselage aerodynamics, it is assumed that longitudinal forces and moments are dependent on fuselage angle of attack and lateral forces and moments are dependent on angle of sideslip. The exception is the drag force, which is assumed to have a contribution from both angles of attack and sideslip. The modeling is based on a low- and a high-angle representation of forces and moments, according to Appendix F in [25], with data obtained through a detailed computational fluid dynamics characterization. Phasing between the two approximations is performed by means of cubic spline interpolation, with improved performance with respect to the proposed linear transition.

The modeling of the two vertical empennages and of the horizontal stabilizer also follows the approach in [25]. The aerodynamics of the MR hub and parachute pod are assessed by the equivalent flat plate area model. As an example, the force vector generated by MRH is expressed as:

$$\boldsymbol{F}_{MRH} = -\frac{1}{2}\rho\big([A_{xMRH}, A_{yMRH}, A_{zMRH}]\boldsymbol{V}_{MRH}\big)\boldsymbol{V}_{MRH} \tag{11}$$

where $V_{MRH} = [u_{MRH}, v_{MRH}, w_{MRH}]^T$ is the velocity, relative to the air mass, of the main rotor hub and includes the contribution of MR downwash, according to [31]. A_{xMRH}, A_{yMRH}, and A_{zMRH} are the equivalent flat plate drag areas, respectively orthogonal to x_B, y_B, and z_B. The moment generated by F_{MRH} about CG is given by $M_{MRH} = d_{MRH} \times F_{MRH}$, where

$$d_{MRH} = \begin{bmatrix} STA_{CG} - STA_{MRH} \\ BL_{MRH} - BL_{CG} \\ WL_{CG} - WL_{MRH} \end{bmatrix} \tag{12}$$

is the vector directed from CG to MRH position, assumed to be coincident with its center of pressure, with constant components expressed in $\mathcal{F}_B$.

4. Trim and Stability Analysis

The nonlinear model described in Section 3 is implemented in the Matlab/Simulink environment, where differential equations are solved by the Dormand-Prince ode8 method with a frequency of 1000 Hz [32]. In what follows, (1) the trim conditions are determined for different cruise speeds, (2) a linearization procedure is applied to the complete model about such equilibria, and (3) an open-loop dynamic analysis is performed to investigate the helicopter control and stability properties.

4.1. Trim Analysis

The helicopter model is numerically trimmed for straight-and-level flight at $h = 50$ m in standard atmospheric conditions. Different values of forward speed are considered, ranging from 0 km/h (hover) to 180 km/h (approximately the never-exceed speed), with steps of 5 km/h. For the sake of brevity, the results of both the static and the following dynamic analysis are summarized only for the hovering condition, for which dedicated flight tests were performed for validation purposes.

The main results of trim analysis for the hovering condition are given in Table 3 and compared with the data available from flight tests performed with the same vehicle configuration (deviations with respect to measured data are reported in terms of absolute values of percentage errors). To this end, the helicopter was equipped with a set of sensors, including: (a) potentiometers for cockpit command acquisition and blade pitch measurement, (b) torque–meters for MR and TR torque analysis, and (c) a AHRS providing rigid body attitude, angular rate, acceleration, speed, and position information.

According to Table 3, good agreement is found between predicted and measured values, showing the validity of the modeling approach. A major difference characterizes the longitudinal cyclic pitch, with a 42% error. It must be noted that a degree of uncertainty characterizes the knowledge of CG position (especially the STA_{CG} parameter) in the actual flight configuration, which is estimated by means of CAD analysis and suspension techniques. Uncertainty also characterizes the aerodynamics of the fuselage, especially in the case of hovering and low-speed forward flight, where MR wake envelops a large portion of the fuselage. For the aim of the present analysis, the model adopted for both MR inflow and fuselage aerodynamics necessarily represents a compromise solution, which allows for satisfactory accuracy in terms of the dynamic characterization of rotorcraft without the cost of excessively-complex aerodynamic models.

The match expected at hover between MR cyclic pitch angles and flapping coefficients, namely $A_{1s} = b_1$ and $B_{1s} = -a_1$, holds almost exactly in Table 3. Slight differences occur for the simulated hover condition, which is actually obtained by flying the helicopter at a residual forward speed of 0.1 m/s. With respect to the experimental campaign, effective environmental conditions were also monitored, provided the helicopter was maintained in upwind hover while estimating a maximum wind speed of 20 km/h.

Table 3. Trim analysis for the hovering flight.

Parameter	Symbol	Value	Units	Est. Error $|\cdot|$
Main Rotor				
Long. first–harmonic flapping coeff.	a_1	2.92	deg	N/A
Lat. first–harmonic flapping coeff.	b_1	-1.10	deg	N/A
Induced speed	v_i	7.79	m/s	N/A
Aerodynamic torque	Q	1 579.5	Nm	3.1%
Tail Rotor				
Long. first–harmonic flapping coeff.	a_{1TR}	0.24	deg	N/A
Lat. first–harmonic flapping coeff.	b_{1TR}	-0.24	deg	N/A
Induced speed	v_{iTR}	11.64	m/s	N/A
Aerodynamic torque	Q_{TR}	26.1	Nm	4.4%
Fuselage				
Roll angle	ϕ	-2	deg	17.6%
Pitch angle	θ	-1.96	deg	6.7%
Control Pitch Angles				
MR lat. cyclic pitch	A_{1s}	-1.10	deg	4.8%
MR lon. cyclic pitch	B_{1s}	-2.91	deg	42.0%
MR collective pitch	θ_0	12.62	deg	2.1%
TR collective pitch	θ_{0TR}	8.28	deg	2.1%

4.2. Dynamic Analysis

Consider the equations of motion introduced in Section 3 and detailed in [25]. In nonlinear form, it is

$$\dot{x} = f(x, u, t) \tag{13}$$

provided x is rigid–body state vector, namely

$$x = [u,\ w,\ q,\ \theta,\ v,\ p,\ \phi,\ r]^T \tag{14}$$

while time evolution of x_E, y_E, z_E, and ψ is not accounted in the framework of system linearization. Control vector u has four components, expressed in terms of pilot commands as:

$$u = [\delta_c,\ \delta_e,\ \delta_a,\ \delta_p]^T \tag{15}$$

Using small perturbation theory [29], helicopter motion is described in terms of perturbation from the equilibrium condition, $x_e = [U_e, W_e, Q_e, \Theta_e, V_e, P_e, \Phi_e, R_e]^T$ and $u_e = [U_{1e}, U_{2e}, U_{3e}, U_{4e}]^T$, written in the form $x = x_e + \delta x$ and $u = u_e + \delta u$. By following the approach and the nomenclature of [29], given the trim conditions in Section 4.1, the model in Equation (13) is linearized at all considered speeds to respectively obtain system and input matrices

$$A = \left(\frac{\partial f}{\partial x}\right)_{x_e, u_e}, \quad B = \left(\frac{\partial f}{\partial u}\right)_{x_e, u_e} \tag{16}$$

as a function of aerodynamic derivatives. The latter are estimated by numerical differencing in the Matlab/Simulink environment [29]. To this end, aerodynamic forces and moments are positively perturbed by each of the state and input vector components in turn, with amplitude equal to 0.02 (respectively intended in terms of m/s for u, v, w, rad/s for p, q, r, rad for ϕ, θ, and non-dimensional units for control inputs). State and control derivatives are written in the form:

$$X_u = \frac{1}{m}\frac{\partial X}{\partial u} \tag{17}$$

and

$$L'_p = \frac{J_{zz}}{J_{xx}J_{zz} - J_{xz}^2}\frac{\partial L}{\partial p} + \frac{J_{xz}}{J_{xx}J_{zz} - J_{xz}^2}\frac{\partial N}{\partial p} \tag{18}$$

$$N'_r = \frac{J_{xz}}{J_{xx}J_{zz} - J^2_{xz}}\frac{\partial L}{\partial r} + \frac{J_{xx}}{J_{xx}J_{zz} - J^2_{xz}}\frac{\partial N}{\partial r} \tag{19}$$

A total of 36 stability derivatives and 24 control derivatives are determined in the standard 6DOF representation for each flight condition. Due to the highly-classified nature of the data involved in the project, only one sample derivative is analyzed in the present paper at hover. A qualitative discussion about the behavior of the most significant derivatives is provided in what follows.

The effect of linear velocity on aerodynamic forces is principally taken into account by X_u, Y_v, and Z_w. The force damping derivatives $X_u < 0$ and $Y_v < 0$, which respectively reflect the drag and side force on rotor–fuselage combination, steadily increase in absolute value and are practically linear with speed beyond 50 km/h. At low speed, the effect of disc tilt following perturbations in u and v becomes predominant. Similar considerations hold for the heave damping derivative Z_w, which is mostly influenced by the fuselage and horizontal empennage in high-speed flight. At low speed, the MR tends to dominate Z_w through a reduction in C_T determined by a vertical speed perturbation. In order to validate the numerical linearization routine, a comparison is performed with the analytical results obtained for the stability and control derivatives according to formulas available in the literature. As an example, the MR contribution only to Z_w can be analytically estimated as [29]

$$Z_w = -\frac{\rho(\Omega R)\pi R^2}{m}\frac{\partial C_T}{\partial \mu_z} \tag{20}$$

where $\mu_z = w/(\Omega R)$ is MR climb ratio and

$$\frac{\partial C_T}{\partial \mu_z} \approx \frac{2 a \sigma |\lambda|}{16|\lambda| + a\sigma} \tag{21}$$

Based on the data in Tables 1 and 3, it is $\lambda = -0.0371$ at hover, such that $\partial C_T/\partial \mu_z \approx 0.018$. It follows $Z_w \approx -0.317$ 1/s, which is close to the value numerically obtained in the same condition for the full helicopter, namely -0.345 1/s. In such a case, the estimation error obtained according to literature results is -8.1%, provided that fuselage and appendages contributions are disregarded.

The speed stability effect is observed in $M_u > 0$ and $L'_v < 0$, the latter showing a practically linear behavior with speed. $M_w > 0$ is representative of the incidence static stability effect, which increases non-monotonically with speed and approximately tracks M_u, being influenced by MR inflow on helicopter components. Finally, $N'_v > 0$ accounts for the weathercock effect by means of TR and vertical fins (stabilizing with speed) and the fuselage (destabilizing).

The damping derivatives $L'_p < 0$, $M_q < 0$, and $N'_r < 0$ reflect short-term, small, and moderate-amplitude handling characteristics. If, on the one hand, L'_p and M_q principally account for MR flapping motion in the presence of roll and pitch rate perturbations, N'_r is dominated by loads on TR and vertical fins, with a stronger yaw-damping effect at high forward speeds.

Given the stability and control derivatives obtained above, the complete system and input matrices A and B are generated according to the structure provided on page 277 in [29]. Note that, with the idea of designing closed-loop control systems, the input matrix B is configured for application to the non-dimensional pilot commands. The formulation in terms of blade pitch control angles is however possible by means of the mapping functions C_1, C_2, C_3, and C_4 introduced in Section 3.3.1.

For the aim of the present work, however, the results of a decoupled analysis are first discussed. Based on the approximate separation between the longitudinal and the lateral-directional dynamics, the decoupled representation is available in ref. [29], where input matrices are applied to blade pitch control angles. The longitudinal dynamics are described by the forced system:

$$\frac{d}{dt}\begin{bmatrix} u \\ w \\ q \\ \theta \end{bmatrix} = \underbrace{\begin{bmatrix} X_u & X_w & X_q - W_e & -g\cos\Theta_e \\ Z_u & Z_w & Z_q + U_e & -g\sin\Theta_e \\ M_u & M_w & M_q & 0 \\ 0 & 0 & 1 & 0 \end{bmatrix}}_{\boldsymbol{A}_{lon}}\begin{bmatrix} u \\ w \\ q \\ \theta \end{bmatrix} + \underbrace{\begin{bmatrix} X_{\theta_0} & X_{B_{1s}} \\ Z_{\theta_0} & Z_{B_{1s}} \\ M_{\theta_0} & M_{B_{1s}} \\ 0 & 0 \end{bmatrix}}_{\boldsymbol{B}_{lon}}\begin{bmatrix} \theta_0 \\ B_{1s} \end{bmatrix}$$

(22)

A pair of complex-conjugate poles is determined from $\boldsymbol{A}_{lon}$, which is related to an unstable phugoid mode with natural frequency ω_{ph} and time constant τ_{ph} (calculated as the reciprocal of the real part of the poles in its absolute value). Two real stable modes are also evaluated, namely heave and pitch subsidence effects. The first pole, identified by $p_{hv} < 0$, is practically determined by the vertical damping derivative Z_w. The second pole, $p_{ps} < 0$, accounts for the fundamental contribution of both Z_w and M_q and is characterized by a time constant approximately estimated as $\tau_{ps} \approx -1/(Z_w + M_q)$ [29]. The decoupled lateral-directional dynamics are defined by the system:

$$\frac{d}{dt}\begin{bmatrix} v \\ p \\ r \\ \phi \end{bmatrix} = \underbrace{\begin{bmatrix} Y_v & Y_p + W_e & Y_r - U_e & g\cos\Phi_e\cos\Theta_e \\ L'_v & L'_p & L'_r & 0 \\ N'_v & N'_p & N'_r & 0 \\ 0 & 1 & \cos\Phi_e\tan\Theta_e & 0 \end{bmatrix}}_{\boldsymbol{A}_{lat}}\begin{bmatrix} v \\ p \\ r \\ \phi \end{bmatrix} + \underbrace{\begin{bmatrix} Y_{A_{1s}} & Y_{\theta_{0TR}} \\ L'_{A_{1s}} & L'_{\theta_{0TR}} \\ N'_{A_{1s}} & N'_{\theta_{0TR}} \\ 0 & 0 \end{bmatrix}}_{\boldsymbol{B}_{lat}}\begin{bmatrix} A_{1s} \\ \theta_{0TR} \end{bmatrix}$$

(23)

A pair of complex-conjugate poles is derived from $\boldsymbol{A}_{lat}$ with the natural frequency ω_{dr}. Such poles characterize the dutch-roll mode, which is unstable but slowly develops with a time constant τ_{dr}. The roll subsidence mode, mostly determined by the damping derivative L'_p, is related to the real pole $p_{roll} < 0$. The spiral subsidence mode at hover is stable, $p_{spiral} < 0$, and dampens with a time constant τ_{spiral}.

The analysis of coupled representation behind state matrix A is also considered, and the obtained poles are marked by a superscript 'c'. A comparison with the corresponding values derived through the decoupled analysis is provided where possible. Two real poles are first extracted. The roll subsidence effect is recognized in the first pole, $p_1 = 0.85 \cdot p_{roll}$, provided $p_1 \approx L'_p$. The same consideration holds for vertical damping mode, identified by $p_8^{(c)} \approx p_{hv} \approx Z_w$. Three pairs of complex-conjugate poles complete the analysis. The first pair, $p_{2,3}^{(c)}$, is stable with a real part proportional to $M_q + Z_w$. It is representative of a damped oscillation with a natural frequency $\omega_{py}^{(c)}$ and time constant $\tau_{py}^{(c)}$, determined by the coupling of pitch and yaw subsidence modes. The second pair, $p_{4,5}^{(c)}$, characterizes the unstable phugoid mode, which develops with a time constant $\tau_{ph}^{(c)} = 0.75 \cdot \tau_{ph}$ and shows natural frequency $\omega_{ph}^{(c)} = 1.38 \cdot \omega_{ph}$. The last pair of complex poles, $p_{6,7}$ characterizes the dutch roll motion, which is unstable and develops with natural frequency $\omega_{dr}^{(c)} = 0.92 \cdot \omega_{dr}$ and time constant $\tau_{dr}^{(c)} = 1.07 \cdot \tau_{dr}$.

4.3. Model Validation

In Section 4, a comparison is provided between simulated and measured variables regarding the static characterization of hovering conditions. In what follows, predicted dynamic properties about the same equilibrium are validated through identification methods [33]. To this end, flight data are collected and eventually filtered after performing frequency sweep maneuvers about the hover, according to the approach described in [34]. The frequency response for each selected input–output pair is then identified during an optimization process driven by the difference between the computed and the predicted frequency responses. The fidelity of the model is finally established using time domain verification, according to which time response predicted by the identified model is compared with the response recorded during flight tests.

Different maneuvers and data pairs are considered for the identification of transfer functions, such as $p(s)/A_{1s}(s)$ and $q(s)/B_{1s}(s)$, with the aim of validating the predicted dynamic information. For the sake of brevity, the adopted procedure is detailed for the characterization of the heave subsidence mode, whose dominant derivative Z_w is discussed above. In particular, the first input-output data pair describes the effect of MR collective pitch angle θ_0 on vertical acceleration, $a_z = \dot{w}$, expressed in a body-fixed frame. A detail of the data taken into account for such an identification procedure is reported in Figure 5. The predicted transfer function as obtained from the state-space representation in Equation (22) is:

$$\left.\frac{a_z(s)}{\theta_0(s)}\right|_{mdl} = \frac{Z_{\theta_0}\, s\,(s - z_1)(s - z_2)(s - z_3)}{(s - p_{hv})(s - p_{ps})\left(s^2 - 2/\tau_{ph}\, s + \omega_{ph}^2\right)} \tag{24}$$

where a set of 4 zeros is determined. The first one is located at the origin, $z_1 = 0.9983\, p_{ps}$ is real negative, and z_2, z_3 are a complex–conjugate pair such that $z_2\, z_3 = 1.008\, \omega_{ph}^2$ and $z_2 + z_3 = 1.059 \cdot 2/\tau_{ph}$. Heave subsidence mode evidently dominates the motion along z_B, provided that almost perfect pole-zero cancellation characterizes the terms depicted in gray color. It follows:

$$\left.\frac{a_z(s)}{\theta_0(s)}\right|_{mdl} \approx \frac{Z_{\theta_0}\, s}{s - p_{hv}} \tag{25}$$

Numerical identification is performed by using a Prediction Error Minimization (PEM) method focused on simulation [35], provided the transfer function in Equation (24) is assumed as the initial guess model. The identified transfer function is:

$$\begin{aligned}
\left.\frac{a_z(s)}{\theta_0(s)}\right|_{id} &= \frac{0.9964\, Z_{\theta_0}(s - \pi_1)(s - \pi_2)}{(s - 1.0421\, p_{hv})(s - 0.9912\, \pi_2)} \\
&\quad \cdot \frac{\left(s^2 + \pi_3\, s + \pi_4\right)}{\left(s^2 + 1.0090\, \pi_3\, s + 0.9964\, \pi_4\right)} \\
&\approx \frac{0.9964\, Z_{\theta_0}\, s}{(s - 1.0421\, p_{hv})}
\end{aligned} \tag{26}$$

where pole–zero cancellation can evidently be performed for the gray terms. It must be noted that $\pi_1 \approx 0$, such that the zero at the origin is also recovered. The estimation error between model-predicted and identified parameters is provided in the second line of Equation (26), where the updated values of $\left|Z_{\theta_0}\right|$ and $\left|p_{hv}\right|$, respectively, result in being 0.36% smaller and 4.21% bigger than the model-predicted ones in Equation (25). A sample comparison between measured and refined-simulation data after heave subsidence mode characterization is finally provided in Figure 6 for the acceleration.

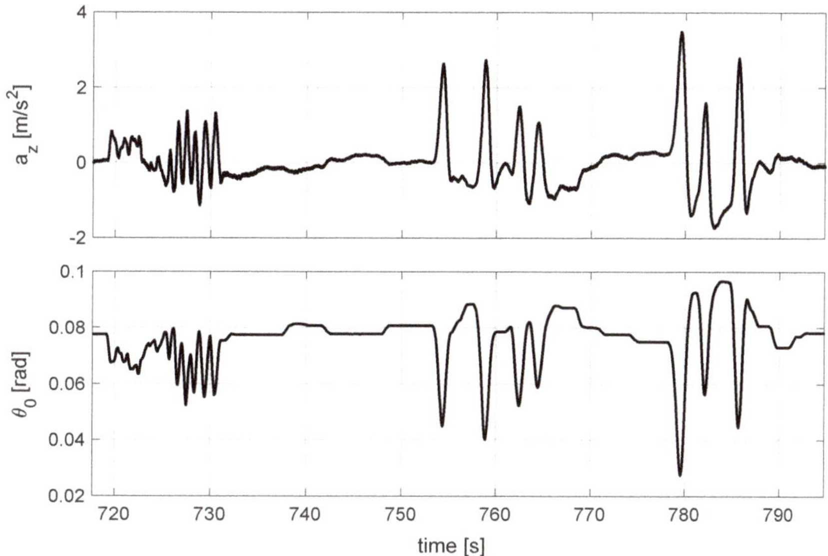

Figure 5. The input–output measured data used for heave subsidence mode characterization near hover (detail).

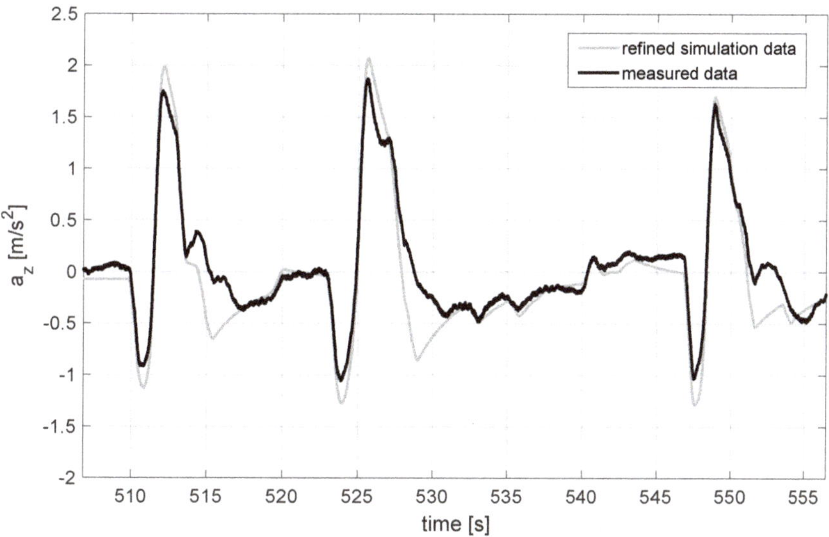

Figure 6. Measured and simulated data after heave subsidence mode characterization at hover (detail).

Encouraging results are indeed obtained for other input-output pairs, thus validating the modeling approach. In all cases, in fact, very good agreement is found between the dynamic properties obtained through numerical simulation and identification techniques.

5. Control System Design and Test

In what follows, the control system design phase is described based on the mathematical model in Section 3 and the analysis performed in Section 4. The closed-loop system is first analyzed by Model-In-the-Loop (MIL) simulations, where linear controllers are directly designed and validated in the nonlinear framework by means of an extensive campaign of simulations performed in collaboration with the candidate pilot. HIL tests are then performed to refine the control gains and validate the software/hardware setup [36].

5.1. Model–in–the–Loop Validation

Pilot commands, here named $\delta_a^{(pilot)}$, $\delta_e^{(pilot)}$, $\delta_c^{(pilot)}$, and $\delta_p^{(pilot)}$ are an input to the control system and follow the same convention described in Section 3.3.1. According to the given requirements, no closed-loop control is designed for MR collective pitch, such that $\delta_c \equiv \delta_c^{(pilot)}$.

In the framework of control system design, simulation models for selected AHRS and actuators are also developed. Modeling parameters in terms of accuracy and performance are obtained from both datasheets and dedicate experiments performed in laboratory facilities.

The first controller is designed to stabilize yaw motion by the actuation of TR collective pitch angle θ_{0TR} through the closed-loop feedback of yaw rate r. Let $e_r = \xi_r \delta_p^{(pilot)} - r$ be the error between the desired and the measured angular rate, provided that $\xi_r > 0$ is a prescribed constant that transforms the non-dimensional command provided by the remote pilot into the desired yaw rate. The control scheme is described by the equation:

$$\delta_p = k_p^{(r)} e_r + k_i^{(r)} \int_0^t e_r(s)\, ds \tag{27}$$

where $k_p^{(r)} > 0$ and $k_i^{(r)} > 0$ are control gains, respectively, providing proportional and integral contributions related to the error signal $e_r(t)$.

The second controller is used to stabilize the fuselage's attitude in terms of roll and pitch angles by the actuation of MR lateral and longitudinal cyclic control angles, respectively. With respect to roll angle stabilization, it is:

$$\delta_a = k_p^{(\phi)} e_\phi + k_i^{(\phi)} \int_0^t e_\phi(s)\, ds + k_d^{(\phi)} p \tag{28}$$

where $k_p^{(\phi)} > 0$ and $k_i^{(\phi)} > 0$. A derivative–like contribution is also provided by the direct feedback of roll rate p through the gain $k_d^{(\phi)} < 0$. The error between desired and measured roll angle is calculated as $e_\phi = \xi_\phi \delta_a^{(pilot)} - \phi$, where $\xi_\phi > 0$ is a prescribed constant. Controller structure for the stabilization of pitch angle follows the same approach, namely:

$$\delta_e = k_p^{(\theta)} e_\theta + k_i^{(\theta)} \int_0^t e_\theta(s)\, ds + k_d^{(\theta)} q \tag{29}$$

where $k_p^{(\theta)} > 0$, $k_i^{(\theta)} > 0$, and $k_d^{(\theta)} < 0$. The error between desired and measured pitch angle is $e_\theta = \xi_\theta \delta_e^{(pilot)} - \theta$, where $\xi_\theta > 0$.

In Figures 7 and 8 the results of a sample maneuver are reported. Simulation is started at $h = 50$ m with null attitude of the helicopter ($\phi_0 = \theta_0 = \psi_0 = 0$ deg) and an initial angular rate about the yaw axis, such that $p_0 = q_0 = 0$ deg/s and $r_0 = -5$ deg/s. MR collective pitch angle is kept constant and equal to the value obtained in Table 3 for the hovering condition, namely $\theta_0 = 12.62$ deg, corresponding to $\delta_c^{(pilot)} = 0.495$. Let $\xi_r = 40 \cdot \pi/180$ rad/s, $\xi_\phi = 25 \cdot \pi/180$ rad, and $\xi_\theta = 12 \cdot \pi/180$ rad. Input values to the controllers are $\delta_p^{(pilot)} = 0$, $\delta_a^{(pilot)} = -0.08$, and $\delta_e^{(pilot)} = -0.163$, which respectively provide the desired values $\phi = -2$ deg, $\theta = -1.96$, and $r = 0$ deg/s necessary to hover. In Figure 7, state variables describing fuselage attitude are plotted as a function of time, showing the stabilizing effect of implemented controllers. The corresponding control pitch angles are depicted in Figure 8, where the hover trim variables reported in Table 3 are retrieved. Given the highly-classified nature of the data involved during the dronization process, the adopted first-guess controller gains are omitted.

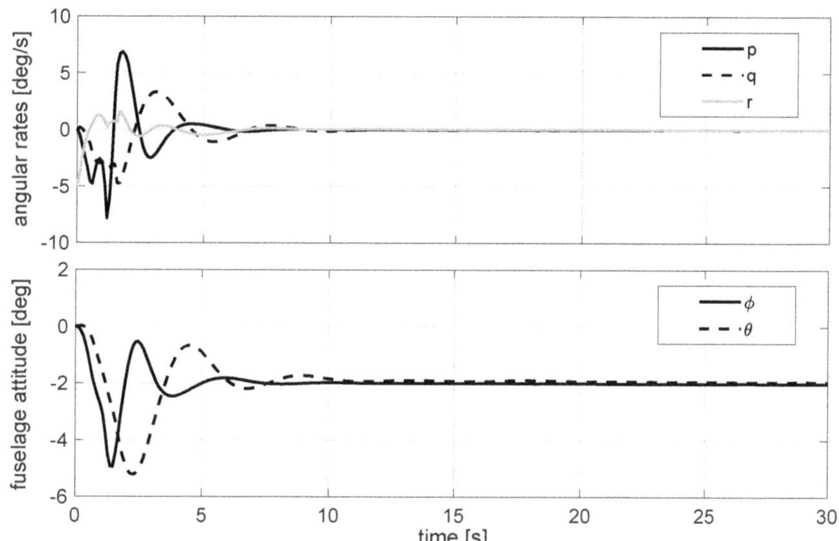

Figure 7. MIL stabilization of attitude variables.

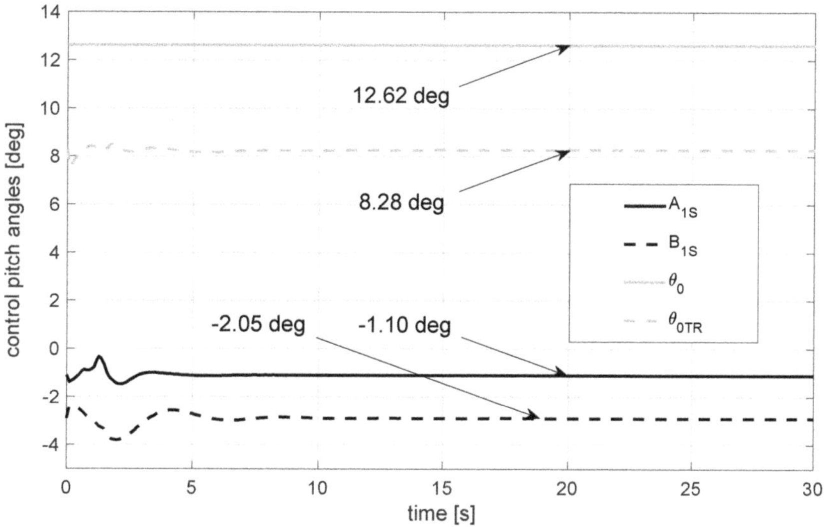

Figure 8. MIL stabilization of control pitch angles to the hovering condition.

5.2. Hardware–in–the–Loop Validation

The simulation setup described above is deployed to a HIL laboratory facility according to the scheme outlined in Figure 9. System components are set up as follows:

- The software developed in Matlab/Simulink for the mathematical modeling of helicopter dynamics and AHRS devices is automatically coded and deployed to a high-performance Real-Time Target Machine (RTTM) by Simulink Real-Time^{TM} tools. Solver frequency is set at 20 kHz, while AHRS model data are generated at 100 Hz. Software coding and deployment are performed through a host desktop PC, where the FlightGear open-source application is used to represent simulation data through a 3D graphical interface.

- The output of RTTM is provided via a dedicated standard industrial bus I/O module with two isolated ports. The first port is used to output the emulated AHRS data. The second port is used to generate repeatable control commands for HIL validation only, as if they were provided by the pilot on the ground.

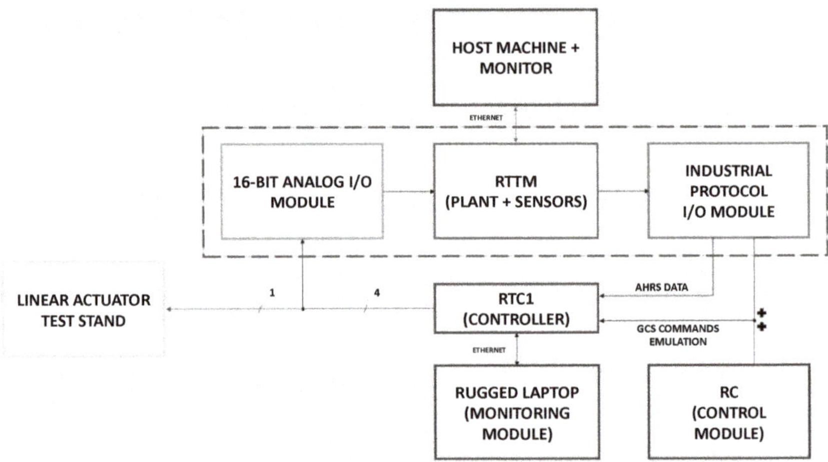

Figure 9. Sketch of HIL simulation setup.

- AHRS data and pilot commands from the RC device are the inputs to the onboard computer. At the time of the HIL experiments the RTC2, was already mounted on the helicopter as a pure acquisition device for an extensive campaign of manned flight tests. Hence, laboratory HIL tests were performed by arranging the RTC1 as an onboard computer. To emulate the presence of the radio modem, signals from the RC are converted from serial to standard industrial protocol by a micro-controller board equipped with a dedicated conversion shield. The control laws designed in Matlab/Simulink are coded and deployed to the RTC1 through the rugged laptop. The code developed for the onboard computer makes use of proprietary libraries for PID control implementation, acquisition and processing of input signals (including the application of Butterworth filters with order 1 and a cut-off frequency of 5 Hz to measured data), and real-time monitoring of selected variables.
- Control signals are acquired through a terminal board by a dedicated I/O module, a 16 bit analog input device selected to close the control loop. An ad hoc test bench is also provided where 1 EMA is controlled, in turn, by a voltage signal. Information about the linear motion are acquired and made available to evaluate the actuation performance.

Different maneuvers are performed during HIL simulations to validate the control strategy in Section 5.1 and the hardware implementation. A sample case is reported in what follows. Starting from a hovering condition, a step input $\delta_a^{(pilot)} = 0.08$ is generated via the RTTM in order to reach a desired roll angle of 2 deg while keeping the other inputs unaltered. In Figures 10 and 11, the results of the maneuver are reported in terms of variation with respect to the hover trim variables.

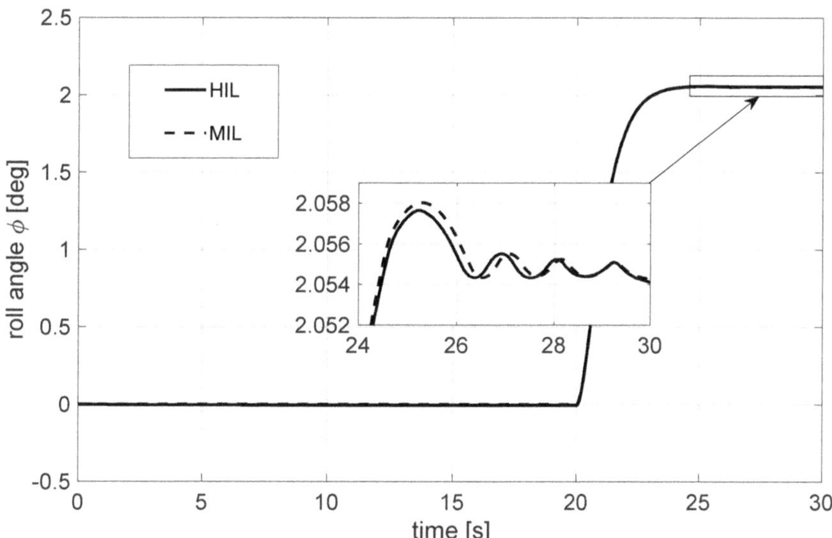

Figure 10. Roll angle stabilization maneuver: comparison between MIL and HIL simulations (roll angle, variation with respect to the hover condition).

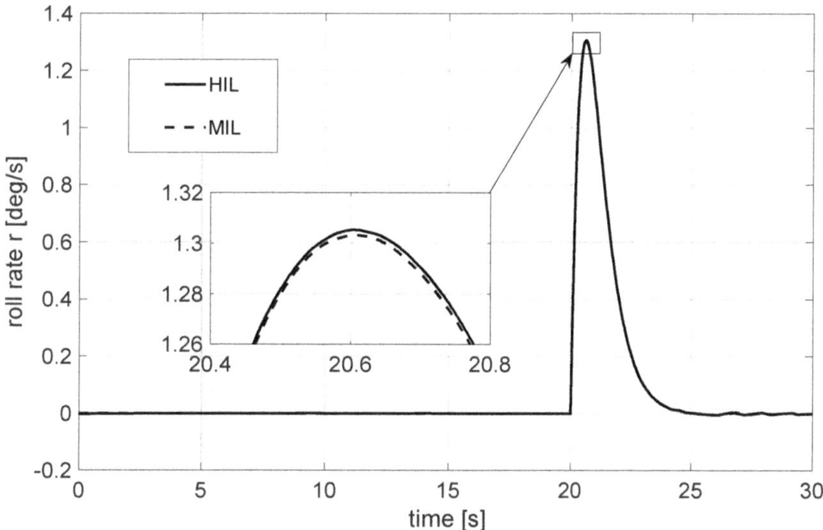

Figure 11. Roll angle stabilization maneuver: comparison between MIL and HIL simulations (roll rate).

It can be noted that, for the same maneuver, the error between HIL and MIL simulations always remains bounded and smaller than 0.001 deg (roll angle) and 0.005 deg/s (roll rate). Furthermore, discretization and quantization effects of signals are investigated, which, however, do not affect controller efficacy. This and many other simulation tests validate the quality of the simulation software and the correct implementation of acquisition, actuation, and control system protocols in the presence of real flight hardware in a controlled environment prior to flight.

6. Flight Tests with the Unmanned Helicopter

After an extensive campaign of HIL simulations aiming at the fine-tuning of controller gains and the correct setup of hardware implementation, the helicopter is finally configured for unmanned flight tests and equipped with the ballistic parachute canopy. In order to simulate the presence of on board passengers, sandbags are put on the two seats, thus replicating the inertial configuration analyzed in Section 4, with the exception of the canopy. The campaign, performed in June 2018 at the airport of Oristano–Fenosu (Sardinia, Italy) in 4 days, is organized according to the following steps:

1. Step 1. Direct control of onboard actuators by remote pilot commands, such that $\delta_a = \delta_a^{(pilot)}$, $\delta_e = \delta_e^{(pilot)}$, $\delta_c = \delta_c^{(pilot)}$, and $\delta_p = \delta_p^{(pilot)}$. This piloting configuration allows for validation of the overall actuation setup and represents a reversion mode in case of AHRS failure (manual mode).
2. Step 2. The controller in Equation (27) is activated in order to stabilize the yaw rate. Different flight tests are performed and control gains are refined according to remote pilot recommendations, such that $k_p^{(r)}$ and $k_i^{(r)}$ are respectively increased by about 15% and 14% with respect to the first-guess values in Section 5.1.
3. Step 3. Before activating the controllers in Equations (28) and (29), an intermediate test is performed in order to evaluate the damping contribution only provided by gains $k_d^{(\phi)}$ and $k_d^{(\theta)}$ to the flying qualities about the roll and the pitch axis, respectively. To this end, the yaw rate is stabilized as in Step 2, while the direct control action of the pilot on lateral and longitudinal cyclic commands is supported by roll and pitch damper controllers, configured as follows:

$$\delta_a = \delta_a^{(pilot)} + k_d^{(\phi)}\, p \tag{30}$$

$$\delta_e = \delta_e^{(pilot)} + k_d^{(\theta)}\, q \tag{31}$$

At the end of Step 3, control gains are fine-tuned such that $k_d^{(\phi)}$ and $k_d^{(\theta)}$ are respectively increased by about 2% and 13% with respect to the first-guess values.
4. Step 4. The attitude controllers in Equations (28) and (29) are investigated, leaving the pilot with direct control of MR collective pitch only. Control gains are corrected such that $k_p^{(\theta)}$ and $k_i^{(\theta)}$ are respectively increased by 25% and 60% with respect to the precautionary small values proposed in Section 5.1. Finally, $k_p^{(\phi)}$ and $k_i^{(\phi)}$ are left unaltered.

Some flight data is reported, which describes the tests performed after Step 4 with the unmanned system in its definitive mission configuration.

In Figure 12, the commanded value of yaw rate, calculated as $\xi_r\, \delta_p^{(pilot)}$ (black line), is compared with the corresponding value measured by the AHRS (gray line). The data are expressed in deg/s and show the correlation between the desired and achieved attitude motion while the pilot performs oscillatory yawing maneuvers.

In Figure 13a,b the stabilization of roll and pitch angles is also analyzed over the same time period (80 s). In particular, roll angle oscillates with a standard deviation of 0.78 deg about the mean value of -1.91 deg. Similar considerations hold for the pitch angle, characterized by a standard deviation of 0.74 deg and a mean value of 0.35 deg. If, on the one hand, the roll angle is consistent with the simulation results obtained in Table 3, the pitch angle shows major difference. This is caused by the presence of light tail wind and the fact that the inertial and aerodynamic configuration of the unmanned helicopter differs because of the presence of the parachute canopy over MRH. Collective command, characterized by a standard deviation of 0.01, remains almost constant and equal to 0.66 (corresponding to 13.67 deg pitch angle).

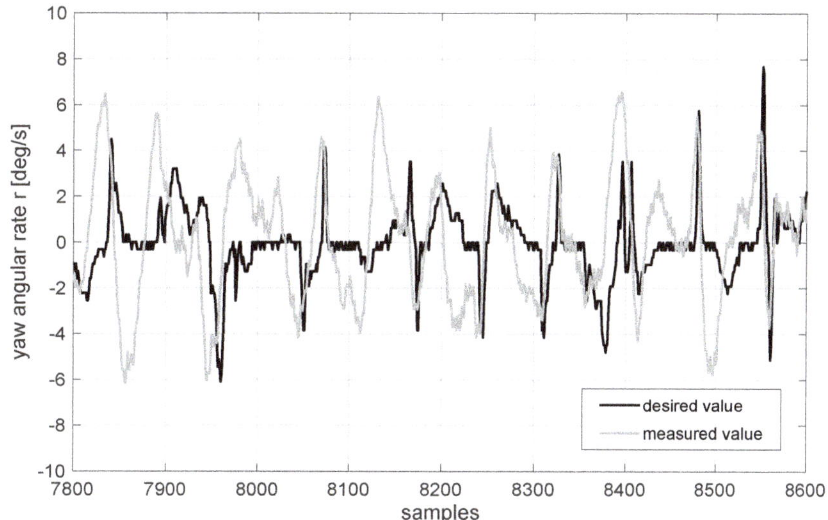

Figure 12. Yaw rate stabilization in a near–hover condition (flight tests, 10 Hz sampling).

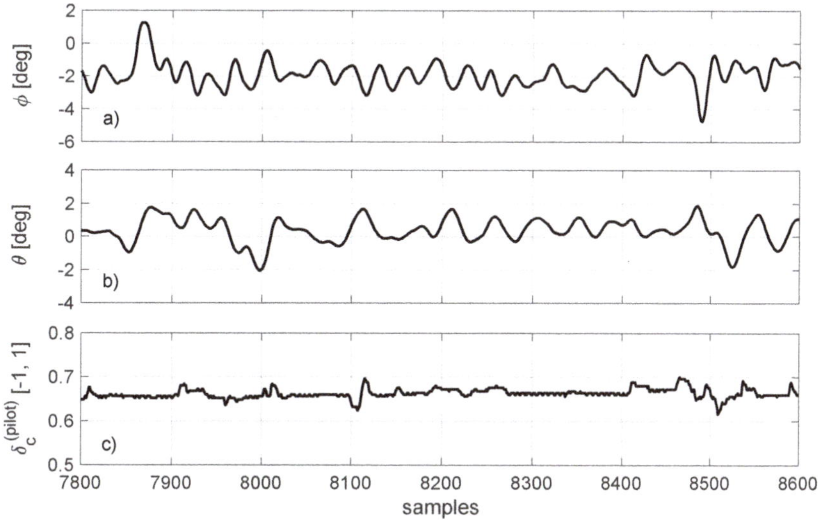

Figure 13. (**a**,**b**) Attitude stabilization and (**c**) collective pitch command in a near–hover condition (flight tests, 10 Hz sampling).

The final experiment, performed on 22 June, is described in Figure 14, where helicopter trajectory is plotted in a 3D environment. Position data are obtained from GPS measurements provided by the AHRS and recorded by the RTC2. After the initial phase required for pre–flight checks and turbine engine warm up, the take–off occurs at time t_0. The climb phase to the height $h_1 = 330$ m is performed in $t_1 - t_0 = 97$ s in the presence of South–West (SW) wind, with an average climb rate of about 3.4 m/s. In particular, during the first 40 s the climb rate is stabilized at 2 m/s by pilot's action, and then pushed to 4.5 m/s until reaching the maximum height. At time t_1, the prescribed flight termination procedure is activated by switching-off the engine and commanding parachute ejection at time $t_2 = t_1 + 4$ s. Complete parachute deployment is performed in about 5 s, at time $t_3 = t_2 + 5$ s (see Figure 15). During the helicopter accelerated free fall the total height loss is $h_3 - h_2 = -146$ m, with an average vertical speed of -16.2 m/s. After $t = t_3$ the rate of descent stabilizes to a practically constant value of 7.5 m/s until the helicopter safely lands at $t_4 = t_3 + 27$ s. The effect of wind is visible in Figure 14, where helicopter trajectory deviates in the North–East direction and stops near the runway at about 285 m from the take–off point. Upon impact with the ground, acceleration peaks are recorded that fall within the parameters of crash tests in both the aeronautical and automotive sectors. Test data show that the system is likely to achieve its goal of saving lives, even at a lower altitude of just 150 m.

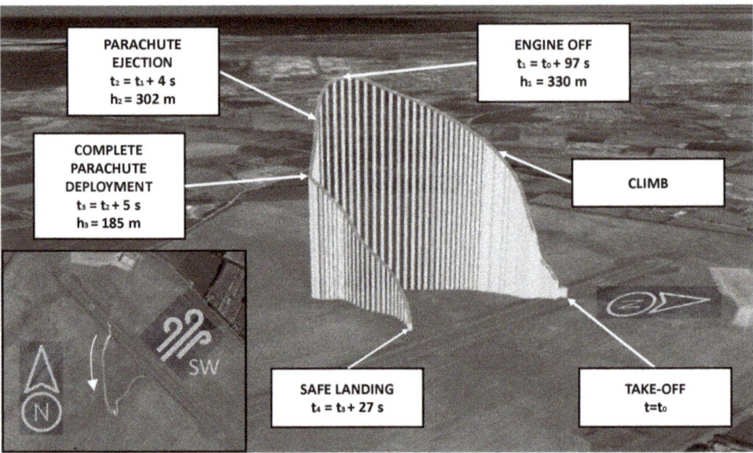

Figure 14. Trajectory followed during the final mission with parachute ejection (Maps Data: Google Earth © 2020 TerraMetrics).

Figure 15. Parachute ejection phases (courtesy of Curti Aerospace Division).

7. Conclusions

In the present paper, the complete procedure adopted to transform a light helicopter into an unmanned rotorcraft is described. By adopting the MBD approach, mission requirements were first outlined, and the design of the control system was addressed in terms of system architecture definition. Particular attention was devoted to the mathematical model of the helicopter and its subsystems, made on the basis of geometric, inertial, and aerodynamic data provided by the manufacturer and refined by identification techniques.

With the purpose of validating an innovative ballistic parachute rescue system, a closed-loop controller was developed to allow stable maneuvering in the field of view of a remote pilot. To this end, attitude stabilization algorithms were first tested in a Model-In-the-Loop environment. Furthermore, laboratory experiments allowed for (1) Hardware-In-the-Loop validation of involved equipment and (2) control software deployment on real-time target machines. Dedicated flight tests were performed to prove the effectiveness of the approach and the achievement of the desired closed–loop flying qualities. The final mission successfully showed the feasibility of the proposed termination procedure by securing a safe helicopter landing in the event of engine failure. The experiment allowed researchers to focus on the design and experimental validation of technologies at the core of future UAM, envisaging a more efficient, safe, and possibly sustainable exploitation of the vertical dimension.

Author Contributions: Conceptualization, E.L.d.A.; methodology, E.L.d.A., F.G., M.T., G.R. and C.A.; software, E.L.d.A. and M.T.; validation, E.L.d.A., F.G., M.T., G.R. and C.A.; formal analysis, E.L.d.A., F.G. and C.A.; investigation, E.L.d.A., F.G., M.T. and C.A.; resources, E.L.d.A., F.G., M.T. and C.A.; data curation, E.L.d.A. and C.A.; writing—original draft preparation, E.L.d.A. and F.G.; writing—review and editing, E.L.d.A., F.G. and C.A.; visualization, E.L.d.A. and F.G.; supervision, F.G. and C.A.; project administration, F.G. and C.A.; funding acquisition, F.G. and C.A. All authors have read and agreed to the published version of the manuscript.

Funding: This research was funded by European Commission in the framework of H2020–EU.2 and H2020–EU.3 DISRUPT project (ID: 691436).

Data Availability Statement: Not applicable.

Acknowledgments: The authors wish to express their sincerest gratitude to Hypertech Solution S.r.l., for the effective cooperation that allowed the successful achievement of the goal.

Conflicts of Interest: The authors declare no conflict of interest.

References

1. Garrow, L.A.; German, B.J.; Leonard, C.E. Urban air mobility: A comprehensive review and comparative analysis with autonomous and electric ground transportation for informing future research. *Transp. Res. Part. C Emerg. Technol.* **2021**, *132*, 1–31. [CrossRef]
2. de Angelis, E.L.; Giulietti, F. Dynamic stability and control of rotorcraft for suspended load transportation: An analytical approach. In Proceedings of the 48th European Rotorcraft Forum, Winterthur, Switzerland, 6–8 September 2022; pp. 1–14.
3. Schneider, D. The delivery drones are coming. *IEEE Spectrum* **2020**, *57*, 28–29. [CrossRef]
4. de Angelis, E.L.; Giulietti, F.; Pipeleers, G. Swing angle estimation for multicopter slung load applications. *Aerosp. Sci. Technol.* **2019**, *89*, 264–274. [CrossRef]
5. Causa, F.; Fasano, G. Multiple UAVs trajectory generation and waypoint assignment in urban environment based on DOP maps. *Aerosp. Sci. Technol.* **2021**, *110*, 1–15. [CrossRef]
6. Dai, W.; Pang, B.; Low, K.H. Conflict-free four-dimensional path planning for urban air mobility considering airspace occupancy. *Aerosp. Sci. Technol.* **2021**, *119*, 1–17. [CrossRef]
7. Schweiger, K.; Knabe, F.; Korn, B. An exemplary definition of a vertidrome's airside concept of operations. *Aerosp. Sci. Technol.* **2022**, *125*, 107144. [CrossRef]
8. Le Tallec, C.; Joulia, A.; Harel, M. A personal plane air transportation system-The PPlane Project. *SAE Int. J. Aerosp.* **2011**, *4*, 1281–1292. [CrossRef]
9. Hlinka, J.; Trefilova, H. Identification of major safety issues for a futuristic personal plane concept. *Aviation* **2014**, *18*, 120–128. [CrossRef]
10. Shihab, S.A.M.; Wei, P.; Ramirez, D.S.J.; Mesa-Arango, R.; Bloebaum, C. By Schedule or on Demand?—A Hybrid Operation Concept for Urban Air Mobility. In Proceedings of the AIAA Aviation 2019 Forum, Dallas, TX, USA, 17–21 June 2019; pp. 1–13. [CrossRef]
11. Boelens, J.H. The Road-Map to Scalable Urban Air Mobility, Volocopter White Paper 2021; Chapter 4. Available online: https://www.volocopter.com/content/uploads/Volocopter-WhitePaper-2-0.pdf (accessed on 23 November 2021).

12. de Angelis, E.L.; Giulietti, F.; Pipeleers, G.; Rossetti, G.; Van Parys, R. Optimal autonomous multirotor motion planning in an obstructed environment. *Aerosp. Sci. Technol.* **2019**, *87*, 379–388. [CrossRef]

13. de Angelis, E.L.; Giulietti, F.; Avanzini, G. Optimal cruise performance of a conventional helicopter. *Proc. Inst. Mech. Eng. Part G J. Aerosp. Eng.* **2021**, *263*, 865–878. [CrossRef]

14. Hardesty, M.; Guthrie, D.; Cerchie, D. Unmanned Little Bird Testing Approach. In Proceedings of the 2009 American Helicopter Society International Technical Specialists Meeting on Unmanned Rotorcraft Systems, Phoenix, AZ, USA, 20–22 January 2009; pp. 1–12.

15. Downs, J.; Prentice, R.; Dalzell, S.; Besachio, A.; Ivler, C.M.; Tischler, M.B.; Mansur, M.H. Control System Development and Flight Test Experience with the MQ–8B Fire Scout Vertical Take-Off Unmanned Aerial Vehicle (VTUAV). In Proceedings of the American Helicopter Society 63rd Annual Forum, Virginia Beach, VA, USA, 1–3 May 2007; pp. 1–27.

16. Mansur, M.H.; Tischler, M.B.; Bielefield, M.D.; Bacon, J.W.; Cheung, K.K.; Berrios, M.G.; Rothman, K.E. Full flight envelope inner–loop control law development for the unmanned K–MAX. In Proceedings of the American Helicopter Society 67th Annual Forum, Virginia Beach, VA, USA, 3–5 May 2011; pp. 1–17.

17. Perry, D. Eurocopter Demonstrates Unmanned EC145. Flight Global. 2013. Available online: https://www.flightglobal.com/video-eurocopter-demonstrates-unmanned-ec145/109550.article (accessed on 23 November 2021).

18. Adams, E. A Tap–to–Fly Helicopter Shows How Flying Cars Might Take Off. Wired. 2019. Available online: https://www.wired.com/story/sikorsky-sara-helicopter-autonomous-flying-car-air-taxi-tech/ (accessed on 23 November 2021).

19. Curti Costruzioni Meccaniche, The New Zefhir Helicopter by Curti Aerospace Division. Available online: https://www.zefhir.eu (accessed on 23 November 2021).

20. Bertolani, G.; Ryals, A.D.; Pollini, L.; Giulietti, F. L1 adaptive speed control of a helicopter. In Proceedings of the 48th European Rotorcraft Forum, Winterthur, Switzerland, 6–8 September 2022; pp. 1–11.

21. Pavel, M.D.; Bertolani, G.; Giulietti, F. Non–linear (incremental) backstepping control applied to helicopter flight. In Proceedings of the 48th European Rotorcraft Forum, Winterthur, Switzerland, 6–8 September 2022; pp. 1–14.

22. Huber, M. Zefhir Successfully Tests Whole Helicopter Parachute. AINonline. 2018. Available online: https://www.ainonline.com/aviation-news/general-aviation/2018-09-28/zefhir-successfully-tests-whole-helicopter-parachute (accessed on 23 November 2021).

23. Head, E. How Curti Put a Parachute on Its Zefhir Helicopter. Vertical. 2019. Available online: https://verticalmag.com/news/curti-parachute-zefhir-helicopter/ (accessed on 23 November 2021).

24. Abdelmaksoud, S.I.; Mailah, M.; Abdallah, A.M. Control Strategies and Novel Techniques for Autonomous Rotorcraft Unmanned Aerial Vehicles: A Review. *IEEE Access* **2020**, *8*, 195142–195169. [CrossRef]

25. Talbot, P.D.; Tinling, B.E.; Decker, W.A.; Chen, R.T. *A Mathematical Model of a Single Main Rotor Helicopter for Piloted Simulation*; NASA TM 84281; NASA: Washington, DC, USA, 1982; pp. 1–55.

26. Department of Defense. *World Geodetic System 1984—Its Definition and Relationship with Local Geodetic Systems*; DMA TR 8350.2; Department of Defense: Arlington, VA, USA, 1991; Chapter 5.

27. U.S. Government Printing Office. *U.S. Standard Atmosphere*; NOAA-S/T 76-1562; U.S. Government Printing Office: Washington, DC, USA, 1976; Chapter 2.1.

28. Leishman, J.G. *Principles of Helicopter Aerodynamics*, 2nd ed.; Cambridge University Press: New York, NY, USA, 2006; Chapters 2 and 5.

29. Padfield, G.D. *Helicopter Flight Dynamics*, 2nd ed.; Blackwell Pub.: Oxford, UK, 2007; Chapter 3.

30. Scavo, T.R.; Thoo, J.B. On the geometry of Halley's method. *Am. Math. Mon.* **1995**, *102*, 417–426. [CrossRef]

31. Jewel, J.W.; Heyson, H.H. *Charts of the Induced Velocities Near a Lifting Rotor*; NASA TM 4-15-59L; NASA: Washington, DC, USA, 1959; pp. 1–68.

32. Dormand, J.R.; Prince, P.J. A family of embedded Runge-Kutta formulae. *J. Comput. Appl. Math.* **1980**, *6*, 19–26. [CrossRef]

33. Tischler, M.B. *Advances in Aircraft Flight Control*, 1st ed.; Taylor & Francis: London, UK, 1996; Chapter 2. [CrossRef]

34. Tischler, M.B.; Fletcher, J.W.; Diekmann, V.L.; Williams, R.A.; Cason, R.W. *Demonstration of Frequency Sweep Testing Technique Using a Bell 214–ST Helicopter*; NASA TM 89422; NASA: Washington, DC, USA, 1987; Chapters 5 and 6.

35. Ljung, L. Prediction error estimation methods. *Circuits Syst. Signal Process.* **2002**, *21*, 11–21. [CrossRef]

36. Cai, G. Design and implementation of a hardware–in–the–loop simulation system for small–scale UAV helicopters. In Proceedings of the 2008 IEEE International Conference on Automation and Logistics, Qingdao, China, 1–3 September 2008; pp. 29–34. [CrossRef]

drones

Article

Software Aging Effects on Kubernetes in Container Orchestration Systems for Digital Twin Cloud Infrastructures of Urban Air Mobility

Jackson Costa [1], Rubens Matos [1,2,*], Jean Araujo [1,3], Jueying Li [4], Eunmi Choi [5], Tuan Anh Nguyen [4,6,*], Jae-Woo Lee [6,*] and Dugki Min [4,*]

1   Departamento de Computação, Universidade Federal de Sergipe, São Cristóvão 49100-000, Brazil
2   Coordenação de Redes de Computadores, Instituto Federal de Sergipe, Lagarto 49400-000, Brazil
3   UNAME Group, Universidade Federal do Agreste de Pernambuco, Garanhuns 55292-270, Brazil
4   Department of Computer Science and Engineering, College of Engineering, Konkuk University, Seoul 05029, Republic of Korea
5   School of Software, College of Computer Science, Kookmin University, Seoul 02707, Republic of Korea
6   Konkuk Aerospace Design-Airworthiness Institute (KADA), Konkuk University, Seoul 05029, Republic of Korea
*   Correspondence: rubens.junior@ifs.edu.br (R.M.); anhnt2407@konkuk.ac.kr (T.A.N.); jwlee@konkuk.ac.kr (J.-W.L.); dkmin@konkuk.ac.kr (D.M.)

Citation: Costa, J.; Matos, R.; Araujo, J.; Li, J.; Choi, E.; Nguyen, T.A.; Lee, J.-W.; Min, D. Software Aging Effects on Kubernetes in Container Orchestration Systems for Digital Twin Cloud Infrastructures of Urban Air Mobility. *Drones* **2023**, *7*, 35. https://doi.org/10.3390/drones7010035

Academic Editors: Ivana Semanjski, Antonio Pratelli, Massimiliano Pieraccini, Silvio Semanjski, Massimiliano Petri and Sidharta Gautama

Received: 19 October 2022
Revised: 11 December 2022
Accepted: 22 December 2022
Published: 3 January 2023

Abstract: It is necessary to develop a vehicle digital twin (DT) for urban air mobility (UAM) that uses an accurate, physics-based emulator to model the statics and dynamics of a vehicle. This is because the use of digital twins in the operation and control of UAM vehicles is essential for the UAM operational digital twin infrastructure (UAM-ODT). There are several issues that need to be addressed in this process: (i) the lack of digital twin engines for the digitalization (twinization) of the dynamics and control of UAM vehicles at the core of UAM-ODT systems; (ii) the lack of back-end system engineering in the development of UAM vehicle DTs; and (iii) the lack of fault-tolerant mechanisms for the DT cloud back-end system to run uninterrupted operations 24/7. On the other hand, software aging and rejuvenation are becoming increasingly important in a variety of computing scenarios as the demand for reliable and available services increases. With the increasing use of containerized systems, there is also a need for an orchestrator to support easy management and reduce operational costs. In this paper, an operational digital twin (ODT) of a typical urban air mobility (UAM) infrastructure is developed on a private cloud system based on Kubernetes using a proposed cloud-in-the-loop simulation approach. To ensure the ODT can provide uninterrupted operational control and services in UAM around the clock, we propose a methodology for investigating software aging in Kubernetes-based containerized clouds. We evaluate the behavior of Kubernetes software using the Nginx and K3S tools while they manage pods in an accelerated lifetime experiment. We continuously execute operations for creating and terminating pods, allowing us to observe the utilization of computing resources (e.g., CPU, memory, and I/O), the performance of the Nginx and K3S environments, and the response time of an application hosted in those environments. In some conditions and for specific metrics, such as virtual memory usage, we observed the effects of software aging, including a memory leak that is not fully cleared when the cluster is stopped. These issues could lead to system performance degradation and eventually compromise the reliability and availability of the system when it crashes due to memory space exhaustion or full utilization of swap space on the hard disk. This study helps with the deployment and maintenance of virtualized environments from the standpoint of system dependability in digital twin computing infrastructures where a large number of services are running under strict continuity requirements.

Keywords: operational digital twin; urban air mobility; cloud-in-the-loop simulation; software aging; software rejuvenation; Kubernetes; Nginx; K3S

1. Introduction

Digital twin (DT) technology is a cutting-edge innovation that has the potential to revolutionize various industries. DT involves creating a virtual replica of a physical object or system, and using data-driven analysis and decision-making to continuously update and improve it. The virtual replica, or digital twin, is made up of computational models that evolve and change over time, reflecting the structure, behavior, and environment of the physical object or system they represent [1,2]. Digital twin systems are digital representations of physical systems, such as vehicles, buildings, or manufacturing processes. They are used to simulate the behavior and performance of the physical system, and to predict its behavior or performance under different conditions. This can be useful for a variety of applications, such as planning for maintenance, optimizing the operation of the physical system, or analyzing the impact of changes to the system's design or operation.

The development of an operational vehicle digital twin system for urban air mobility (UAM-ODT) includes the following fundamental modules: (i) neural digital twin dynamic engines (DTDE), (ii) neural digital twin control engines (DTCE), (iii) digital twin control frame (DTCF), and (iv) digital twin cloud infrastructure (DTCI) as shown in Figure 1. The DTDE module is responsible for creating a virtual replica of the aerodynamics of UAM vehicles using learning-based techniques. The DTCE module performs control tasks, such as robust control, optimal control, and adaptive control, to ensure the safety of the vehicle. These two modules digitalize the dynamics and control of the vehicle to ensure that the operations of the vehicle in the digital space are identical to those in the physical space. The DTCF module serves as a bridge between the digital twin and the physical twin of the vehicle. It can provide teleoperation services, fault-tolerant control, or traffic prediction and management, with the belief that if the dynamics and control of the physical vehicle are accurately captured in the digital space along with the digital environment (e.g., city, region, country), the operations in the digital space can be effectively transferred to the physical space. The DTCI module is the common computing platform that hosts the entire UAM-ODT system, running constantly to create a virtual space of the real-world UAM physical infrastructure. Due to the stringent requirements for the high availability of the digital twin system, the DTCI must handle any failures and maintain constant digital operations and services in the long run. Particularly, if a digital twin runs all day and night, it can be subject to a phenomenon known as "software aging". Software aging is the gradual deterioration of the performance and reliability of software over time, due to factors such as changes in the operating environment, errors and defects in the software, or the accumulation of wear and tear on the software. If a digital twin runs continuously, it can experience software aging more quickly than if it were run only intermittently. This can cause the digital twin to become less accurate and less reliable over time, which can affect the quality of the predictions and decisions it makes. In this work, we investigate the software aging problems in the digital twin cloud infrastructure which is developed upon Kubernetes-based cloud environment using a cloud-in-the-loop simulation approach.

Software aging is a phenomenon that occurs when software systems become less reliable and less efficient over time. This can happen for a variety of reasons, such as changes in the environment, changes in the software itself, or the accumulation of errors and defects. When software ages, it can become less accurate and less reliable, which can affect the performance and behavior of the systems that it is used to control or manage. The software aging phenomenon occurs in operating software systems, causing sudden failures such as crashes and continuous performance degradation, which can be circumvented by a proactive strategy such as software rejuvenation to avoid abrupt system interruptions [3,4]. The relevance of such a phenomenon is remarkable, considering that the demand for availability and reliability in the provision of services in practically all areas has increased in order to have quality and competitiveness in each field of activity. Considering high availability requirements, the services of computing, health, security, financial system, geolocation, and routing are examples that can be cited. In order to meet such service demands without the unwanted effects of software aging, it is necessary to use an architecture capable of maintaining its offer without huge operational costs of employing several redundant servers with high computational power, requiring human resources for their handling and management, and also incurring higher energy costs. Using virtual machines in contexts such as these has been an alternative because they provide functionalities of a physical server based on the same traditional computational architecture. Thus, it is possible to create several virtual machines on a single server, and each virtual machine can run different environments allowing the execution of heterogeneous systems [5]. The scalability and flexibility of IT (Information Technology) can be increased through virtualization, in addition to generating significant savings in operational costs. Thus, IT administration becomes easier to manage by obtaining better availability, operability, performance, and greater workload mobility through virtualization [6].

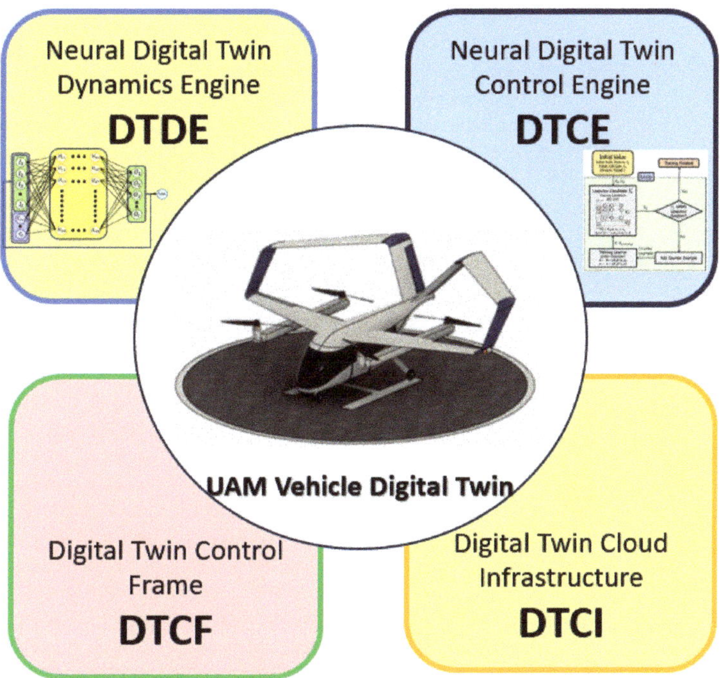

Figure 1. Operational Digital Twin for Urban Air Mobility (UAM-ODT).

If the flight control software of an unmanned aerial vehicle (UAV) experiences software aging, it can affect the performance and behavior of the UAV. As the software ages, it can become less accurate and less reliable, which can cause the UAV to behave in unexpected or unsafe ways. To address software aging in the flight control software of a UAV, it is important to periodically update and maintain the software. This can involve installing patches and updates, fixing errors and defects, and re-tuning or re-calibrating the software to account for changes in the environment or the UAV itself. Regular maintenance and updates can help to ensure that the flight control software remains accurate and reliable over time, and can help to prevent or mitigate the effects of software aging. In some cases, software aging can cause the flight control software to become unstable or unreliable. If this happens, it may be necessary to take the UAV out of service temporarily in order to perform maintenance or repairs. This can involve replacing or upgrading the flight control software, or making other changes to the UAV in order to improve its performance and reliability. So, software aging in the flight control software of a UAV can affect the performance and behavior of the UAV. To address this problem, it is important to periodically update and maintain the flight control software, and to take the UAV out of service if necessary in order to perform maintenance or repairs. This can help to ensure that the UAV remains safe and reliable over time.

To create a digital twin, a mathematical model of the physical system is created using data about the system's behavior and performance. This model is then used to simulate the behavior of the physical system under different conditions, and to make predictions about its performance. In order to create a reliable and accurate digital twin, it is important to use accurate and reliable software to create the model and simulate the system's behavior. However, software aging can be a problem for digital twin systems. As the software used to create and simulate the digital twin ages, it can become less accurate and less reliable. This can affect the accuracy and reliability of the digital twin, and can cause it to produce incorrect or inconsistent predictions. In some cases, this can lead to incorrect or sub-optimal decisions or actions based on the digital twin's predictions. To address software aging problems in digital twin systems, it is important to periodically update and maintain the software used to create and simulate the digital twin. This can involve installing patches and updates, fixing errors and defects, and re-tuning or re-calibrating the software to account for changes in the environment or the system being modeled. Regular maintenance and updates can help to ensure that the digital twin remains accurate and reliable over time, and can help to prevent or mitigate the effects of software aging.

Digital twin systems can experience a variety of errors, depending on the specific characteristics of the system and the software being used. Some common types of errors that can occur in digital twin systems include:

- *Data errors:* Digital twin systems are typically based on data about the behavior and performance of the physical system being modeled. If the data are incorrect or inconsistent, it can cause errors in the digital twin. For example, if the data contain missing or invalid values, or if the data are not properly pre-processed or cleaned, it can affect the accuracy and reliability of the digital twin.
- *Modeling errors:* Digital twin systems are based on mathematical models of the physical system being modeled. If the model is incorrect or incomplete, it can cause errors in the digital twin. For example, if the model does not accurately represent the underlying physical principles or behaviors of the system, or if the model is not properly calibrated or validated, it can affect the accuracy and reliability of the digital twin.
- *Software errors:* Digital twin systems are implemented using software, and software can contain errors or defects. If the software used to create or simulate the digital twin contains errors, it can cause the digital twin to behave in unexpected or incorrect ways. For example, if the software contains bugs or syntax errors, or if the software is not properly designed or implemented, it can affect the accuracy and reliability of the digital twin.

Overall, digital twin systems can experience a variety of errors, including data errors, modeling errors, and software errors. To address these errors and improve the accuracy and reliability of the digital twin, it is important to carefully collect and pre-process the data, to create accurate and well-calibrated models, and to use high-quality software that is free of errors and defects.

When using virtualization, it is possible to implement many servers in a smaller number of hosts (physical servers), which consequently implies the gain of physical spaces and energy cost reduction. However, once the virtual machine is initialized, all the hardware on which the Operating System (OS) is running is loaded and not just a copy of the OS, resulting in the consumption of many system resources, making virtualization very expensive from a computational point of view [7]. The use of containerization mitigates the operational cost of traditional virtualization, as stated by [5], in which the author addresses host-level virtualization known as container, which is another type of virtualization. This type of virtualization acts on top of the physical server offering support to several independent systems since the physical server already has an OS installed, not needing to load all the host hardware or its copy. Container-based virtualization has recently gained much attention [8,9]. This virtualization makes an application run efficiently in the most varied computing environments through its encapsulation and its dependencies [10]. This virtualization technique is said by the author of [11] to be lightweight, as the system significantly decreases workloads by sharing OS resources from host. Containers provide an isolated environment for system resources such as processes, file systems and networks to run at the host OS level, without having to run a Virtual Machine (VM) with its own OS on top of virtualized hardware. By sharing the same OS kernel, containers start much faster using a small amount of system memory compared to booting an entire virtualized OS like in [10] VMs. Kubernetes is a widely used tool for managing containers, configure, maintain and manage solutions that have containers as an approach to the detriment of VMs. Thus, this work aims to evaluate the effects of software aging and the performance of Kubernetes when undergoing a high-stress load, characterized by creating replicas of pods to maintain service availability in the Nginx and K3S environments. Furthermore, the aging problem on an unmanned vehicle refers to the degradation of the vehicle's performance over time, due to factors such as wear and tear, corrosion, and obsolescence. As an unmanned vehicle ages, its components may become less reliable and less capable of performing their intended functions, which can affect the vehicle's ability to operate safely and effectively. The aging problem can be particularly challenging for unmanned vehicles, as they often operate in harsh or hostile environments, and they may be subjected to high levels of stress and strain. For example, an unmanned aerial vehicle (UAV) may experience high levels of vibration and air turbulence during flight, which can cause its components to wear out faster. Similarly, an unmanned underwater vehicle (UUV) may be exposed to corrosive saltwater, which can cause its components to corrode and deteriorate over time. In this work, our focus is on the investigation of a digital twin cloud infrastructure in which a Kubernetes-based cloud environment is investigated regarding software aging phenomenon of the cloud if hosting the UAM-ODT with no downtime.

The study in this work extends the related research area on software aging in virtualized environment through the following *key contributions:*

- proposed a cloud based simulation platform with provisioning for the development of UAM-ODT infrastructures
- proposed a methodology for measurement and assessment of software aging in a container-based environment with a Kubernetes cluster in the digital twin cloud.
- performed comprehensive test-bed experiments and observations of software aging phenomena along with software rejuvenation in Kubernetes clusters based on Minikube and K3S environments.
- Findings and impacts:
    - It is important to stress that aging events found in test-bed experiments indicate the threats of system failures and performance degradation due to software aging symptoms. However, the time that those events will occur depends on the characteristics and intensity of the workload that the system needs to process, as well as the hardware and software specification of that Kubernetes system.
    - If the system has more resources available or less workload than those employed in this experiment, the aging phenomenon would be slower, and subsequently, the failures due to resource exhaustion would take longer to occur. This fact does not reduce the importance of evaluating software aging in those systems as well as planning actions for their mitigation.

To the best of our knowledge, this work contributes to the practical implementation and maintenance of virtualized environment on the perspectives of system dependability in digital twin computing infrastructures in which a huge amount of services are running with a stringent requirement of continuity. The findings of this study bring about the comprehension of software aging phenomena in digital twin computing infrastructures developed on top of Kubernetes, which is at very early stage of current research on software aging problems for a high level of dependability and fault-tolerance in digital twin computing infrastructures.

In order to facilitate the understanding of this work, the paper is organized as follows. Section 2 addresses the related works that inspired this study on software aging assessment; Section 3 presents the fundamental concepts and system design used in this work; Section 4 deals with the methodology used in the research; the objective and planning, covering the context in which it was produced, the tools selected, variables involved, scripts for reproduction and the hardware used are discussed in Section 5. The results are presented and discussed in Section 6. In Section 7 are the remarks arising from our research results.

2. Related Work

The work described in [10] analyzes the performance of running containers with services hosted on them, carrying out experiments with containers monitoring system resources, including network, memory, disk, and CPU. The testbed environment consists of a Kubernetes cluster manually deployed to carry out the evaluation, considering the Microsoft Azure Kubernetes Service (AKS), Google Kubernetes Engine (GKE), or Amazon Elastic Container Service for Kubernetes (EKS).

The authors in [12] evaluated the memory utilization, network overhead of containers, storage, and CPU using Docker, comparing them with KVM hypervisors. They exposed in their experiments that the containers obtained, in the worst case, similar or superior performance when compared to the VMs.

The work presented in [13] conducted a similar study, however, comparing the performance obtained from containers when monitoring the number of requests an application server could handle in relation to the same application deployed in a VM and the results showed that the VMs had significantly outperformed the containers.

The research reported in [14] performed application experiments for HPC (high-performance computing), using benchmarking tools to evaluate memory, network, disk, and CPU performance in Linux Container (LXC) related virtualization implementations, along with OpenVZ and Linux VServer, showing that all containerized apps performed similarly to a native system.

The authors of [15,16] showed improvements obtained related to performance isolation for MapReduce workloads. However, when evaluating disk workloads, LXC failed to fully isolate resources, opposite behavior to that of hypervisor-based systems.

Through memory, network, and disk metrics, the authors of [17] evaluated the performance of LXC, Docker, and KVM running many benchmarking tools to measure the performance of these components and concluded that the overhead caused by container-based virtualization technologies could have its weight considered irrelevant, despite the performance being compensated by safety.

Our main focus is on software aging investigation on a private cloud system hosting an operational digital twin of an eVTOL vehicle flying in a virtualized urban air mobility. Operational digital twins of vehicles in urban air mobility are digital representations of real-world vehicles that can be used for a variety of purposes. Some potential uses of operational digital twins in urban air mobility include:

- *Performance modeling and simulation:* Operational digital twins can be used to model and simulate the performance of vehicles in urban air mobility systems, including their flight dynamics, propulsion systems, and control systems. This can help to optimize the design and operation of vehicles, to improve their performance and efficiency, and to identify potential issues or risks.
- *Fleet management and maintenance:* Operational digital twins can be used to monitor the condition and performance of vehicles in real time, and to provide information about their current state and status. This can be used to support fleet management and maintenance operations, by providing timely and accurate data about the health and safety of vehicles, and by enabling proactive maintenance and repair.
- *Traffic management and control:* Operational digital twins can be used to support traffic management and control in urban air mobility systems, by providing information about the location, orientation, and velocity of vehicles. This can help to coordinate

the movement of vehicles, to avoid collisions and other hazards, and to optimize the flow of traffic in urban airspace.

- *Emergency response and rescue:* Operational digital twins can be used to support emergency response and rescue operations in urban air mobility systems, by providing real-time information about the location and status of vehicles. This can help to quickly and accurately identify the location and condition of vehicles in distress, and to coordinate rescue and recovery efforts.

Operational digital twins of vehicles in urban air mobility can be used for a variety of purposes, including performance modeling and simulation, fleet management and maintenance, traffic management and control, and emergency response and rescue. Due to such constant operational services, the UAM-ODT cloud system is inevitable to suffer software aging problems. In this study, we specifically investigate the software aging problems of a UAM-ODT cloud system based on Kubernetes virtualization environment.

3. System Design

3.1. Cloud in the Loop Simulation (CILS):

The ability to simulate a wide range of heterogeneous personal aerial vehicles (PAV) in the same virtual environment is critical and required to verify a variety of AI control algorithms even before their practical implementation on physical twin vehicles in digital twin infrastructures of future urban air mobility (UAM). One may deploy the AI control algorithms on actual vehicles and train them through practical flight testing in actual surroundings in order to improve the precision and calibre of neural network-based AI control algorithms (for example, neural Lyapunov control [18], deep reinforcement learning [19]). However, it frequently takes a lot of work and a considerable amount of time to collect enough flight test data for developing a competent AI control model for autonomous PAVs. As a result, one of the popular approaches is to develop an operational digital twin system for UAM (abbreviated as UAM-ODT) to replicate the actions of UAM vehicles in real-world settings within a shared virtual environment [20].

Inspired by the idea, we propose to adopt cloud-based solutions to develop a UAM-ODT system for a specific eVTOL PAV, called eVTOL KADA-UAM vehicle, under development by Konkuk Aerospace Design-Airworthiness Research Institute (KADA), Konkuk University, Seoul, Republic of Korea as shown in Figure 2 and its virtual environment of UAM-ODT as shown in Figure 3.

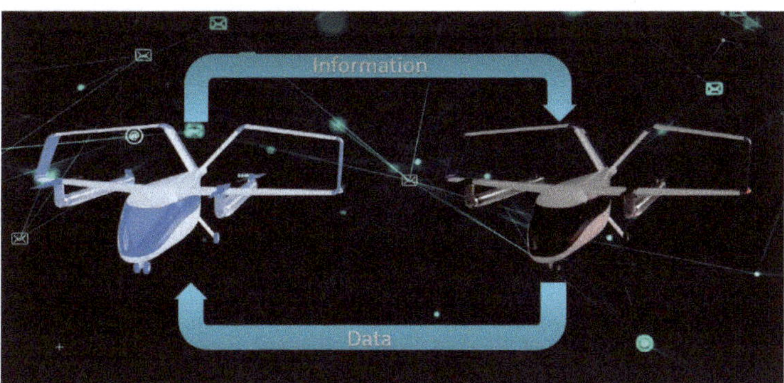

Figure 2. A digital replica of UAM vehicle in UAM-ODT infrastructure.

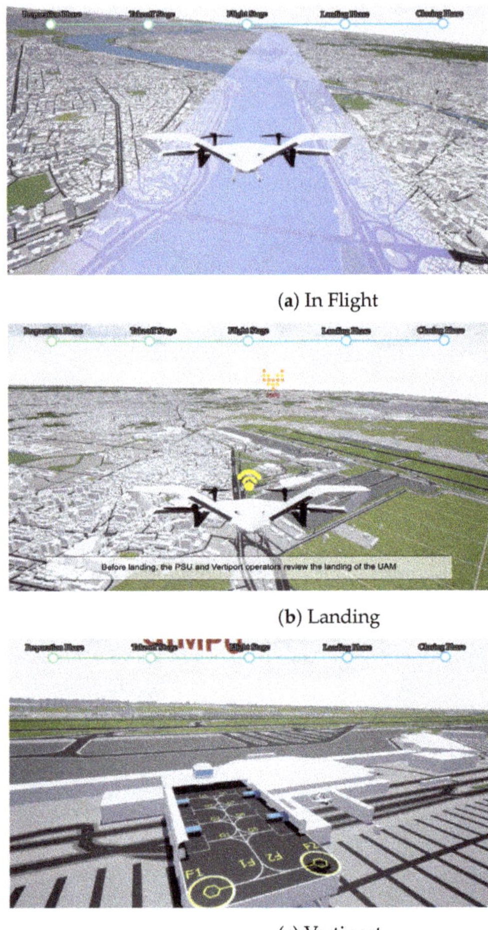

(a) In Flight

(b) Landing

(c) Vertiport

Figure 3. A visualization of UAM-ODT infrastructure. (The figures are excerpts from a video at https://blog.naver.com/yy8661 provided by Hyeon Jun Lee, Konkuk Aerospace Design-Trustworthiness Institute, Konkuk University, Seoul, Republic of Korea (rain9138@gmail.com)).

The proposal is the overall cloud-in-the-loop simulation (CILS) framework that can simulate the operations of a multitude of heterogeneous UAM vehicles with completely different aerodynamics in a UAM-ODT system, and thus can be used for verification and training of AI control algorithms in virtual world before practical implementation. The overall conceptual CILS architecture is designed as in Figure 4. To simulate multi-mode operations of heterogeneous environment which consists of multi-vehicles with different dynamics and configuration, we adopted the virtualization concept in cloud computing paradigm to separate a multitude of SILS processes onto different VMs. A single SILS process encompasses a PX4-based autopilot multi-mode AI control module (abbreviated as PX4) and a JSBsim based aerodynamic module called Konkuk Flight Simulation-Digital Twin Dynamics module (KFS-DT). The encapsulation of these two modules is called a dynamics-control SILS package, which is deployed on a multitude of VMs. The KFS-DT module guarantees the concept of digital twin framework for K-UAM vehicles in which it captures a high-fidelity CFD dynamics model of each physical vehicle. The autopilot control PX4 module transfer controls of each vehicle **u** to the dynamic module KFS-DT. The KFS-DT module computes new vehicle states **s** and returns them to the PX4 module.

The PX4 updates the current states **s** to a VM controller module. On the other hand, the PX4 receives updated sensor data from the VM controller module to generate new controls **u**. On each VM, there is a VM controller module to handle the data transactions between the VMs with a physical server for operational control management and for the visualization of the virtual environment (called environment control center). The VM controller module in each VM transmits vehicle states s receiving from a control PX4 module in the same VM and receives sensor data or mission data to/from the visualization center using Airsim [21] for AI application and Unity™for visualization. While an environment controller module in the control center is designed to handle the operations of all vehicles in the simulated environment upon the ground control module, AI module, Airsim server and Unity client module. The scalability of this cloud-based simulation framework is guaranteed by an auto-provisioning cloud system.

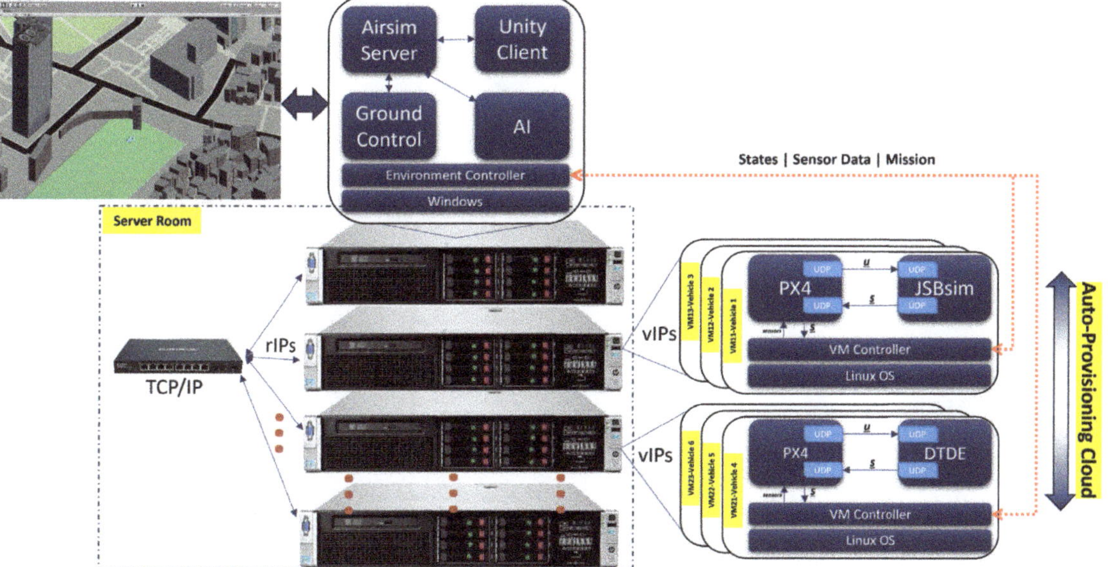

Figure 4. Cloud in the loop simulation framework.

In this work, our focus is on the digital twin version of an eVTOL vehicle to capture flight operations in an urban air mobility for further studies on performance modeling and simulation of vehicle, fleet management and maintenance, traffic management and control, and emergency response and rescue. Thus, the detailed design of the overall cloud in the loop simulation framework running on a private cloud platform is presented in a comprehensive manner while the detailed design of the eVTOL vehicle in consideration regarding circuit and IT problem in the individual engine control systems is out of scope of the study. We consider how the dynamics of the vehicle and its corresponding control are simulated in a virtualized environment of urban air mobility to mimic the real-world flight operations for air traffic managements in urban areas.

3.2. Cloud Provisioning Hardware System:

The hardware infrastructure of CILS is designed as shown in Figure 5, consists of main two components: one is virtual cluster (VC) disk image provisioning, and the other is auto provisioning virtual instance creator. Virtual machines existing in the same VC subgroup can be used by sharing the virtual disk image with homogeneous S/W as read-only. The numerous existing cloud systems provides GUI where users can build instances. It seems that this function simplifies the manipulation of creating instances on the cloud system whereas it just repeats useless operation to prepare requisites and build VMs. The auto provisioning virtual instance creator based on infrastructure as a code (IaC) provides a consistent CLI workflow to manage hundreds of cloud services and customize simulation environments.

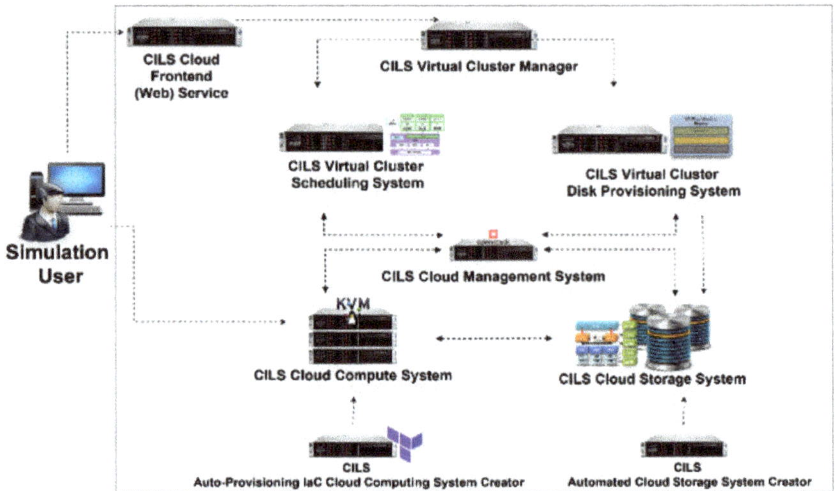

Figure 5. Cloud provisioning hardware system architecture.

Figure 6 shows the provisioning technology for virtual cluster images. The VC disk image provisioning based on union mounting technique can integrates one instance with diversified simulation virtual disk images depending on the simulation requirement such as PX4-Autopilot, JSBSim, Airsim, FlightGear and so on. Due to the orchestration of the cloud management, it can implement the communication via layer 2 or layer 3 between instances. Thanks to the capabilities of cloud provisioning and orchestration as designed in Figures 5 and 6, the UAM management can be maintained and stored on CILS Cloud Compute System complying the CILS framework in Figure 3.

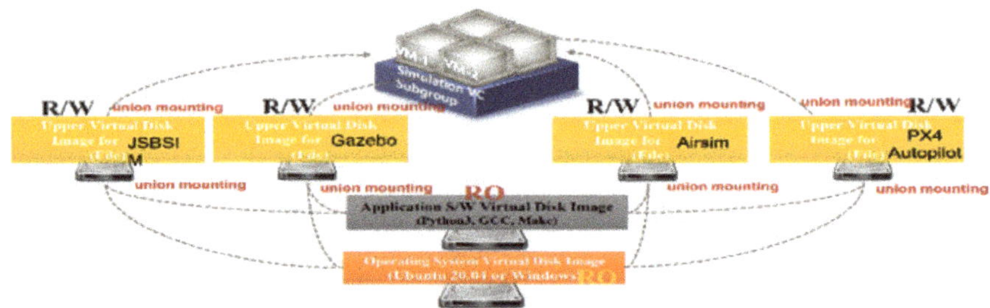

Figure 6. Virtual cluster image provisioning technology.

3.3. Software Aging and Rejuvenation

For urban air mobility (UAM), it is necessary to create a vehicle digital twin (DT) that uses a precise, physics-based emulator to characterise a vehicle's statics and dynamics. As a result, the UAM operational digital twin infrastructures need the deployment of the digital twin in vehicle operations and control (UAM-ODT). The problems are, (i) the absence of digital twin engines for the digitalization (twinization) of dynamics and control of UAM vehicles running at the core of UAM-ODT systems; (ii) the absence of back-end system engineering in the development of UAM vehicles; and (iii) the absence of fault-tolerant mechanisms for the DT cloud back-end systems running 24/7 uninterrupted operations.

Unmanned vehicles, also known as drones, are a relatively new technology and there are still many challenges and limitations associated with their use. One of the main causes of errors in the management of unmanned vehicles is the lack of a reliable and robust communication system. This can make it difficult for the operator to control the drone and receive accurate information about its location and status. Additionally, the complex algorithms used to control the drone's movements can sometimes produce unexpected or unpredictable behavior, leading to errors in the management of the vehicle. Other potential causes of errors in the management of unmanned vehicles include software bugs, hardware malfunctions, and interference from external sources such as radio waves or other electromagnetic signals. Software aging refers to the gradual degradation of a software system's performance over time. In the context of unmanned vehicles, software aging can be caused by a number of factors, including the accumulation of data, changes in the operating environment, and the introduction of new features or updates. As the software continues to be used, it may become slower, less reliable, and more prone to errors. This can affect the performance of the unmanned vehicle and make it more difficult to manage. Other potential causes of software aging in the management of unmanned vehicles include the use of outdated or inefficient algorithms, inadequate testing and debugging, and the lack of proper maintenance and support.

Software aging and rejuvenation has been an active line of research since 1995 when it was proposed by Huang et al., then at AT&T Bell Labs [22]. The reasons that lead to software aging include data loss, accumulated operating system error, resource consumption, and sudden crashes, for example. These phenomena, which accumulate gradually over time, can lead to software performance degradation, which can lead to a sudden crash or shutdown of software systems [23]. A fault tolerance prevention strategy, called software rejuvenation, aims at circumventing the negative effects of software aging, thus making it an important issue for systems reliability by avoiding sudden system failures caused by software aging, providing security and availability [24,25]. Companies such as Amazon and Google have increased interest in adopting technology architecture based on microservices (which usually rely on containerization) [26]. The reason for such an adoption is that application systems based on microservices architecture have the advantage of being easier to develop, deploy, and scale compared to monolithic architecture systems [27]. Containerization systems allow the configuration of the environment for software deployment in the shortest possible time, solving problems of integration of the most diverse applications [28].

When using containers, the application code in any offered service is involved in containerization along with its libraries, all dependencies, and configuration files necessary for its execution in the most diverse types of environments. This containerization becomes autonomous and portable, as it is abstracted from the host OS and can be executed on any computing platform [29]. This approach has been widely applied in the computing industry and demonstrated in several studies. Refs. [12,17] report overall cost reduction and overall application performance optimization in containers. The applications' microservices are generated by dividing them into small units and independently, increasing the scalability and portability of the services and the containers [10]. Nonetheless, the increase in the use of containers implies the need for tools capable of managing, through the control of tasks such as the operation of applications in containers throughout the infrastructure, scaling, and automation of application deployment [30]. An example of a container orchestration tool that is increasingly needed and widespread is Kubernetes, open-source and made available by Google. Container management tools are at the peak of expectations in the Hype Cycle for Cloud Computing from Gartner [31].

Such expectations give signs that the field in container orchestration technologies is on the rise, attractive, and very competitive, and should continue at an increasing pace as several organizations consider adopting the container-based approach [9,32,33]. A fair amount of tools as solutions for the execution and orchestration of containers emerged and quickly became solution standards in this context, among them Docker and Kubernetes.

They enable the creation of new containers and pods (a computational unit in Kubernetes comprising one or more containers) with their deployments in an agile way when an increase in application workload is detected, or even a pod drops due to excessive consumption of its resources, such as memory or CPU (Central Processing Unit), through monitoring [34]. However, its various components and related complexity have a very costly learning curve, which may not be easy to manage even with its proven efficiency in scaling, configuring, and maintaining services. Therefore, managing a Kubernetes infrastructure is a complex task. This has given rise to a new market for managing Containers, such as hosted Kubernetes solutions.

4. Methodology

The methodology adopted in this work followed the flow shown in Figure 7, which in summary is based on an experimental evaluation applied by measuring the use of system resources and performance in a container-based environment with a Kubernetes cluster. The evaluation was carried out in different scenarios using the Nginx or K3S tool to manage the cluster.

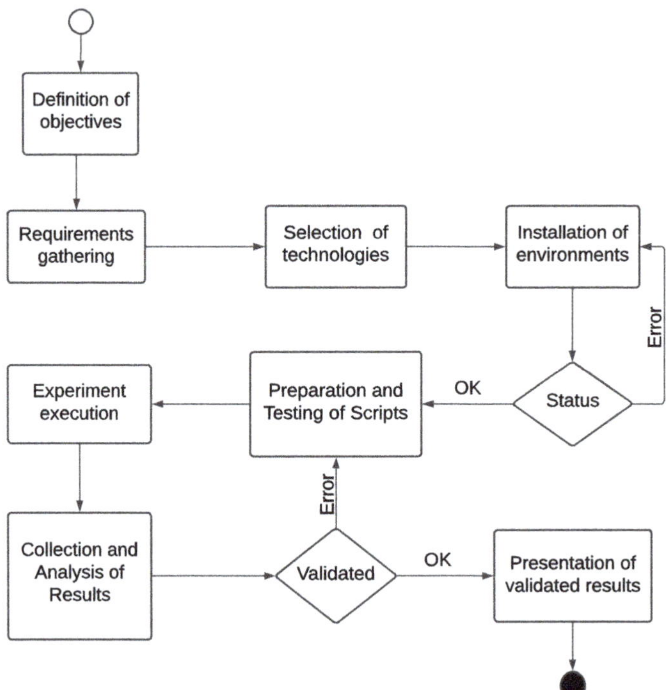

Figure 7. Methodology of the software aging measurement and assessment in Kubernetes environment.

Kubernetes is an open-source platform for managing and orchestrating containerized applications. It allows you to deploy and manage multiple containers, such as those created with Docker, across a cluster of machines, and it provides many features and tools to help you automate, scale, and manage applications and their dependencies. One of the main benefits of Kubernetes is that it helps to simplify and automate the process of deploying and managing applications in a distributed environment. This can save time and effort, and it can help to improve the reliability and scalability of applications. Additionally, Kubernetes provides many features and tools that can help you manage and monitor applications, such as: (i) Service discovery and load balancing: Kubernetes can automatically assign unique IP addresses to each of containers, and it can automatically distribute incoming traffic across the containers in cluster. (ii) Configuration management: Kubernetes allows you to define application's configuration in a declarative manner, using YAML files or other configuration formats. This can make it easier to manage and update the configuration of applications. (iii) Health checking: Kubernetes can monitor the health of containers and applications, and it can automatically restart or replace containers that are not functioning properly. (iv) Self-healing: Kubernetes can automatically detect and recover from failures in application, such as when a container crashes or when a node in cluster goes down. Therefore, he benefits of Kubernetes include improved automation, scalability, and reliability for applications, as well as a rich set of features and tools for managing and monitoring applications in a distributed environment [35].

In this research, the experiments were carried out using scripts developed for these scenarios to simulate a service's distribution using a Kubernetes cluster, which can be accessed externally through the Internet, receiving a high-stress load by performing requests. We have also developed scripts to monitor software aging metrics, such as CPU utilization, memory consumption, and disk utilization, among others, in order to measure the performance of a service hosted in Kubernetes by checking the time of requests correctly fulfilled.

The environments of Nginx and K3S adopted in this experimental evaluation are composed of a cluster containing 5 Pods and 1 Service—that allows communication between the Pods. One of the Pods was configured as a Deployment of an Nginx web server, which enabled testing the performance of an application hosted in Kubernetes, responding to user requests from anywhere connected to the Internet.

CPU utilization, memory consumption, disk utilization, and total response time were some of the metrics used for this study, based on the metrics used in [10,14]. The results of these measures were captured by scripts developed for this purpose and, finally, evaluated through analysis of their behavior.

The proposed methodology actually can be applied for typical operational digital twin version of heterogeneous UAM vehicles including rotary aircrafts such as drones or helicopters, fixed-wing aircrafts or hybrid aircrafts such as eVTOL vehicles. Since the UAM-ODT platform is designed to run on a private cloud computing system based on cloud-in-the-loop simulation paradigm with heterogeneous digital twin modules of dynamics and controls as shown in Figure 4.

5. Experimental Planning

5.1. Goal Definition

To guarantee the mitigation of software aging emergence in the UAM-ODT platform with proper operational management and maintenance, this work presents a developed methodology for the investigation of software aging phenomenon by measuring the use of system resources and performance. We use different tools for the measurement in different experiments. The objective of this work was formally defined when using the Goal Question Metric (GQM) method [36] to verify the emergence of the effects of Software Aging as well as the performance of the Kubernetes Cluster in Minikube and K3S through the response time of the service when responding to requests.

5.2. Planning

The independent variables in the experiment are the number of simultaneous pod replication requests made to the service, overloading the Nginx application server, and emulating Kubernetes' autoscaling so that the service continues to be available under high workloads. The dependent variables were: CPU utilization, memory consumption, disk utilization, and average response time to requests.

Following the GQM method, the following research questions were designed to broadly cover the scope of this work:

- Q1?: Have indicators of software aging been found?

- Q2?: Which environment had the best performance in controlling the consumption of resources related to software aging?
- Q3?: Was there a similarity in behavior between the results obtained with Minikube and K3S?

In order to answer Q1 and Q2, the following metrics were evaluated: CPU utilization, memory consumption, and disk utilization. In order to answer Q3, we conducted a comparative analysis of the results obtained from metrics that have been used to answer Q1 and Q2.

5.3. Object Selection

Samples of 125 h of monitoring in the Minikube environment and 95 h in the K3S environment were considered to evaluate the system's performance and verify the aging effects.

5.4. Experimental Design

The following steps were developed for the execution of the experiment:

- Step 1: Survey of the requirements for its realization in the Minikube and K3S environment.
- Step 2: Development and analysis of monitoring scripts, execution of the environment and its stress.
- Step 3: The experiment execution script, both in Minikube and in K3S, followed the following general script:
  - Step 3.a: Execute the monitoring script for 2 h without any workload before the cluster is started.
  - Step 3.b: Run script that starts the cluster with the container orchestrator and keeps monitoring for initial 2 h without stress.
  - Step 3.c: Run the high workload emulating the auto-scaling 420 times in a loop.
  - Step 3.d: After the end of stress, wait 2 h and execute a script that ends the container orchestrator as a possible software rejuvenation action.
  - Repeat steps 3.a, 3.b, 3.c, and 3.d until completing five cycles.
- Step 4: Generate graphs of the results obtained and analyze them.

To reinforce the understanding of our experiment, Figure 8 depicts a diagram that represents the sequence of operations performed by the general script we just described.

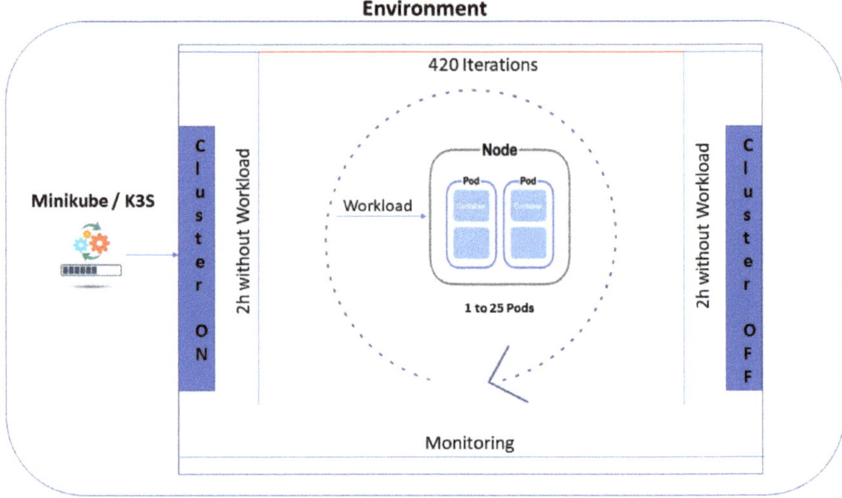

Figure 8. Diagram for cycles of operations performed by the experiment script.

Throughout all routines in `step 3`, another script sends client requests to the service that is hosted in the cluster. Those requests are effectively serviced by any pods that might have been created throughout the stress workload. Figure 9 illustrates the interaction between a client and the service in the `Kubernetes` cluster both in the `Minikube` environment and in the `K3S` environment, in both, the infrastructure architecture is configured as in Figure 2, which defines a logical set of Pods and enables exposure external traffic, load balancing and service discovery for these Pods, which have `Nginx` as a lightweight HTTP server, which is represented with Other App.

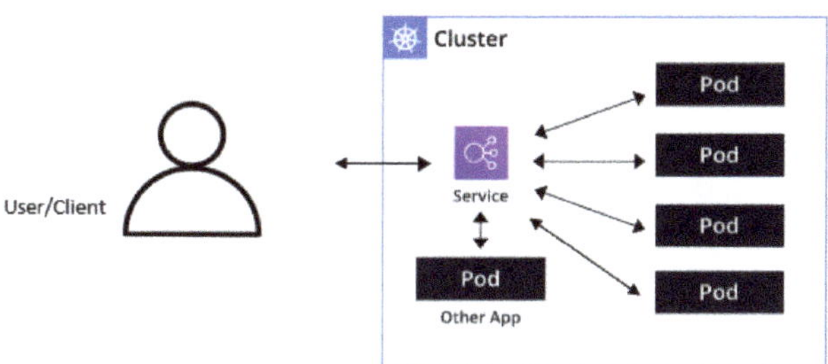

Figure 9. Cluster and Client Interaction Overview.

5.5. Instrumentation

The hardware used in this experiment was: a host with 8 `GB` of RAM, a Core i3 processor with a 3.1 `GHz` clock, a WiFi module, and Ubuntu Linux OS version 20.04 64 bits. The software used were: Shell script, for experiment implementation, monitoring and collection of data generated as results through the bash command interpreter; K3S version `v1.22.5+k3s1`; Minikube version `1.15.1` commit: `23f40a012abb52eff365ff99a709501a61 ac5876`; Kubernetes `v1.19.4` on Docker `19.03.13` for running the Kubernetes Cluster and Pods.

Metrics were collected with an interval of 60 s for monitoring CPU utilization and memory consumption, while for the disk usage metric the interval was 5 s, which we considered necessary to have enough samples, avoiding interference from monitoring activity. on actual system performance.

6. Experimental Results

In this section, the results collected from the experiment will be presented for both `Minikube` and `K3S` environments, considering the metrics of CPU utilization, memory consumption, disk utilization, and, finally, the requests made to the service. Each metric result is described in the following subsections. These are metrics for continuity and performance of the UAM-ODT cloud infrastructure. The data were collected from the cloud infrastructure rather than from the vehicle. The reason is we are investigating the software aging problems in a private cloud to host 24/7/365 operational digital twin services for UAM management.

It is worth highlighting that the experiments' total time differs in `Minikube` and `K3S` due to a difference in the average time to restart the pods within the auto-scaling process. This information was also measured and is presented in Table 1, showing the fastest execution of this action in the `K3S` environment, 25.4% faster than `Minikube`, evidencing an improved efficiency in auto-scaling of `K3S` when compared to `Minikube`.

Table 1. Average Pod Reset Time.

Environment	Time (s)
Minikube	97.56
K3S	72.80

6.1. CPU Utilization

In the CPU utilization evaluation, data were collected from the following specific metrics: USR, which is the percentage of CPU used by the task during execution at the user level; SYS, which is the percentage of CPU used by the task during execution at the kernel level of the OS; WAIT is the percentage of CPU spent by the task while waiting to be executed; and finally, the CPU_TOTAL, which is the total percentage of CPU time used by the task monitored by Pidstat tool, which provides statistics report for the tasks on GNU/Linux systems.

Figure 10 shows a peak of 180% of CPU_TOTAL during the initialization of the Cluster, but with an average slightly above 100% during the entire experiment in the Minikube environment. It is also possible to notice in the graph a controlled behavior within Minikube about the metrics limits since the limit is only exceeded when starting the environment. Notice also that values of utilization higher than 100% in this context are related to the usage of more than one core of the processor by this process.

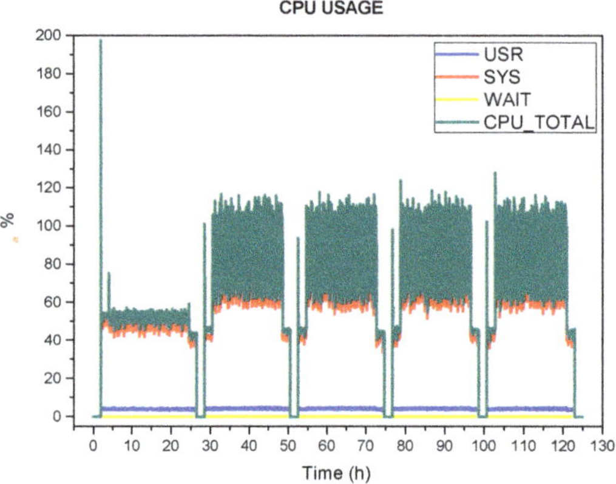

Figure 10. CPU utilization in Minikube.

Figure 11 shows a different behavior of K3S about Minikube regarding CPU utilization when we look at the CPU_TOTAL metric, which, unlike Minikube, it shows an increase in CPU_TOTAL utilization together with the USR metric over time, being interrupted when applying the cluster termination, which seems to act as a software rejuvenation technique for this situation. Although, during the entire experiment, the CPU_TOTAL did not exceed 60%.

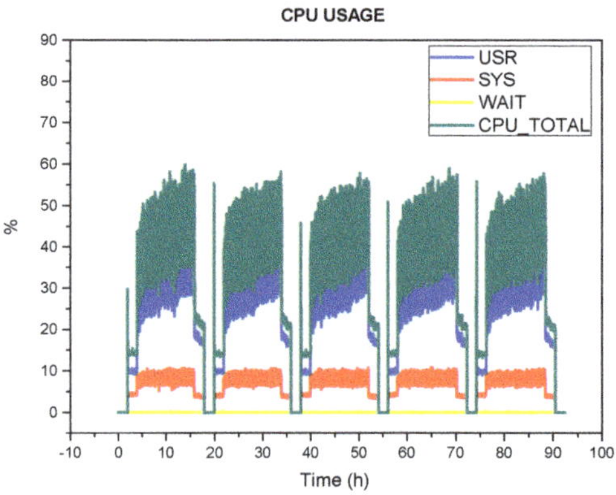

Figure 11. CPU utilization on K3S.

6.2. Disk-Related Metrics

In the evaluation of disk-related metrics, data were collected for the following metrics: READ, which represents the amount of kilobytes per second that the task took to be read; WRITE, which is the amount of kilobytes per second that the task sent to be written to the disk; and finally CANCELLED, which is the amount of kilobytes per second whose disk writing was canceled by the task, that can also occur when the task truncates some dirty page cache. All these metrics were monitored by the Pidstat tool in both Minikube and K3S.

In Figure 12, the WRITE and CANCELLED metrics have their behavior unchanged throughout the experiment, always walking close to 0 KB/s. Although, the READ metric had a distinct behavior, holding the same value throughout a single cycle of the cluster stress, and presenting a linear growth among cycles until the fourth execution cycle, being interrupted abruptly when reaching about 4,000,000 KB/s due to the limiting factor of the Minikube environment. Such behavior may be indicative of the software aging phenomenon in this environment.

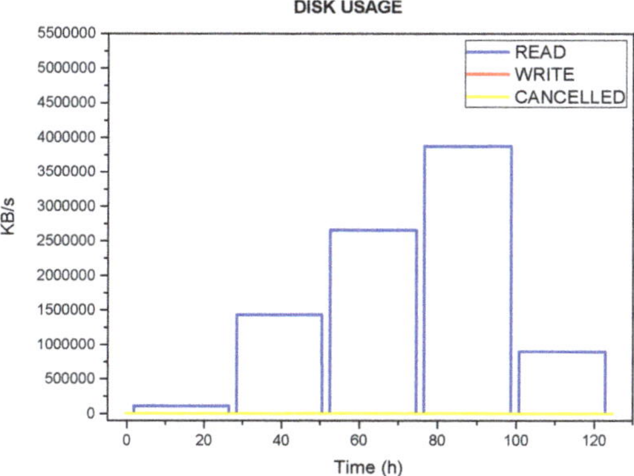

Figure 12. Disk-related metrics in Minikube.

In Figure 13, the WRITE and CANCELLED metrics also remain close to 0 KB/s throughout the experiment, similar to the execution in the Minikube environment. However, the behavior is different in the READ metric, which was not interrupted abruptly and had a linear growth from one cycle to another until the end of the experiment execution in K3S. It is important to mention that K3S presented smaller values of bytes read per second than Minikube, which might have prevented it from the abrupt fall observed there.

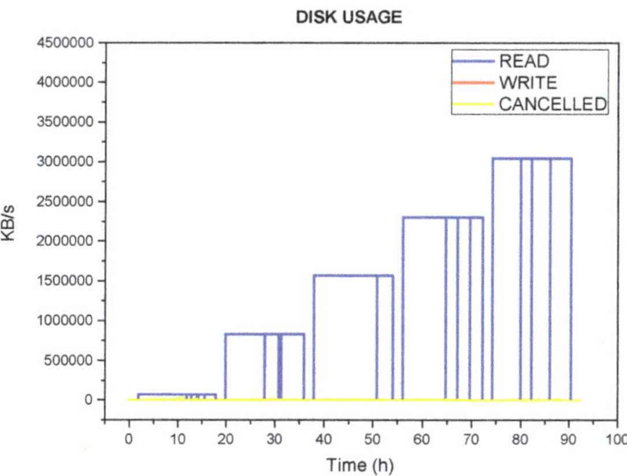

Figure 13. Disk usage in K3S.

6.3. Memory Usage

In the evaluation of memory consumption, data were collected for the metrics: MEM_USED, which represents the calculation of the total memory used; MEM_FREE, which is the memory that is not being used; MEM_AVAILABLE, which estimates how much memory is available to start new applications without swapping (it may include memory space that is being used for buffers or cache); MEM_SHARED, which is the memory mainly used by TMPFS which is the file system that keeps all files in virtual memory; MEM_BUFFERS_CACHED, which is the sum of the memory buffers and cache; SWAP_USED and SWAP_FREE metric, which represent respectively the used and free amount of virtual memory's swap space, that allows the system to use a part of the hard disk as physical memory. All these metrics were monitored using the "free" tool in both the Minikube and K3S environments.

In the evaluation of memory utilization in Minikube, the MEM_USED metric in Figure 14 has its behavior mirrored with that of the MEM_AVAILABLE metric, while the MEM_USED increases throughout the experiment, the MEM_AVAILABLE decreases in an inversely proportional trend. MEM_USED has a consumption increase of around 70% at the end of the experiment, even applying rejuvenation (i.e., cluster termination and restart between cycles). Such an action drops the memory usage temporarily, but when the cluster is started again, the system restores the same memory usage level observed at the end of the previous cycle. The MEM_FREE metric has a drop close to 48%. The MEM_BUFFERS_CACHED metric has a drop of around 41%. The SWAP_USED metric also behaves inversely to the SWAP_FREE metric, while the SWAP_USED has a 20% increase at the end of the experiment and SWAP_FREE a drop of 11%. The MEM_SHARED metric in both Minikube and K3S behave similarly, maintaining a regularity between 48 to 179 MB of consumption.

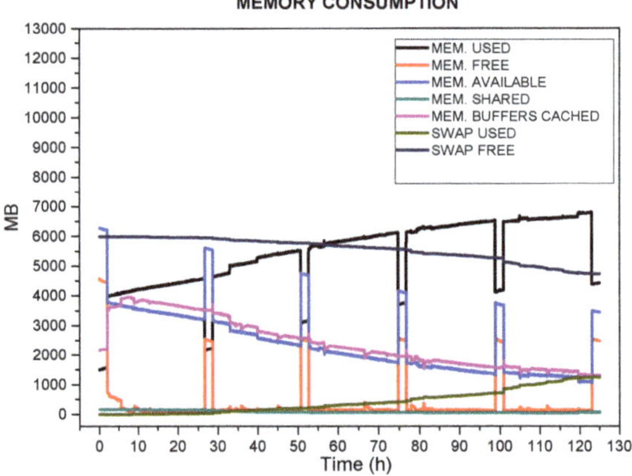

Figure 14. Memory consumption in Minikube.

In the evaluation of memory utilization in K3S, the MEM_USED metric in Figure 15 showed behavior similar to that observed in Minikube. MEM_USED has a consumption increase of around 61% at the end of the experiment, even applying rejuvenation. The MEM_FREE metric has a decrease of close to 79%. MEM_BUFFERS_CACHED has an increase of around 12%, which differs from the behavior in Minikube. The SWAP_USED has an increase of 8% when it reaches the end of the experiment and the SWAP_FREE a decrease of 8.5%.

For these memory consumption metrics, both in Minikube and in K3S, linear regression calculations on MEM_USED were performed to estimate the moment when the system would reach its upper limit for RAM usage, which in these cases is 8 GB. To confirm that estimate, we also computed the linear regression for MEM_FREE, which is another way to indicate the exhaustion of the resource, leading to system downtime and, consequently, the interruption of service provision. Similar regression estimates were carried out for the swap space usage.

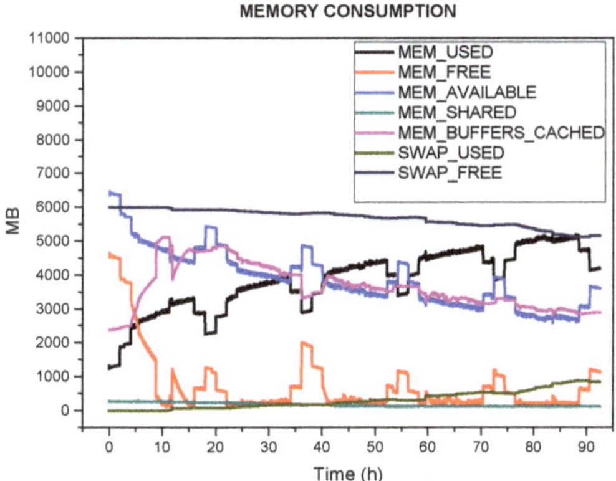

Figure 15. Memory consumption on K3S.

Equation (1) of the linear regression was obtained for the MEM_USED metric in the Minikube environment (MU_Minikube), shown in Figure 14. From this equation, it is possible to observe, as a function of MU_Minikube, that the 8 GB limit is reached after 170 h (i.e., 7 days and 2 h) of continuous execution of the workload used in the experiment. For the SWAP_USED metric, also exposed in Figure 14 for the Minikube environment, the linear regression Equation (2) was obtained.

$$MU_{Minikube} = 3900.84 + 23.98072 \times T_{stress} \tag{1}$$

In Equation (2), it is possible to observe that the upper limit of the SWAP_USED metric of the Nginx environment (SU_Minikube), which in this case is 5.8 GB, is reached after approximately 551 h of experiment, or 22 days of the same, so that the resource was completely exhausted.

$$SU_{Minikube} = -221.43413 + 10.37255 \times T_{stress} \tag{2}$$

For the MEM_USED metric of the K3S environment (MU_K3S), the linear regression Equation (3) was obtained which, through it, it is possible to observe that the upper limit of 8GB of resource for the MEM_USED metric is reached after 187 h (i.e., 7 days and 8 h) of workload execution.

$$MU_{K3S} = 2482.70 + 29.67105 \times T_{stress} \tag{3}$$

Finally, the SWAP_USED metric of the K3S environment (SU_K3S) had the linear regression Equation (4) obtained, which allows the visualization of resource exhaustion, which has a total of 5.8GB, after 603 h (i.e., 25 days and 1 h) of workload performed in the experiment.

$$SU_{K3S} = -120.24857 + 9.30782 \times T_{stress} \tag{4}$$

6.4. Evaluation and Discussions

When evaluating the results presented in Figures 10 and 11, it can be seen that most of the CPU consumption happens through the USR metric in the K3S environment, while the SYS metric does the highest consumption in the Nginx environment. This growth behavior of the USR metric in K3S was recurrent even after applying Software Rejuvenation every cycle, unlike the Nginx environment that maintains stability in the consumption of its CPU utilization metrics.

The results presented in Figures 12 and 13 show similar behavior in the use of disk usage metrics in the K3S and Nginx environment, differing only that in Nginx, the READ metric presents an interruption when it reaches 4,000,000 KB/s, returning in the fifth cycle with a total utilization close to 10%. In the K3S environment, this READ metric does not suffer an interruption but presents a linear growth from one workload cycle to another. In both scenarios, the READ metric generally presents a linear growth representing a greater need for reading from disk at each cycle.

Figures 14 and 15 show similar behavior related to memory consumption metrics in the Nginx and K3S environments, respectively. In both, there is a linear growth of the MEM_USED metric and in the SWAP_USED metric, and the opposite behavior of the MEM_AVAILABLE and SWAP_FREE metrics. Thus, with Equations (1) and (2) obtained from linear regression, it is possible to glimpse the effects arising from software aging related to memory consumption even after the application of a potential software rejuvenation action, that is, the cluster termination.

The results presented in this section are the observation of software aging phenomenon for a private cloud system hosting a UAM-ODT platform for UAM management. It is crucial to emphasise that those findings point to the dangers of system breakdowns and performance declines brought on by signs of software ageing. However, the timing of those events relies on the nature and volume of the workload that the system must handle, in addition to the hardware and software requirements of particular Kubernetes system. The ageing phenomena would be delayed and the failures caused by resource exhaustion would follow if the system had more resources available or a lighter burden than that used in this experiment. This reality does not lessen the significance of assessing the software ageing in those systems and organising countermeasures. Evaluating these scenarios using other software rejuvenation approaches and complementary metrics related to software aging are the most promising steps that could be taken in future work.

Regarding how to avoid the observed software aging phenomenon in the UAM-ODT infrastructure, in general, there are several strategies that can be used to avoid or mitigate software aging. These can include: (i) regularly updating and patching the software to fix bugs and security vulnerabilities; (ii) monitoring the performance of the software and identifying potential problems before they occur; (iii) implementing automation and management tools to help manage the software and its dependencies; (iv) using modular, microservice-based architectures to make it easier to update and maintain individual components of the system; (v) using containerization technologies, such as Docker, to package the software and its dependencies into a self-contained environment that can be easily deployed and managed. These are just some examples of strategies that can be used to avoid software aging in a cloud system. Currently, the technique to avoid software aging is monitoring the performance of the software and identifying potential problems before they occur. Further investigation on how to adopt the software rejuvenation techniques in optimal and automatic manner will be an interesting extension for research into the UAM-ODT system in which the services for UAM management using ODT are constant and at zero downtime.

7. Conclusions

This paper presented a comprehensive study on the effects of software aging problems on Kubernetes in container orchestration system in a digital twin cloud infrastructure for UAM-ODT systems. The behaviours of Kubernetes software were analysed in an accelerated lifespan experiment utilising both Nginx and K3S tools. The operations for establishing and terminating pods were carried out in real time, allowing us to monitor the usage of computational resources (such as CPU, memory, and I/O), the performance of the Nginx and K3S environments, and the response time of an application hosted in those environments. In particular settings and for specific metrics, such as virtual memory utilisation, software ageing effects were detected, indicating a memory leak that is not entirely cleansed when the cluster is halted. The study's findings help to understand the phenomenon of software ageing in digital twin computing infrastructures built on Kubernetes, which is at the very beginning of current research on software ageing issues for highly reliable and fault-tolerant digital twin computing infrastructures.

Author Contributions: Conceptualization, methodology, and supervision, R.M.; project administration, formal analysis and investigation, R.M., J.A., J.-W.L., D.M. and T.A.N.; resources, funding acquisition and investigation, E.C., J.-W.L. and D.M.; validation, J.C., J.L.; writing—original draft, J.C., R.M. and J.A.; writing—review and editing, J.A., R.M., J.L., E.C., J.-W.L., D.M. and T.A.N.; All authors have read and agreed to the published version of the manuscript.

Funding: This research was supported by the Basic Science Research Program through the National Research Foundation of Korea (NRF) funded by the Ministry of Education (2021R1A2C2094943). This research was supported by the Basic Science Research Program through the National Research Foundation of Korea (NRF), funded by the Ministry of Education (2020R1A6A1A03046811). This work is supported by the Korea Agency for Infrastructure Technology Advancement(KAIA) grant funded by the Ministry of Land, Infrastructure and Transport (Grant RS-2022-00143965).

Institutional Review Board Statement: Not applicable.

Informed Consent Statement: Not applicable.

Data Availability Statement: Not applicable.

Conflicts of Interest: The authors declare no conflict of interest.

References

1. Nguyen, T.A.; Jeon, S.; Maw, A.A.; Min, D.; Lee, J.W. Toward dependable blockchain and AI engines of digital twin systems for urban air mobility. In Proceedings of the KSAS 2021 Spring Conference, The Korean Society for Aeronautical & Space Sciences, Samcheok-si, Gangwon-do, Republic of Korea, 7–9 July 2021; pp. 408–409.
2. Nguyen, T.A.; Li, J.; Jang, M.; Maw, A.A.; Pham, V.; Lee, J.W. Cloud-in-the-loop simulation: A cloud-based digital twin HW/SW framework for multi-mode AI control simulation of eVTOL KADA-UAM Personal Aerial Vehicles. In Proceedings of the KSAS 2022 Spring Conference, The Korean Society for Aeronautical & Space Sciences, Goseong-gun, Gangwon Province, Republic of Korea, 20–24 April 2022; pp. 138–139.
3. Jiang, L.; Peng, X.; Xu, G. Time and Prediction based Software Rejuvenation Policy. In Proceedings of the 2010 Second International Conference on Information Technology and Computer Science, Kiev, Ukraine, 24–25 July 2010; pp. 114–117. [CrossRef]
4. Nguyen, T.A.; Min, D.; Park, J.S. A Comprehensive Sensitivity Analysis of a Data Center Network with Server Virtualization for Business Continuity. *Contin. Math. Probl. Eng.* **2015**, *2015*, 521289. [CrossRef]
5. Dewi, L.P.; Noertjahyana, A.; Palit, H.N.; Yedutun, K. Server Scalability Using Kubernetes. In Proceedings of the 2019 4th Technology Innovation Management and Engineering Science International Conference (TIMES-iCON), Bangkok, Thailand, 11–13 December 2019; pp. 1–4. [CrossRef]
6. Chuang, C.F.; Chen, S.S. To Implement Server Virtualization and Consolidation Using 2P-Cloud Architecture. *J. Appl. Sci. Eng.* **2017**, *20*, 121–130.
7. Vaughan-Nichols, S.J. Containers vs. Virtual Machines: How to Tell Which Is the Right Choice for Your Enterprise. 2016. Available online: https://www.networkworld.com/article/3068392/containers-vs-virtual-machines-how-to-tell-which-is-the-right-choice-for-your-enterprise.html (accessed on 10 September 2022).
8. Hat, R. Automation, Cloud, Security Lead Funding Priorities. 2018. Available online: https://www.redhat.com/en/blog/redhat-global-customer-tech-outlook-2019-automation-cloud-securitylead-funding-priorities?source=bloglisting (accessed on 11 June 2019).
9. Portworx. 2018 Container Adoption Survey. 2018. Available online: https://portworx.com/wpcontent/uploads/2018/12/Portworx-Container-Adoption-Survey-Report-2018.pdf (accessed on 11 June 2019).
10. Pereira Ferreira, A.; Sinnott, R. A Performance Evaluation of Containers Running on Managed Kubernetes Services. In Proceedings of the 2019 IEEE International Conference on Cloud Computing Technology and Science (CloudCom), Sydney, NSW, Australia, 11–13 December 2019; pp. 199–208. [CrossRef]
11. Education, I.C. Containerization. 2019. Available online: https://www.ibm.com/cloud/learn/containerization (accessed on 20 September 2022).
12. Felter, W.; Ferreira, A.; Rajamony, R.; Rubio, J. An updated performance comparison of virtual machines and Linux containers. In Proceedings of the 2015 IEEE International Symposium on Performance Analysis of Systems and Software (ISPASS), Philadelphia, PA, USA, 29–31 March 2015; pp. 171–172. [CrossRef]
13. Joy, A.M. Performance comparison between Linux containers and virtual machines. In Proceedings of the 2015 International Conference on Advances in Computer Engineering and Applications, Ghaziabad, India, 19–20 March 2015; pp. 342–346. [CrossRef]
14. Xavier, M.G.; Neves, M.V.; Rossi, F.D.; Ferreto, T.C.; Lange, T.; De Rose, C.A.F. Performance Evaluation of Container-Based Virtualization for High Performance Computing Environments. In Proceedings of the 2013 21st Euromicro International Conference on Parallel, Distributed, and Network-Based Processing, Belfast, UK, 27 February–1 March 2013; pp. 233–240. [CrossRef]

15. Xavier, M.G.; Neves, M.V.; Rose, C.A.F.D. A Performance Comparison of Container-Based Virtualization Systems for MapReduce Clusters. In Proceedings of the 2014 22nd Euromicro International Conference on Parallel, Distributed, and Network-Based Processing, Turin, Italy, 12–14 February 2014; pp. 299–306. [CrossRef]
16. Xavier, M.G.; De Oliveira, I.C.; Rossi, F.D.; Dos Passos, R.D.; Matteussi, K.J.; Rose, C.A.D. A Performance Isolation Analysis of Disk-Intensive Workloads on Container-Based Clouds. In Proceedings of the 2015 23rd Euromicro International Conference on Parallel, Distributed, and Network-Based Processing, Turku, Finland, 4–6 March 2015; pp. 253–260. [CrossRef]
17. Morabito, R.; Kjällman, J.; Komu, M. Hypervisors vs. Lightweight Virtualization: A Performance Comparison. In Proceedings of the 2015 IEEE International Conference on Cloud Engineering, Tempe, AZ, USA, 9–13 March 2015; pp. 386–393. [CrossRef]
18. Chang, Y.C.; Roohi, N.; Gao, S. Neural Lyapunov Control. In *Proceedings of the Advances in Neural Information Processing Systems*; Wallach, H., Larochelle, H., Beygelzimer, A., d'Alché-Buc, F., Fox, E., Garnett, R., Eds.; Curran Associates, Inc.: Nice, France, 2019; Volume 32, pp. 3245–3254. [CrossRef]
19. Jiang, Z.; Lynch, A.F. Quadrotor Motion Control Using Deep Reinforcement Learning. *J. Unmanned Veh. Syst.* **2021**, *9*, 234–251. [CrossRef]
20. Glaessgen, E.; Stargel, D. The Digital Twin Paradigm for Future NASA and U.S. Air Force Vehicles. In Proceedings of the 53rd AIAA/ASME/ASCE/AHS/ASC Structures, Structural Dynamics and Materials Conference 20th AIAA/ASME/AHS Adaptive Structures Conference 14th AIAA, Honolulu, HI, USA, 23–26 April 2012; American Institute of Aeronautics and Astronautics: Reston, VA, USA, 2012. [CrossRef]
21. Shah, S.; Dey, D.; Lovett, C.; Kapoor, A. AirSim: High-Fidelity Visual and Physical Simulation for Autonomous Vehicles. In *Field and Service Robotics*; Springer: Cham, Switzerland, 2018.
22. Huang, Y.; Kintala, C.; Kolettis, N.; Fulton, N. Software rejuvenation: Analysis, module and applications. In Proceedings of the Twenty-Fifth International Symposium on Fault-Tolerant Computing. Digest of Papers, Pasadena, CA, USA, 27–30 June 1995; pp. 381–390. [CrossRef]
23. Xu, J.; You, J.; Zhang, K. A neural-wavelet based methodology for software aging forecasting. In Proceedings of the 2005 IEEE International Conference on Systems, Man and Cybernetics, Waikoloa, HI, USA, 12 October 2005; Volume 1, pp. 59–63. [CrossRef]
24. Avritzer, A.; Weyuker, E.J. Monitoring Smoothly Degrading Systems for Increased Dependability. *Empir. Softw. Eng.* **1997**, *2*, 59–77. [CrossRef]
25. Nguyen, T.A.; Kim, D.S.; Park, J.S. A Comprehensive Availability Modeling and Analysis of a Virtualized Servers System Using Stochastic Reward Nets. *Sci. World J.* **2014**, 1–18. [CrossRef] [PubMed]
26. Singleton, A. The Economics of Microservices. *IEEE Cloud Comput.* **2016**, *3*, 16–20. [CrossRef]
27. Dragoni, N.; Lanese, I.; Larsen, S.; Mazzara, M.; Mustafin, R.; Safina, L. Microservices: How To Make Your Application Scale. In *International Andrei Ershov Memorial Conference on Perspectives of System Informatics*, 11th ed.; Springer: Cham, Switzerland, 2017.
28. Trunov, A.S.; Voronova, L.I.; Voronov, V.I.; Ayrapetov, D.P. Container Cluster Model Development for Legacy Applications Integration in Scientific Software System. In Proceedings of the 2018 IEEE International Conference "Quality Management, Transport and Information Security, Information Technologies" (IT&QM&IS), St. Petersburg, Russia, 24–28 September 2018; pp. 815–819. [CrossRef]
29. Ageyev, D.; Bondarenko, O.; Radivilova, T.; Alfroukh, W. Classification of Existing Virtualization Methods Used in Telecommunication Networks. In Proceedings of the 2018 IEEE 9th International Conference on Dependable Systems, Services and Technologies (DESSERT), Kyiv, UKraine, 24–27 May 2018; pp. 83–86. [CrossRef]
30. Blog, G.C.P. An Update on Container Support on Google Cloud Platform. 2014. Available online: https://cloudplatform.googleblog.com/2014/06/an-update-on-container-support-on-google-cloud-platform.html (accessed on 8 October 2021).
31. Research, G. Hype Cycle for Cloud Computing. 2018. Available online: https://www.https://www.gartner.com/en/documents/3884671 (accessed on 14 October 2021).
32. Diamanti. 2018 Container Adoption Benchmark Survey. 2018. Available online: https://diamanti.com/wp-content/uploads/2018/07/WP_Diamanti_End-User_Survey_072818.pdf (accessed on 12 October 2021).
33. Forrester. The Forrester New waveTM: Enterprise Container Platform Software Suites. 2018. Available online: https://cloud.google.com/containers/ (accessed on 18 October 2021).
34. Kubernetes. Production-Grade Container Orchestration. 2019. Available online: https://kubernetes.io/ (accessed on 18 October 2021).
35. Nguyen, T.A.; Min, D.; Choi, E.; Lee, J.-W. Dependability and Security Quantification of an Internet of Medical Things Infrastructure Based on Cloud-Fog-Edge Continuum for Healthcare Monitoring Using Hierarchical Models *IEEE Internet Things J.* **2021**, *8*, 15704–15748. [CrossRef]
36. Basili, G.; Caldiera, V.R.; Rombach, H.D. The goal question metric approach. In *Encyclopedia of Software Engineering*; John Wiley & Sons, Inc.: Hoboken, NJ, USA, 1994; pp. 528–532.

Article

Gusts Encountered by Flying Vehicles in Proximity to Buildings

Abdulghani Mohamed [1,*], Matthew Marino [1], Simon Watkins [1], Justin Jaworski [2] and Anya Jones [3]

1 School of Engineering, RMIT University, Melbourne, VIC 3001, Australia
2 Mechanical Engineering and Mechanics, Lehigh University, Bethlehem, PA 18015, USA
3 Department of Aerospace Engineering, University of Maryland, College Park, MD 20742, USA
* Correspondence: abdulghani.mohamed@rmit.edu.au

Abstract: There is a growing desire to operate Uncrewed Air Vehicles (UAVs) in urban environments for parcel delivery, and passenger-carrying air taxis for Advanced Air Mobility (AAM). The turbulent flows and gusts around buildings and other urban infrastructure can affect the steadiness and stability of such air vehicles by generating a highly transient relative flow field. Our aim is to review existing gust models, then consider gust encounters in the vicinity of buildings as experienced by flight trajectories over the roof of a nominally cuboid building in a suburban atmospheric boundary layer. Simplified models of fixed- and rotary-wing aircraft are used to illustrate the changes in lift and thrust experienced by flight around the building. The analysis showed that fixed-wing aircraft experienced a substantial increase in angle of attack over a relatively short period of time (<1 s) as they fly through the shear layer at a representative forward velocity, which can be well above typical stall angles. Due to the slow flight speeds required for landing and take-off, significant control authority of rotor systems is required to ensure safe operation due to the high disturbance effects caused by localized gusts from buildings and protruding structures. Currently there appears to be negligible certification or regulation for AAM systems to ensure safe operations when traversing building flow fields under windy conditions and it is hoped that the insights provided in this paper will assist with future certification and regulation.

Keywords: turbulence; gust; UAV; urban; severe; limitation; survey; CFD; city; urban air mobility; buildings; infrastructure; air taxi; advanced air mobility; certification; regulation; vertiports

Citation: Mohamed, A.; Marino, M.; Watkins, S.; Jaworski, J.; Jones, A. Gusts Encountered by Flying Vehicles in Proximity to Buildings. *Drones* **2023**, *7*, 22. https://doi.org/10.3390/drones7010022

Academic Editors: Ivana Semanjski, Antonio Pratelli, Massimiliano Pieraccini, Silvio Semanjski, Massimiliano Petri and Sidharta Gautama

Received: 27 October 2022
Revised: 19 December 2022
Accepted: 23 December 2022
Published: 28 December 2022
Corrected: 31 May 2023

1. Background and Objectives

It is well documented that aircraft of all sizes are adversely affected by turbulence and gusts; as identified by the Federal Aviation Administration (FAA) and the US Transportation Safety Board as a leading cause of accidents—costing over USD 100M p.a. [1]. Severe injuries are reported, such as those in the 2015 Air Canada flight AC088, which injured 21 passengers, including three children [2]; and 2019 Qantas Flight QF108 whereby 3 cabin staff had head and neck injuries [3]. Accidents still continue to occur with more recent ac-cidents that resulted in injured passengers [4] and even a passenger death [5]. As the size, mass and speed of aircraft decrease, the susceptibility to turbulence and gusts increases [6,7]; or in sum, due to lower wing loading [8]. Smaller general aviation aircraft and helicop-ters also tend to fly more at lower altitudes within the Atmospheric Boundary Layer (ABL) which is dominated by high turbulence intensities from ground protruding structures [7,9]. This has led to reported accidents directly relating to turbulence [10–13]. Even the tran-sition through the ABL can be detrimental to aircraft that are designed to fly at very high altitudes such as Facebook's Aquila Uncrewed Air Vehicle (UAV) and Airbus' Zephyr UAV, whereby both had fatal crashes due to turbulence and/or gusts [14,15].

The advent of Advanced Air Mobility (AAM) vehicles involves operating fleets of UAVs in urban environments far more frequently than we have ever anticipated, for the purpose of transporting parcels and passengers. This exposes the fleet of aircraft to a wide range of challenging flow conditions; specifically large-scale gusts induced by urban infrastructure which can persist up to several kilometers away from the source and interact in complex ways. AAM will more than often involve operation in close-proximity to physical structures (e.g., inspection of infrastructure, or take-off and landing operations from building rooftops). In the presence of large-scale gusts, significant flight path deviations can occur, increasing risk of collision with objects. Aircraft collisions with high-rise buildings is not unheard of [16], and the routine operation of UAVs in cities further increases the risk of collisions. There is a need for both research and regulation efforts to enhance safety and minimize the risk through considering vertiport and vehicular design.

The most relevant aspect of aviation to AAM is the operation of helicopters which also fly in urban environments, albeit less frequently and with a human pilot onboard. Landing on buildings poses a specific challenge in some cases, warranting further aerodynamic studies and field wind measurements being prudent [17]. From a vehicular design standpoint, the AAM vehicles' design and flight dynamics are different from the conventional helicopter and airplane design which warrants an exploration into novel design features and technologies that enable lower sensitivity to turbulence and precise maneuvering [1]. From a vertiport standpoint, the existing heliport infrastructure can potentially support AAM; however there is a need for purpose-built buildings (for ease of public access and to account for the autonomy of UAVs). The characterization of the flow fields for different wind conditions around vertiports is warranted, similar to those conducted for heliports [18–21]. New research is, thus, required to characterize the temporal and spatial variation in the flow fields around buildings and vertiports. This will inform vertiport design and site selection to minimize the risk imposed by the local wake of the building from affecting flight safety as well as passenger ride quality.

In recent years, considerable attention has focused on measurements in ground-test facilities or computations that replicate some idealized flow unsteadiness such as a pitching and/or plunging maneuver or an imposed well-characterized gust [22–29]. However, perhaps the most obvious gust problem for UAV flight is steady level operation, or at least, intended steady level operation through the atmospheric boundary layer (ABL), where no discrete obstacle (or associated wake) is present. Previous studies on UAV flight through the ABL [30,31] have shown that three-dimensional (3-D) turbulent structures induce particularly strong disturbances in UAV roll response owing to variation in effective angle of attack along the wingspan. This disturbance in roll was also noted in comments from pilots attempting to hold steady level flight in well-mixed turbulence [32]. Roll disturbances not only degrade payload performance (particularly the blurring of images from optical sensors) but may also lead to undesired flight path deviations. The most critical parts of UAV urban operations entail flight in very close proximity to buildings and may include entering buildings through windows or air vents or landing on their rooftops (see Figure 1). Whilst the flow field around buildings has been extensively studied from a fixed reference frame (e.g., by wind engineers for the purposes of structural loadings [33,34], dispersion of pollutants [35,36], pedestrian wind comfort [37,38], etc.), there appear to be very few studies from the reference frame of the moving aircraft and at the relevant frequencies [39]. We therefore examine this relative flow field with an overall aim to reveal the characteristics of a "severe" gust for UAVs in close proximity to buildings.

In this paper we first review turbulence in the ABL to frame a taxonomy of gusts and consider their relevance to UAVs. The more challenging flight environment for vehicles passing through the local wakes of buildings is then considered and compared to flight in the ABL. Flight in the urban environment is expected to yield gusts of high severity (frequency and/or amplitude), most likely leading to unwanted, severe force spikes and flow separation about the aircraft wing. While the problem is inherently 3D, we first investigate a 2D longitudinal-only case by examining the relative flow near the centerplane of the building. The outcome of this work is an assessment of the most basic research question to characterize the urban environment: What are the disturbances in effective angle of attack and relative flight speed magnitude in a flight-relevant urban gust encounter?

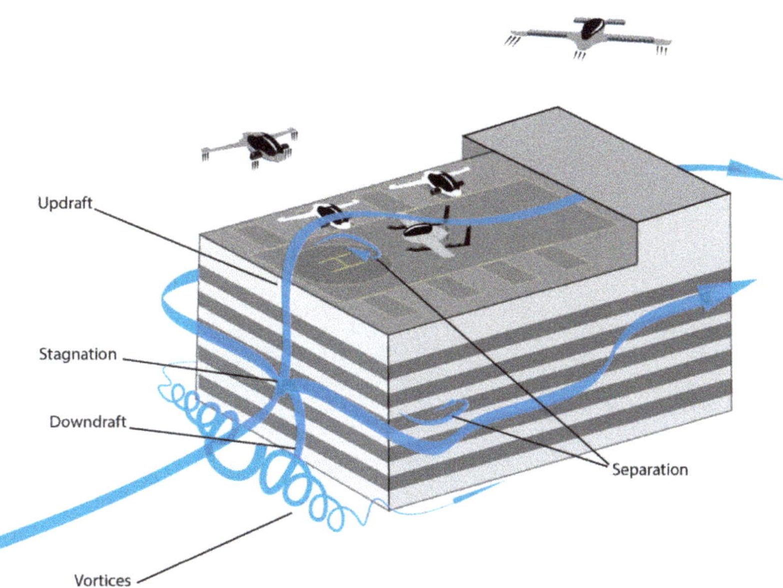

Figure 1. Notional flow field about a building generated by atmospheric winds.

2. Turbulence

Turbulence is defined as a chaotic, random, highly nonlinear and unpredictable flow [40]. In the atmosphere, the characteristics of turbulence are influenced by the thermal stability of the ABL (adiabatic, or various degrees of stability). However, under strong winds mechanical mixing tends to dominate the turbulence generation mechanisms and thermal stability plays a smaller role. Thus, in the current work we ignore thermally driven turbulent flows, as they only tend to dominate under light winds, which are unlikely to generate severe gusts. The ABL extends from the Earth's surface up to an altitude where the wind is no longer influenced by the roughness of the ground, which may include geological or civil structures. The mean wind speeds increase from zero at the Earth's surface up to the "gradient" wind speed, i.e., that which occurs at the gradient height, typically 1–2 km depending upon terrain roughness. Above this height the air is generally smooth, except for bursts of "clear air turbulence," which are not considered here. The ABL is well documented from stationary measurements for various purposes, including meteorological and wind engineering studies (e.g., [41–43]). The interaction of the ABL with obstacles such as buildings, bridges and other infrastructure will generate coherent turbulence structures with length scales of a similar size to the obstacle, as depicted in a 3-D computational fluid dynamics (CFD) simulation shown in Figure 2, from [44]. The building shown is nominally a cuboid of dimension 43 m, and the simulation includes a representation of the velocity and intensity profiles in the approaching ABL. Figure 3 further illustrates the decaying nature of turbulence in an urban scenario, whereby the coherent structures dissipate downstream of obstacles, and a well-mixed turbulent wake then develops (as can be seen downstream of the building in the figure). These flow features yield a velocity field with a broad spectral content that contains a wide range of length and time scales.

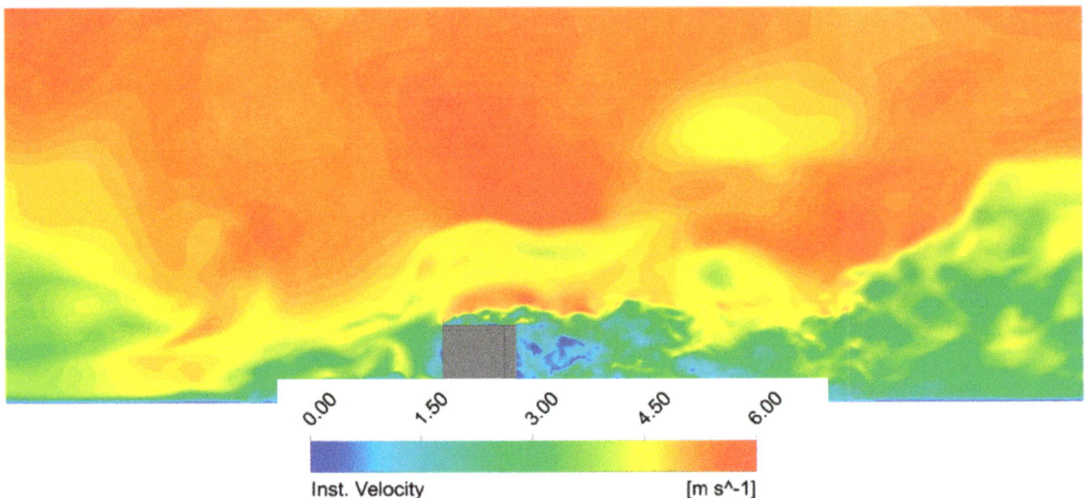

Figure 2. Instantaneous velocities in the atmosphere in an urban environment. Flow travels from left to right. With the reference height and velocity as $U_\infty = 3$ m/s and $y_\infty = 10$ m, this results in a domain (average) Re of approximately 2.05×10^6 [44].

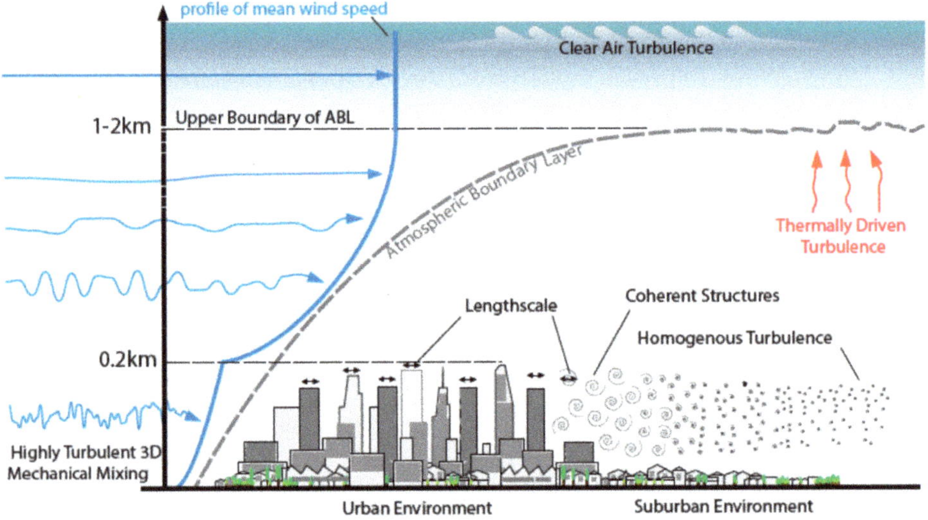

Figure 3. The atmospheric environment in an urban location.

3. Prior Gust Models

Aircraft encounter different types of turbulence while flying through the ABL, and there exists a significant body of knowledge relevant to manned flight focused on the temporal and spatial characteristics of the flow environment that is well-removed from local effects and (usually) from the influence of the ground. These prior works include continuous gust models that represent the structure of the statistically random flow fluctuations in the atmosphere as power spectral functions. These spectra allow for predictions of the mean-square values of the flight vehicle and aeroelastic responses, provided that a transfer function between the gust and response can be established from deterministic or other means [45,46].

The most common continuous gust spectra of Von Karman [47], as well as those of Diederich and Drischler [48], Dryden [49]) are one-dimensional, i.e., they yield three orthogonal velocity components at a single point, a restriction that neglects gradients in the gust across the aircraft as well as any altitude-dependent wind shear effects. These gust models are built up from the statistical theory and measurements of isotropic turbulence. The von Kármán model form interpolates between the isotropic scaling results of Heisenberg [50] at low frequency and the higher-frequency scaling of Kolmogorov [51] in the inertial subrange. The Dryden model instead assumes a functional form that fits experimental measurements of the isotropic turbulent energy spectrum in the early stages of decay; see Liepmann, Laufer [52] for further discussion and comparison of these gust models. The choice of the simpler Dryden form over the more theoretically-grounded von Kármán model is largely a matter of engineering convenience; the correctness advantage of the von Kármán model is important only if significant spectral content relevant also to the flight and aeroelastic dynamics is centered in the microscale range, a decade or more above the integral scale break frequency where the isotropic inertial subrange begins [53]. The isotropic turbulence assumption, central to both models, is valid for turbulence at high altitude. However, at lower altitudes relevant to UAVs/AAM (less than about 2000 ft), anisotropic effects of the ABL without the influence of urban structures may be modeled by adjusting the turbulence intensity and turbulence length scales in the isotropic models according to empirical design specifications. Such specifications at low altitude for the von Kármán and Dryden models, as well as a discussion of more sophisticated gust models, are organized by Standard [53]. Continuous gust models may be compared with traditional discrete gust models including the sharp-edged gust and "1-cosine" gust used to establish severe aeroelastic scenarios. However, if desired, one may readily construct a continuous gust from a known series of discrete gusts [54], and the continuous and discrete models may be superposed provided that the flow disturbances and resulting structural motions are sufficiently small to retain linearity.

Flows within an urban environment are generally inhomogeneous, anisotropic, and time-varying and, therefore, violate many of the core assumptions of traditional gust models. Near the ground, turbulence length scales and intensities vary rapidly with altitude and depend strongly on the terrain [55]; there is a lack of viable models to describe the broad range of general turbulent flows possible in this environment. The introduction of AAM and UAVs further complicates the modelling challenge of the urban environment. Wind shears from the terrain and from multi-scale arrays of buildings produce longitudinal and vertical gusts that generate significant roll and yaw moments, which must be characterized and accounted for in the gust and vehicle dynamics models [56]. In the absence of buildings and terrain, the length scales of the most energetic eddies in the ABL are much larger than the UAV feature lengths, and the high-frequency content of the turbulence spectrum is therefore expected to play a more significant role in the vehicle gust response. However, the urban landscape affects this turbulent flow and can introduce gust length scales pertinent to the air vehicle response. Furthermore, the gusts encountered by UAVs near buildings may be large relative to the local background flow and can lead to catastrophic nonlinear effects, such as stall-induced pitch-up. In light of these challenges, the next sections survey experimental measurements and computational simulations to characterize the three-dimensional gust fields of canonical urban landscapes and investigate scenarios of vehicle trajectories in this environment.

4. Turbulence Experienced by Moving Vehicles (Relative Turbulence)

Turbulence Intensity (*Ti*) is defined as the standard deviation of the fluctuating component of wind velocity (u') divided by the mean wind velocity ($\overline{U}$),

$$Ti = \frac{\sqrt{(u')^2}}{\overline{U}} = \frac{\sigma_{(u')}}{\overline{U}} \tag{1}$$

The variation in the intensities and scales with height from the ground from a stationary perspective (i.e., with reference to the ground) is described in Watkins, Thompson [30], and a database compiled from a wide range of measurements can be found in ESDU 85020 [57]. Movement through the turbulence field at different speeds and directions changes how the turbulence is perceived by moving vehicles. The effect of a moving measurement reference frame has been explored by Watkins and Cooper [58] for ground-based vehicles, where two-component data (in the horizontal plane) obtained from hot-wire anemometers mounted above a vehicle were compared for fixed and moving vehicle frameworks. Turbulence intensities measured from the moving vehicle were found to be in good agreement with those predicted from the measured vehicle-fixed data in relatively smooth domains, well-removed from local wakes such as buildings. However, when data were obtained in rougher terrains, which included traversing local wakes, a significant increase in turbulence intensity was found in the data from the moving vehicle. The lateral intensities were considerably higher than values predicted from ground-fixed data, whereas only slight increases in longitudinal intensities were noted. This result was attributed to the fact that turbulence from a stationary perspective (referenced to the ground) was measured at locations specifically chosen to be removed from local wakes.

Watkins, Milbank [6] extended this work to include three-component data obtained from four laterally spaced, dynamically calibrated, multi-hole Cobra probes. This extension was carried out to understand the turbulent flow environment of UAVs, whereby the lateral separation between the probes could be altered to document the flow impinging at different spanwise locations on a UAV wing. Data were collected over various types of terrain, and under a range of wind speeds and vehicle speeds that included some data closer to buildings than in earlier hot-wire measurements. The closest that the measurement tracks came to buildings was about 5 m due to the vehicle being driven on public roads. The study provided data relating the measured turbulence intensities to relative flight velocity (Figure 4), demonstrating a reduction with increasing freestream speed. In the moving case, the denominator in the turbulence intensity (Equation (1)), $\overline{U}$, becomes V_r, which is the vehicle speed relative to the air (i.e., the wind speed). Figure 5 illustrates the vector addition used to compute V_r,

$$V_r = \sqrt{V_w^2 + V_w^2 - 2\,V_w\,V_v\cos\theta} \tag{2}$$

It is therefore important to differentiate between *Ti* and the Relative Turbulence Intensity (*J*), which takes into account the relative velocity, V_r:

$$J = \frac{\sqrt{(V_W')^2}}{\overline{V_r}} = \frac{\sigma_{V_W'}}{\overline{V_r}} \tag{3}$$

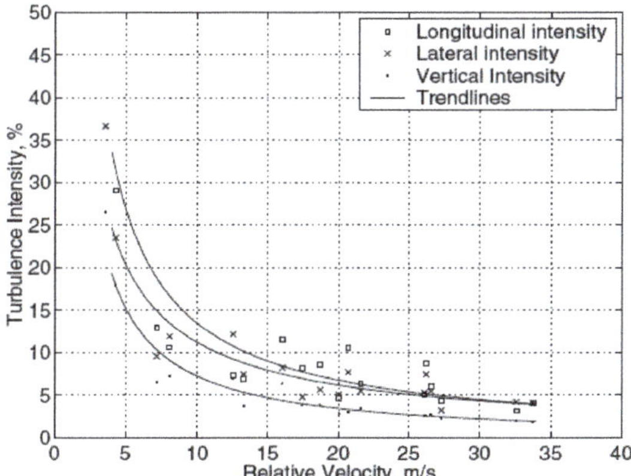

Figure 4. The relationship between relative turbulence intensity J and flight velocity V_V [6].

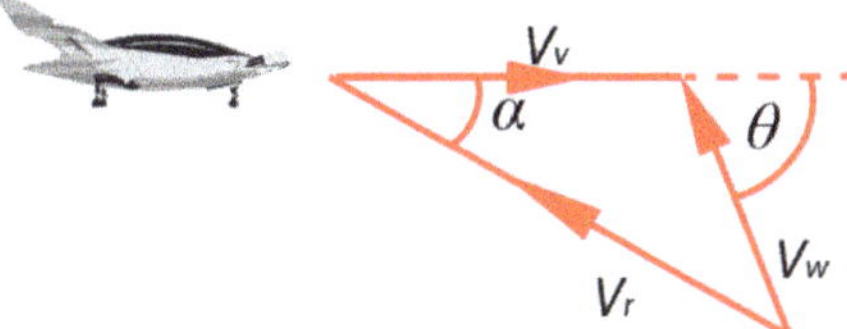

Figure 5. Aircraft and wind velocity vectors.

5. Relevant Gust Characteristics

Excessively large gusts (i.e., those with length scales significantly larger than the vehicle's characteristic dimension) can often be considered quasi-steady, and their effects are relatively easily compensated for [6]. Gust scales equivalent to or smaller than the characteristic length are more deleterious and introduce significant asymmetrical forces and moments. As a gust impacts the leading edge of an aerodynamic surface such as a wing, the flow angle and velocity are altered, inducing variations in the load distribution as illustrated in Figure 6. Gusts of a 3-D nature that are smaller than the wing span will lead to uneven lift distribution over the wings, inducing a rolling motion. Lissaman [59] demonstrated that a sinusoidal load distribution with a period relating to a dimension that is slightly larger than the span of the aircraft results in the maximum roll moment.

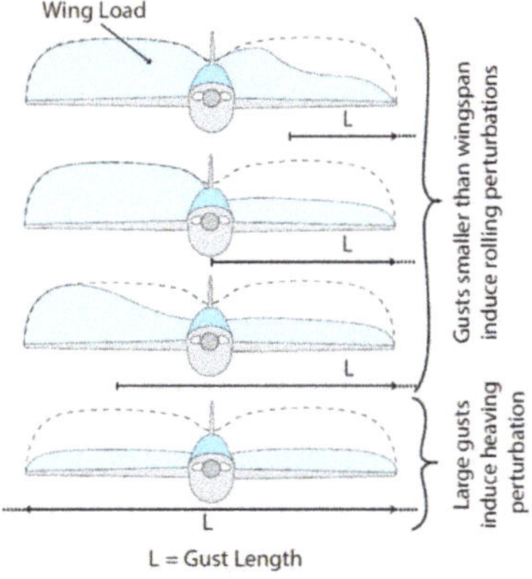

Figure 6. Effect of gust length scale on wing loading. Adapted from Lissaman [59].

Gusts in well-mixed turbulence are highly three-dimensional in nature and it has been shown that out of the possible six degrees of freedom, rolling motion is the most significant disturbing factor for UAVs [6]. Atmospheric measurements in well-mixed turbulence removed from building wakes illustrate the three-dimensionality of gusts, whereby significant flow pitch variations are evident across typical UAV wingspans or rotor diameters. Figure 7a shows a typical time record of the angle of attack, α, recorded by four laterally separated probes during a two-second sampling time, showing large fluctuations of the order of $\pm 10°$. At first, it might seem that there is a strong correlation between the pitch angles measured from the four probes. However, closer examination of the data presented in Figure 7b reveals that there are considerable differences, and at some instances the variation is $\approx 15°$ across probes with a lateral separation of 150 mm.

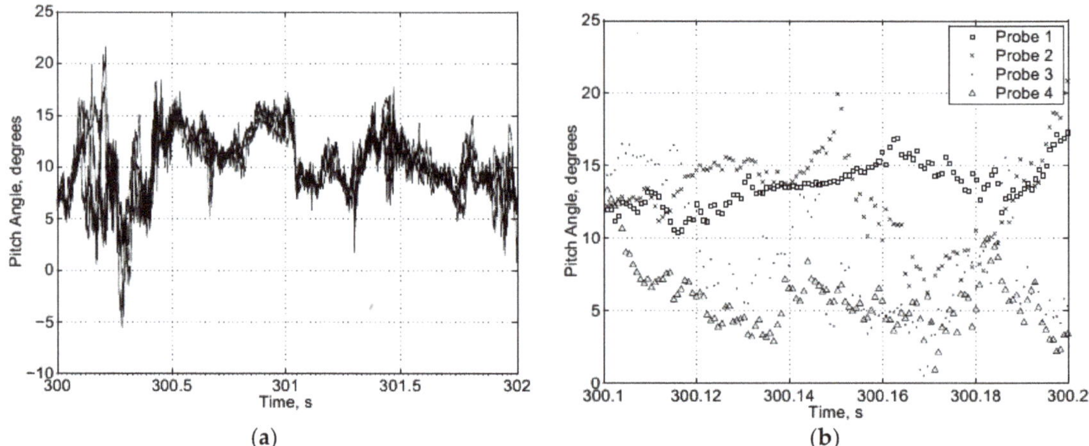

Figure 7. Pitch angle variation: (**a**) 2-s sample (**b**) 0.2-s sample [6].

For fixed wings, Thompson, Watkins [60] showed that typical lateral variations in α are more significant than the associated velocity magnitude variations in generating potential rolling moments (using data from measurements of well-mixed atmospheric turbulence close to the ground applied to simple wing strip theory). The experimental work by Mohamed, Watkins [31] confirmed the high sensitivity of the roll axis to α variation. For rotary wings, among the most relevant work was that conducted by Wang, Dai [61] in which it was found that a variable pitch helicopter blade encountering a downward gust experiences a significant reduction in thrust force. It was also found that the sharper the gust, the more adverse the response is with respect to aerodynamic forces and structural deflection. This behavior is particularly relevant when travelling through shear layers at higher speeds, causing the relative encountered gust front to be perceived as a sharp gust front.

6. Gust Taxonomy

It is desirable to approximate gusts as quasi 1-D or 2-D (see Figure 8) for fundamental studies on the transient flow field around airfoils through, for example, pitch and/or plunge motions in fundamental experiments. However, the reality of well-mixed atmospheric turbulence is intrinsically three-dimensional in nature. Discrete gusts can be categorized as either 1-D or 2-D in the streamwise or transverse directions. Streamwise 1D gusts involve a momentary change in streamwise velocity. For example, as streamwise velocity increases, the corresponding lift over an airfoil also increases, which if not corrected, will result in a translation of the airfoil upwards (due to lift) and backwards (due to the increased drag). Non-symmetric velocity changes along the span of a wing will result in a rolling and yawing motion if not taken into consideration. It is worth noting that Thompson,

Watkins [60], using a simple strip theory model, found that angular flow changes typically have a tenfold greater effect on lift compared to the magnitude changes in atmospheric turbulence. This behavior implies that travelling through a transverse gust will result in a stronger generation of lift than from a streamwise gust.

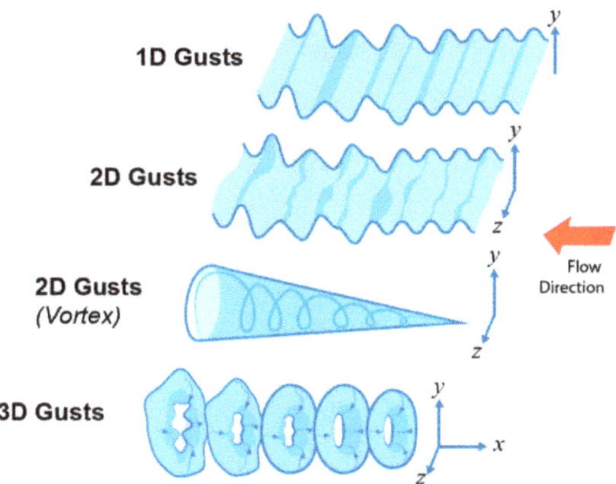

Figure 8. Dimensionality of gusts (modified from Diederich [62]).

7. Severe Gusts around Buildings: Case Studies

Let us now consider the flow field around a nominally cuboid building in a suburban environment. At the juncture between the building and the ground plane, there is the usual horseshoe vortex, perhaps with associated finer structures [63]. Near the building rooftop, there is expected to be a separated flow with meandering shear layers of time-varying position, width, and intensity. Depending on the building's geometry and wind direction, vortices may also be present near the rooftop. Using the taxonomy discussed in the previous section, possible gust encounters by UAV flight in urban environments are illustrated in Figure 9. Given that the angular flow changes typically have a greater effect on sectional lift coefficient in contrast to magnitude changes [31], the most detrimental case in this set is likely to be a transverse gust given the rapidity of the encounter with respect to the flight trajectory. The latter scenario will therefore be the focus of a case study presented in the remainder of this paper, whereby we use the flow field around a representative cuboid building computed by Mohamed, Carrese [44] (see Figure 2) to estimate variations in the lift and rolling moment coefficients of representative UAVs. The CFD simulation representing an urban environment uses an Improved Delayed Detached Eddy Simulation (IDDES) turbulence model. Mohamed, Carrese [44] validated the simulation by demonstrating excellent agreement of the solution strategy with the experimental and large eddy simulation (LES) data of similar but simpler cases. The validation cases examined were: (1) developed channel flow, (2) flow over a backward-facing step, (3) flow over periodic 2D hills, (4) wall-mounted hump flow, and (5) trailing-edge separation over a hydrofoil. Full details of the basis of these simulations can be found in [44] and comparison with point-probe atmospheric measurements is carried out in [39,64].

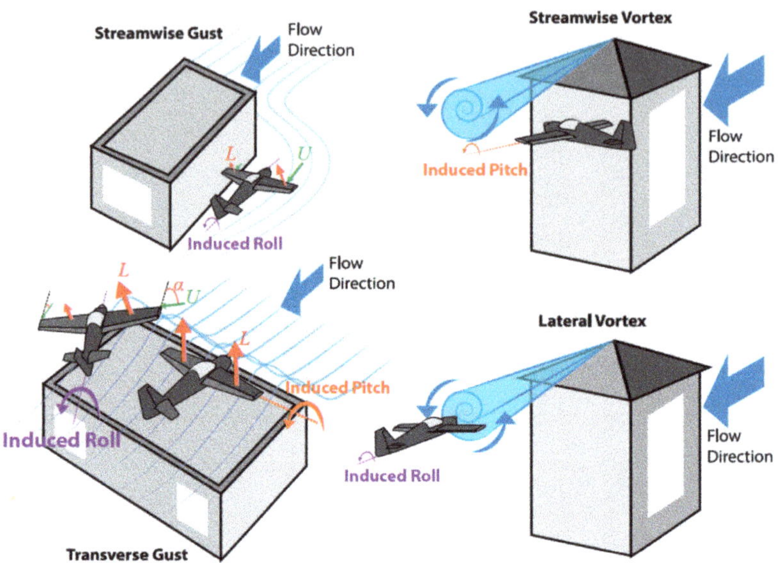

Figure 9. Schematics of possible gust encounters by a fixed wing UAV flying in the vicinity of buildings. Note drawn to scale for illustration purposes.

In the simulation, inflow boundary conditions replicate the relevant velocity and intensity profiles of a suburban ABL. Due to the mesh resolution near the building (~0.05 m), the scales of the resolved turbulence are suitable for the UAV spans discussed in this paper. Initialization of the IDDES simulation was provided using a steady-state k—ω SST model. The RANS momentum field is converted to an instantaneous momentum field before commencing the transient run. The pressure-based Non-Iterative Time-Advancement (NITA) fractional-step solver is utilized, with bounded second-order temporal discretization. The time step is normalized by the ratio of (l_∞/U_∞) with a non-dimensional time-step of Δt^* = 0.003 for the total time of the simulation $t^*_T = 600$ with sampling statistics collected from $t^* > 200$. An average wind speed of 3 m/s at a height of 10 m was used in the upstream boundary condition representing the ABL, and the mean wind direction was normal to the southerly face (i.e., along the x-axis in the figures below). The modelling requirements and profiles for the ABL were obtained from the work of Blocken, Stathopoulos [65], and the ABL velocity profile $U(y)$ was estimated using

$$U(y) = \frac{u^*}{\kappa} \cdot ln\left(\frac{y}{y_0}\right) \tag{4}$$

where u^* is the friction velocity, U_∞ and y_∞ are the reference velocity and height, κ is the von Kármán constant, and y_0 is the equivalent aerodynamic roughness height. The profiles for the turbulence kinetic energy k and specific dissipation ω were estimated using:

$$k(y) = u^{*2} \cdot C_\mu^{-0.5} \tag{5}$$

$$\omega(y) = u^* \cdot C_\mu^{-1.5} \cdot \frac{\kappa}{y} \tag{6}$$

7.1. Flight Trajectory Modelling

Consider a UAV flying at speed V_V in close proximity to a building. Depending on the flight path and the direction of the wind, a wide range of perturbations may be perceived (i.e., the gusts experienced relative to the moving UAV will vary with flight path and wind). Severe gusts are taken to be those that result in a large step change in aerodynamic forces or moments. Realizing that the atmospheric wind can vary from calm to extreme (i.e., storm) levels, it is necessary to select a single atmospheric wind speed and direction, then investigate flight paths relative to the building flow field that would generate the severe cases.

We consider the flight trajectories outlined in Figure 10, representing two flight paths towards the leading edge of the building (0° and 45° flight path angle), performed at some height above the rooftop, thus encountering the shear layers shed from the building structure. The 0° flight path represents the simpler case where the vehicle encounters the gust head-on, and there are no gust-induced rolling moments. The 45° flight path provides insight into the rolling moment that arises due to lift imbalance as one wing is immersed into the shear flow before the other wing. In reality, the UAV's trajectory will be influenced by the flow field. We ignore these vehicle dynamics and any coupling of the vehicle's flow field with that of the building and assume that the vehicle acts as a massless point-particle UAV. Thus, we assume "frozen" turbulence; that is, the computed wind field is sampled at one instant in time, and the "turbulence" encountered by the UAV is the variations in the relative flow field velocity as the vehicle proceeds in its idealized, steady level flight. While such simplification is unrealistic from the viewpoint of airplane flight mechanics, it is arguably sufficient to define a realistic "severe case" to be studied.

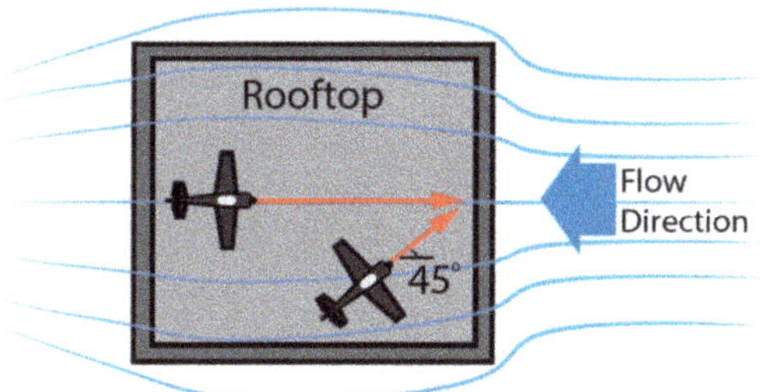

Figure 10. Planform view of flight paths considered in this paper.

The flow fields around a building were extracted from the CFD model (see Figure 11) to identify the gusts encountered as perceived by a moving aircraft. These flow fields were imposed on a simplified model of a fixed wing UAV in a way similar to that by Thompson, Watkins [60] as well as an actuation disk model of a single rotor, in order to extract severe cases during a straight flight path. The chosen aircraft speeds were 5 m/s and 15 m/s with respect to the ground (i.e., typical velocities for UAVs).

The flow extracted from the CFD simulation is presented in this section. The wind along a representative flight path (at fixed points along the flight trajectory) is shown in Figure 12. The flow field for various heights is depicted in Figure 13, and the flow extracted from the CFD simulation is given in Figure 14. The wind velocity is plotted as if it were in polar coordinates following the convention shown in Figure 12. The "flow pitch angle" is the direction of local flow at a h/H value of 0.0023 where h is the height of flight path above the rooftop and H is the building height. The trajectory closest to the roofline is at height ratio of $h/H = 0.0023$, or 10 cm above the roof, which is immersed in a boundary layer of the building itself. This boundary layer is present even at the intermediate trajectory height of $h/H = 0.14$, which is physically 6 m above the building. In this region, from 0 to -1 on the abscissa of Figure 13, the wind speed is low, but highly variable. At $h/H = 0.25$ and 0.33, the flight trajectory is above this building boundary layer and the flow pitch angle variation has settled down to a range within approximately 0–20°. The normalized velocity is the wind speed magnitude normalized by the aforementioned 3 m/s reference velocity. If the wind field were uniform and parallel to the building roof, the flow pitch angle would be zero, and the "normalized velocity" would be a constant. Instead, there are considerable variations in both angle and magnitude. The angle variations are not to be regarded as an angle of attack; at this point in the discussion, the airplane flight has not yet been introduced in the analysis. (Figures 13 and 14 represent the shape of the gust flow independent of the aircraft).

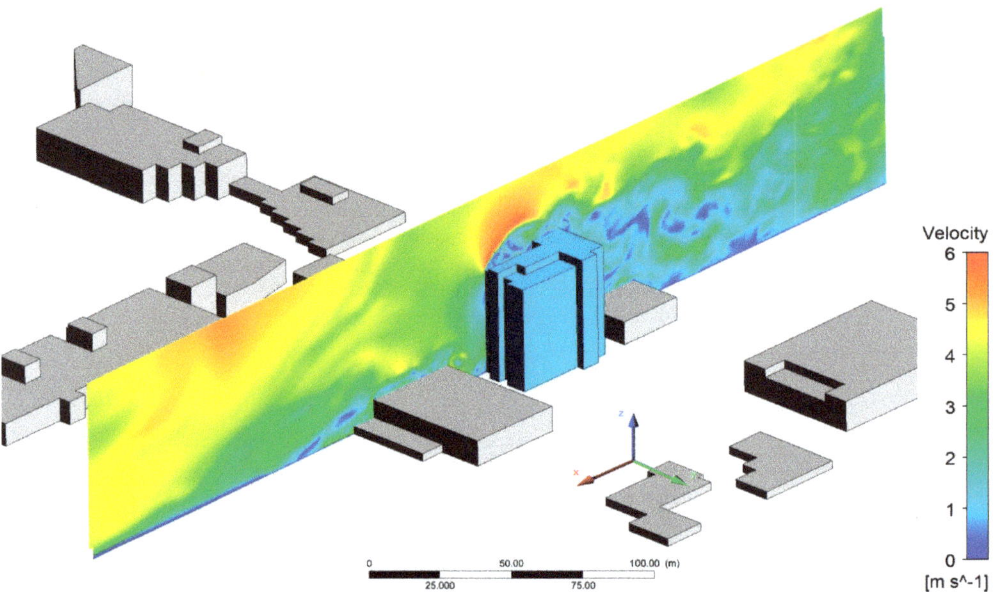

Figure 11. CFD domain, whereby air flows in the positive x-direction. The transient velocity magnitudes are shown in contour plots of the flow around the building located at the same plane of the flight paths (*travelling in the x-direction*) in the vicinity of the rooftop.

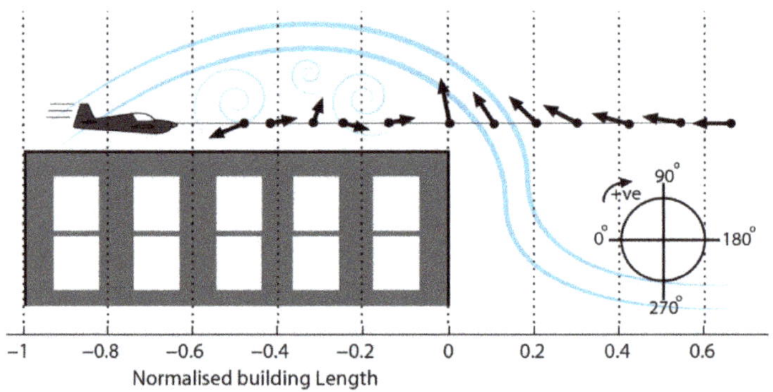

Figure 12. Encountered velocity vectors during proximity flight in the rooftop region of the building.

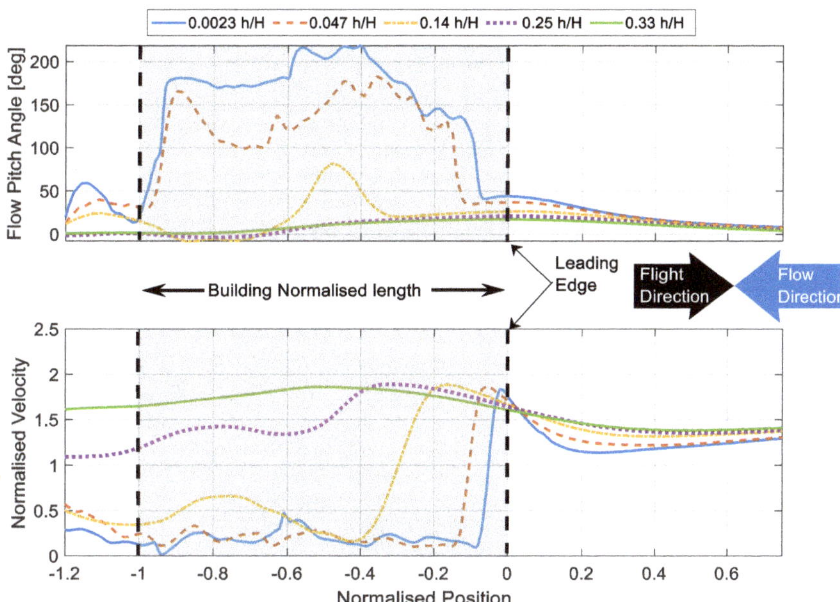

Figure 13. Flow velocity angles and magnitudes at different heights in the vicinity of the building's rooftop. Note that normalized positions −1 and 0 denote the edges of the building.

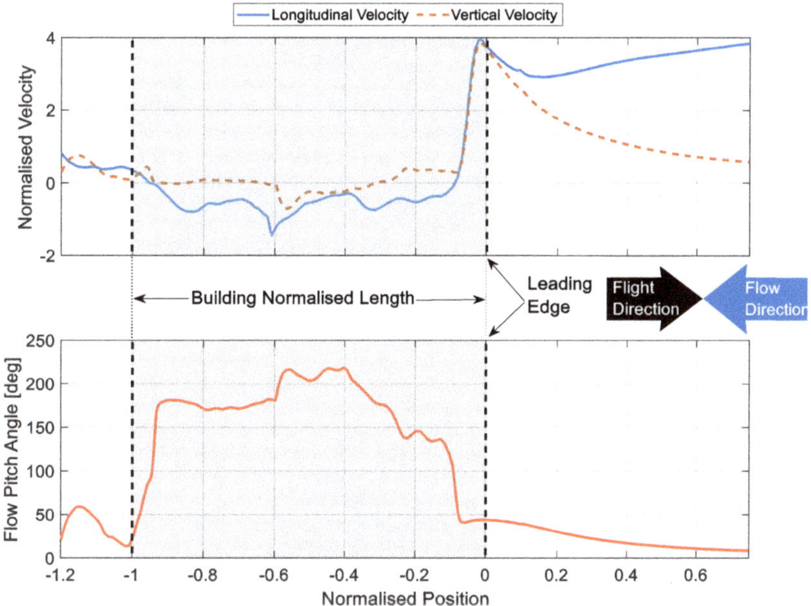

Figure 14. Velocity vectors along a representative flight path ($h/H = 0.0023$) in the rooftop region of the building as extracted from CFD simulation.

The changes in velocity magnitude are greatest very close to the building's top leading edge, whereas the changes in flow pitch angles are smaller. From the results presented in Figures 13 and 14, it is evident that a sharp increase in flow pitch angle at nondimensional position -1 on the abscissa exists at the trailing edge of the building. At the leading edge of the building, 0 on the abscissa, the flow pitch angle drops sharply. The normalized velocity, meanwhile, undergoes no change at the building trailing edge, but rises very sharply at the building leading edge. This rise is closer to the building leading edge for lower h/H, but it is essentially the same in magnitude for all trajectories up to $h/H = 0.14$ (6 m above the roof). This implies that a large change in wind amplitude is experienced by the UAV as it approaches the building edge, even if the desired trajectory is not particularly close to the building itself. In the presentation of Figure 13, wind speeds and angles are the result of the wind field computation, i.e., from a fixed reference frame. We next turn to how the very same results affect candidate UAVs of various kinds, i.e., from the UAVs' frame of reference.

7.2. Estimations of Perceived Gust for Fixed-Wing

Now we consider the flight path of a fixed-wing UAV above the building roof as indicated in Figure 12 at representative speeds of 5 m/s and 15 m/s. Consider how the combined effects of flow angle and magnitude are perceived along one flight path by superimposing the vehicle flight speed V_V onto the vertical speed V_{vert} and horizontal speed V_{horiz} of the wind, V_{vert} and V_{horiz} being the Cartesian analog of the "polar" results given in Figure 13. The superposition of the flight velocity and wind speed enables the relative velocity and angle of attack to be computed. The effective angle of attack, $\alpha(t)$, is calculated using

$$\alpha(t) = \alpha_o + atan\left(\frac{V_{vert}}{V_V + V_{horiz}}\right) \tag{7}$$

The results for two nominal cruise speeds (5 m/s and 15 m/s), converted back into velocity magnitude and angle of attack, are given in Figure 15. The immediately obvious feature of Figure 15 occurs near the building leading edge, "0" of the abscissa. As expected from Figure 13, the shear layer atop the building results in the worst-case perceived gust encounter: at the lower flight velocity (5 ms^{-1}) a $\approx 20°$ change in aircraft relative angle of attack is accompanied by an approximately 50% increase in velocity magnitude, all over a time increment of 0.25 s. At a higher flight velocity of 15 ms^{-1}, the perceived angle of attack is lower ($\approx 10°$), accompanied by a 25% increase in velocity over a time increment of 0.11 s. Using a simple linear relationship between the incident flow changes (angle of attack and relative velocity magnitude) and lift coefficient, and assuming a 2π lift curve slope and an unperturbed flight path (i.e., steady level flight), this gust represents changes in C_L of 8.5 and 2 for a flight velocity of 5 ms^{-1} and 15 ms^{-1}, respectively. For flight paths at 45° (where one wing is immersed into the gust before the other) the roll moment coefficient C_{L_p} presented in Figure 15 is calculated from the lift imbalance between the aircraft's wings:

$$C_{L_p} = \frac{b}{2} * \Delta C_L \tag{8}$$

$$C_{L_p} = M/qSb \tag{9}$$

Taking time lags into consideration, conventional attitude sensing and control systems of a fixed wing UAV travelling at 10 ms^{-1} will typically take 0.52 s to react (from sensing to actuation) [7,66] which can be insufficient to mitigate this gust. The combination of phase-advanced sensors, where flow, forward of the UAV, is measured and used as a control input [67], and novel control techniques may be needed [68] to achieve flight control in this type of environment. Examples of the latter include rotations of the entire wing, leading-edge control surfaces [68], or "fast flaps" at the trailing edge [69], which are intended to deflect faster than one convective time, producing lift transients well beyond what would be considered quasi-steady.

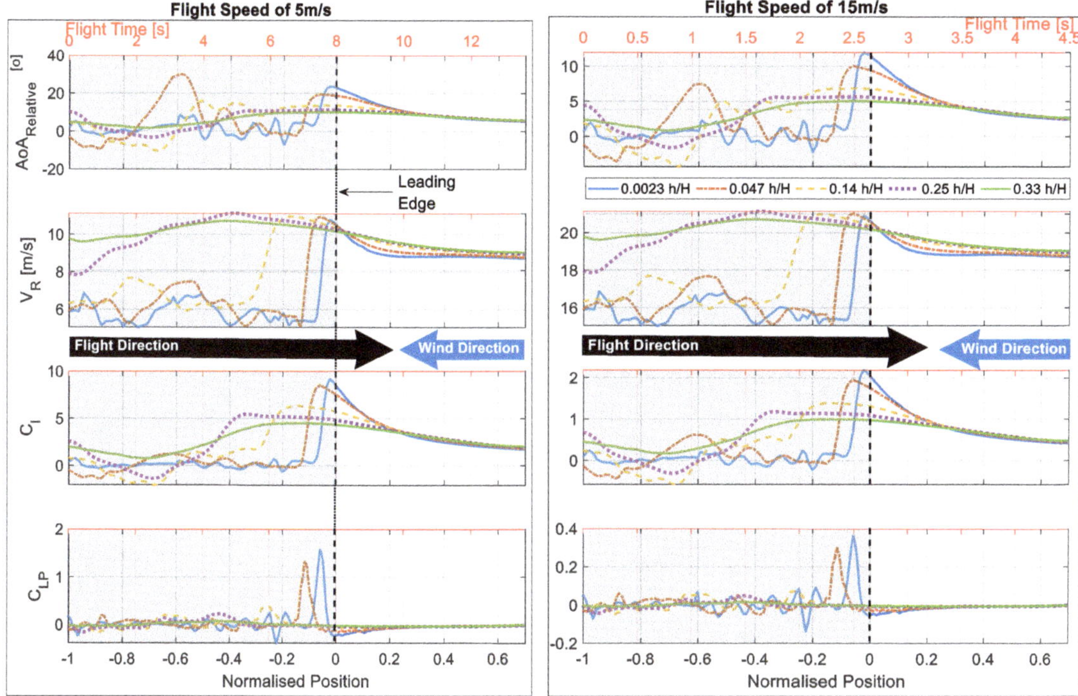

Figure 15. Gust shape as perceived by a moving aircraft. And the resultant C_L in the vicinity of the rooftop (from a simple strip theory model, utilizing transient flow data). Note that position 0 denotes the physical edge of the building.

7.3. Estimations of Perceived Gust for a Rotor

For multirotor aircraft, gust disturbances do not affect the aircraft in the same manner as fixed wing aircraft, especially in forward flight versus hover. This difference is due to the nature in which lift is created via its rotors and the forward motion flight state that requires a multirotor to tilt forward the rotors to generate forward speed. The purpose of this section is to explore the effect of the encountered gust on the total thrust generated while being agnostic about geometrical features of the rotor. This approach is key to making the analysis non-specific to a particular rotor and configuration but more generic and applicable to different multirotor configurations and even hybrid vehicles (i.e., fixed wing with rotors for Vertical Take-Off and Landing, VTOL). We will therefore use momentum disk theory and consider thrust of a single rotor. Aircraft designers can replicate this study and approximate the moments around the center of gravity for any number of rotors they intend to use. There are, however, limitations to this method, as it cannot account for geometric interferences between rotors and/or lifting surfaces, stall conditions, induced downwash effects from forward rotors, and other interactional aerodynamics of the configuration it may be modelling. However, it is sufficient for purposes of analyzing gust response within the context presented.

We consider two types of vehicles as outlined in Table 1 which represent two different scales of rotorcraft. The first vehicle represents a relatively small quadrotor delivery drone while the second is a larger octorotor AAM used for carrying human passengers. The tabulated specifications are generic for purposes of the presented analysis for two configurations which are likely to fly around buildings. The disk loading is determined by the hover weight divided by the total rotor area.

Table 1. Specifications of aircraft used for this analysis.

Parameter	*Specification*	
	Delivery Drone	Advanced Air Mobility Vehicle
MTOW	4 kg	1683 kg
Disk Loading	14.14 kg/m^2	35.68 kg/m^2
Configuration	Quadrotor	Octorotor
Rotor Size (diameter)	0.3 m	2.74 m
Flight Velocities	5 ms^{-1}, 15 ms^{-1}	5 ms^{-1}, 15 ms^{-1}

A single rotor disturbance model is used and is shown in Figure 16. Similar models have previously been used for turbulence and disturbance analysis for small multirotor aircraft with success [70,71].

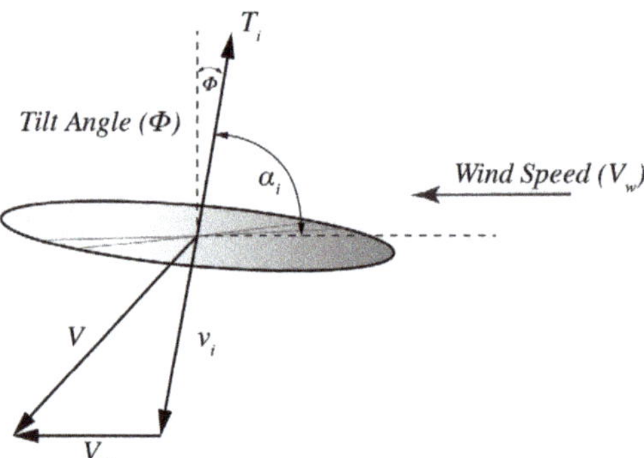

Figure 16. Wind disturbance model for a single disk.

The total induced thrust of the rotor can be represented b

$$T_i = 2\rho A V v_i \tag{10}$$

where the velocity components can be written as the induced velocity of the thrusting disk (v_i), the wind velocity (V_w), and the summation of the two vectors resulting in total induced speed (V).

$$V = V_w + v_i \tag{11}$$

The wind disturbance model allows the oncoming wind vector to be separated into its horizontal and vertical components to resolve the total induced speed vector using

$$V = \sqrt{(V_w cos\alpha_i + v_i)^2 + (V_w sin\alpha_i)^2} \tag{12}$$

where α_i is the induced angle between the rotor disk and the relative oncoming wind vector. As induced angle is influenced by the rotor tilt angle (ϕ) required for forward flight, induced angle can be calculated using

$$\alpha_i = \pi/2 - \phi \tag{13}$$

With no gust disturbance, the relative induced angle between the disk and the oncoming wind vector is completely perpendicular ($\alpha_i = \pi/2$). A purely vertical gust would result in a wind vector at zero or π radians with the thrusting vector parallel to the disk in hover. Vertical disturbances affect the angle of the disk relative to the oncoming wind vector, which allows oncoming gusts to approach the model between the angles of $0 \leq \alpha_i \leq \pi$.

Tilt angles relative to the ongoing wind vector are calculated to be 34° and 63° using forward flight speeds of 5 ms^{-1} and 15 ms^{-1} by resolving the induced angle from Equation (12) assuming no wind gust disturbance. The upper flight velocity of 15 ms^{-1} is regarded as high in terms of the normal flight speeds of multirotor aircraft at this scale; however, we offer this analysis to directly compare to the fixed wing case shown earlier. Referring to Figure 17, large variations are seen in the relative induced angle of the flow relative to the rotor. The most obvious effect can be seen at the buildings edge where larger variations of relative induced flow angle cause significant changes to thrust. The variation in thrust is more significant at the lower flight speed of 5 ms^{-1}. The higher flight speed of 15 ms^{-1} yields lower thrust variance due to a higher relative thrust required to maintaining flight and relatively lower gust vector. In other words, the faster the drone speed, the lower overall effect of the gust on the rotor as the thrusting vector to maintain flight becomes more dominant. Rotor thrust reactions to turbulence are more erratic and greater in magnitude than lift variations seen for the fixed wing aircraft found in the previous fixed wing study featured in this paper. On a rotor disk, turbulent flow vectors from all directions directly influence the aircraft incidence angle, the thrust required for steady level flight, and any perturbations which result in altitude loss or gain. All these directly influence the amount of thrust produced and incidence angle of the rotor significantly.

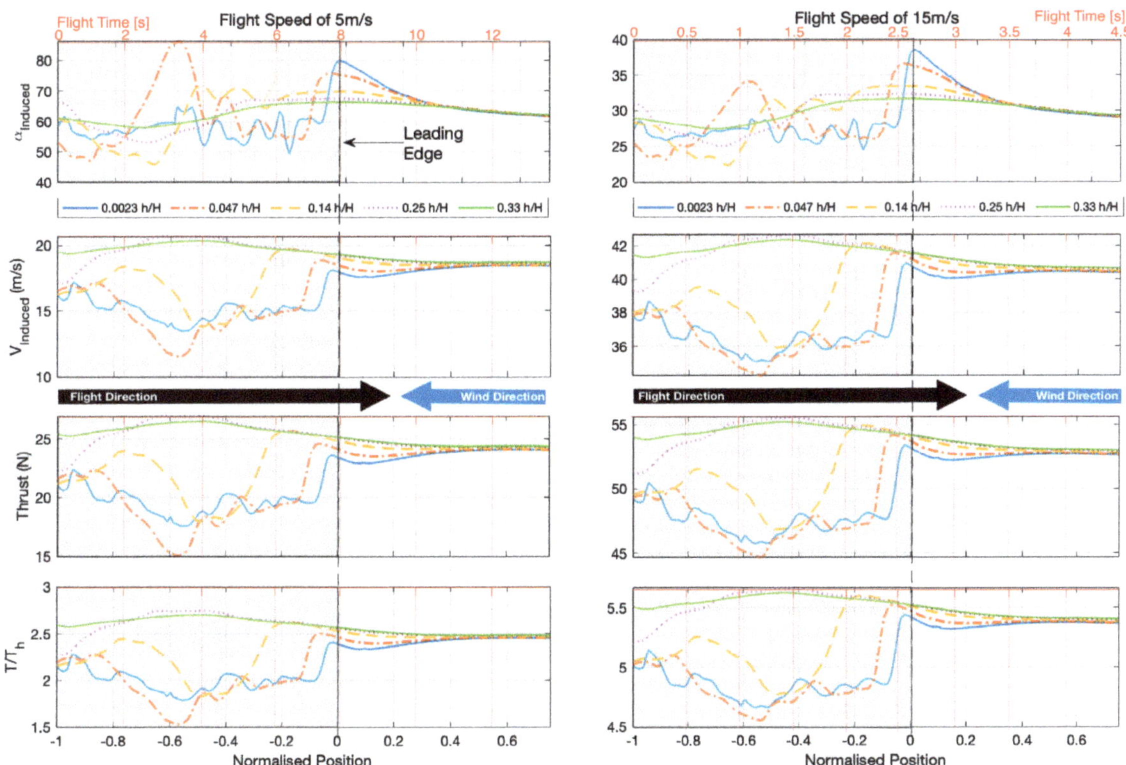

Figure 17. Gust shape as perceived by a moving thrusting disk of a delivery drone and the resultant effects on induced velocity, thrust, and normalized thrust in the vicinity of the rooftop. Note that position 0 denotes the physical leading edge of the building. T/T_h is the thrust required over thrust to hover fraction.

The fixed wing aircraft seems to be more passively tolerant to turbulence (although further experimental studies are required to explore this). Lower tolerance of rotary wing is assumed to be due to the loss in forward flight of the disk when traversing though the gust resulting in increased power and induced velocity while inducing larger rotor tilt angles to maintain attitude and flight speed. As both lift and forward flight is maintained by propulsive means, large variations in power/thrust are required when each are influenced and have a compounding effect when traversing through a gust.

Flight altitudes closest to the building produce the most unsteadiness in rotary wing thrust variation, which is a direct result of the relatively thin shear layer producing a relatively sharp flow vector change. These effects near the building are consistent with the fixed wing aircraft in the previous analysis. We observe that upward gusts result in additional thrusting force required to maintain flight speed and altitude, which is clearly seen when the UAV experiences the large upward gust in Figure 17 at a normalized position between -0.4 and 0. The induced flow vector is altered in this region and results in a higher thrust production in the direction of flight. Flight altitudes of $h/H = [0.0023, 0.047$ and $0.14]$ demonstrate similar trends in thrust. All stabilizes when the disk traverses past the edge of the building to free stream flow which is upwind of the building. Flight altitudes of $h/H = [0.25, 0.33]$ involve flight through a less sharp gust as perceived by the UAV whereby less variations in thrust are observed. This is due to the UAV flying above the shear layer and recirculating flow area caused by the leading edge of the building, where only gradual changes in relative flow angle impinge on the disk. Unlike most fixed wing aircraft, multirotor aircraft are inherently unstable and rely heavily on stabilization through the variation of the thrust of each rotor. The response of the thrusting system (i.e., propellor, motor, and controller) is the limiting factor in correcting for disturbances. Slow-flying multirotor systems traversing through a building-induced gust will experience a relatively high magnitude of thrust variation and will thus require an active stabilization response to maintain steady level flight.

Figure 18 displays the same analysis presented in Figure 17 but for a larger AAM vehicle capable of carrying a human passenger, thus resulting in higher disk loading and thrust. Consequently, the vehicle is relatively less sensitive to the gust, whereby the induced velocity through the rotor, the overall thrust, and the non-dimensional thrust (T/T_h) all show a lower thrust magnitude relative to the delivery drone. These changes are further highlighted in the non-dimensional thrust subplot where the maximum variance near the leading edge of building is $\Delta T/T_h > 0.4$, while the maximum variance for the delivery drone is greater at values of $\Delta T/T_h > 0.8$ at the same flight altitude and speed of 5 m/s. Variance in thrust magnitude is greater in all instances for the delivery drone relative to the urban mobility vehicle. In both flight examples, the variance occurs more so in the locations of highly separated and mixed flow featured at heights between $0.0023 < h/H < 0.14$. Greater heights show smaller flow vector variation, suggesting regions of flow that are out of the turbulence shear layer. For both vehicles, the turbulence effects cannot be neglected and require active turbulence mitigation through autopilot stabilization. The suitability of stabilization systems depends on the actuation speed achievable for the given scale of the aircraft. Higher actuation speeds will reduce the sensing-to-actuation time-lag which thus enables the vehicle to mitigate sharper and higher-amplitude gusts. Light-weight rotors, high-torque motors, greater excess thrust, and power will all contribute to required turbulence reaction speeds.

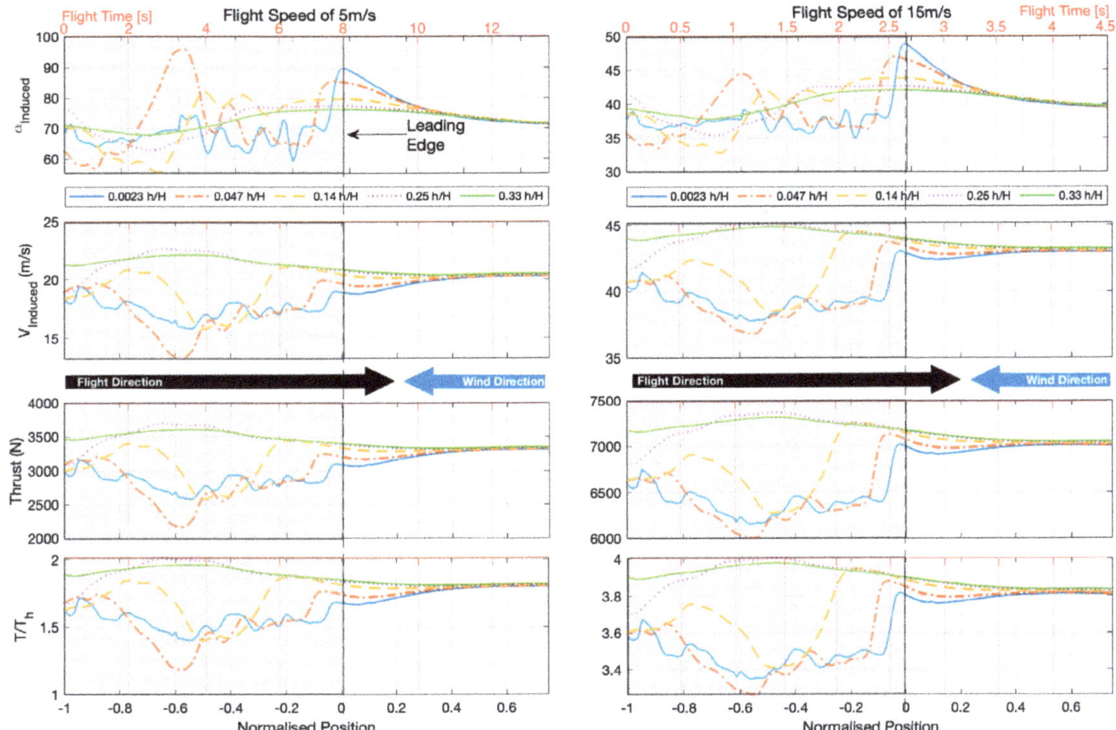

Figure 18. Gust shape as perceived by a moving thrusting disk of an advanced air mobility vehicle, and the resultant effects on induced velocity, thrust, and normalized thrust in the vicinity of the rooftop. Note that position 0 denotes the physical edge of the building. T/T_h is the thrust required over thrust to hover fraction.

7.4. Airframe Design and Certification Considerations

Considerable recent published work has considered the alleviation of gust loads on aircraft [72–76] and in some cases even harvesting it [77–79]. Severe gusts around buildings can pose a major challenge for flight of different vehicular scales and configurations. Smaller UAVs are more sensitive to the disturbances, however larger UAVs are still affected albeit to a lesser extent. The latter will depend on the relative magnitude and scale of a gust with respect to the aircraft's scale. Also, the UAV configuration (rotary vs fixed wing) will respond differently to the disturbances. Hybrid configurations which have a combination of lifting surfaces (i.e., fixed wing) and an array of thrusting disks (i.e., rotary wing) are well suited for close proximity flight to buildings. However, there is a spectrum of design possibilities which require careful design choices to truly alleviate the disadvantages of both fixed and rotary wing. Further research is required to identify the intrinsic aerodynamic deficiencies of these hybrid configurations and what are they particularly susceptible to. For example, fixed-wing craft will stall if flown too slow, while rotary wing craft are susceptible to the vortex ring state and weather cock stability. Some deficiencies may be resolved with hybrid configurations while others may persist or even give rise to new deficiencies especially during hover. Vehicles with large surface areas facing the wind direction (e.g., tilt wings) will experience significant attitude control and flight-path tracking challenges due to the relatively large forces generated by these surfaces. Such designs should be avoided where possible if a UAV is expected to fly at low speeds near buildings and gust-generating infrastructure. The frontal projection area of the UAV regardless of the configuration needs to be minimized most critically during proximity flight. This may be even achieved through active wing area reduction, but the structural and mechanical challenges of an airframe capable of reducing area or changing its wing planform. This design challenge is complex but not impossible. There are also other means of mitigating turbulence and gusts through the control systems [80–82], aerodynamic configuration [68,83–86], and novel sensors [66,67,87]. Counteracting such flow disturbances comes at the cost of increased weight and power demands which will affect range and battery consumption. The question then becomes, how smooth of a flight will the passenger demand? How much control will we need to give to the pilot and/or the system?

As a hybrid UAV flies slow and in proximity to a building, any fixed wing control surface on the airframe become ineffective in controlling attitude due to low speed. The effectiveness or relative force of control surfaces reduces by the square of the flow velocity it is exposed to. In this case the UAV relies mainly on the rotary wings for lift and attitude control. There are opportunities for unconventional fixed wings designs to increase the control authority and rapidity [84,85], however the rotary wings will be required to achieve the majority of the control and lifting work in such scenarios, and therefore require the ability to rapidly adjust thrust to mitigate any gusts encountered. Variable-pitch propellers are effective in generating rapid actuation and more efficient thrust vectoring to enable the vehicle to approach a vertiport at low approach speeds with more control authority and stability.

From a certification standpoint, AAM airframes need to demonstrate the ability to counter attitude disturbances and flight path deviations for a reasonable range of wind speeds and gust conditions to make AAM operational for the majority of the year despite weather. Coping with high wind speeds, certification should include a demonstration of limits on the angular perturbations allowed in the vehicles' three axes during the highest operational wind and gust magnitudes. These angular limits should be selected to ensure that the physical extremities of the vehicle do not collide with the vertiport during touchdown or take-off. Limits should also be imposed on how much flight-path drift occurs for a range of wind and gust speeds to reduce risk of collision with infrastructure. Airframe manufacturers can either conduct physical experimentations to demonstrate compliance with the limitations imposed or utilize numerical-based modelling (with a form of validation) [61].

Helicopter certifications requirements rely on the presence of human pilots on board that can assess hazardous situations. Regulations for autonomous UAV operations in cities (especially large air taxis) will be different and rely on measurable numerical thresholds, which are used by the flight control system for automated decision making and planning, given there is no human-in-the-loop to make such rapid judgments:

- "Operating in close proximity to obstructions can lead to recirculation and loss of performance. Aerodromes, geographically situated in hilly, mountainous areas, including certain coastal regions, can be subject to hazardous turbulent conditions in moderate to strong wind conditions. Pilots should be aware that, in certain cases, aircraft performance can be severely affected. History has shown, in extreme cases, that turbulence has prevented the aircraft from climbing or being controlled near the ground and has also caused structural damage".
- "In winds below 15 kts, the turbulence may be experienced in the lee of an obstruction, vertically to about one third higher than the height of the obstruction. Above 20 kts, turbulence may be experience on the leeward side of an obstruction to a distance of 10–15 times the obstruction height and up to twice the obstruction height above the ground".
- "During take-off or landing in gusty wind situations where wind shear is likely to be present, may require a greater power margin to deal with varying power demands or an unexpected loss of airspeed and accompanying sink. Large anti-torque pedal inputs to maintain directional control also act to reduce the excess power available".

Regulations for autonomous UAV operations in cities (especially large air taxis) will have to be more detailed and require the reliance on measurable numerical thresholds, which are used by the flight control system for automated decision making and planning, given there is no human-in-the-loop to make such rapid judgments.

7.5. Vertiport Design and Certification Considerations

Currently, a small body of knowledge exists around specific heliport requirements that deal with the surrounding turbulence levels from nearby buildings [13,88,89]. There also exists some regulations that can be used as a basis to guide the design and location of vertiport landing infrastructure [20,89]. A turbulence criterion was introduced for helicopters to ensure safe flight is maintained [89]. The criterion sets a threshold on the standard deviation of the vertical flow velocity, which results in a high helicopter pilot workload. Mentzoni and Ertesvåg [88] later suggested the use of turbulence energy instead as a criterion, arguing its benefits over the standard deviation of vertical velocity. Similarly, a new criterion or threshold is needed for the autonomous operation of AAM vehicles, which relies on the limitations of the flight control system instead of the workload of human pilots. The results presented here have implications for vertiport design and a similar analysis can be used to identify thresholds for such a criterion.

Most of the research on building aerodynamics presented in the literature focuses on surface pressure measurements for predicting facade loadings. However, the advent of AAM requires a unique understanding of the velocity field induced by the interaction of the wind with the building on which UAVs will be operating form. Specifically, the shear layers that form and their impact on flight. A thorough characterization of the flow field for different wind directions is essential for each vertiport to be designed since each one will have a unique flow environment. Similar methods and tools, such as those used in the field of wind engineering, can be used.

Vertiport designers will need to avoid design features that generate turbulence or sharp gusts of high amplitude and of length scales that are detrimental to UAVs. A few studies explore this area [13,19,21,90]; however, more research is needed, with full-scale validation. There exists a body of knowledge on designing wind sheltering systems (such as porous fences) for road and rail vehicles which will be relevant. Similarly, building design features, such as round corners and porous deflectors near rooftops, can help reduce the sharpness of the perceived gust, which translates to a lower actuation requirement, thus providing a UAV's flight control system with more time to react and counter the flow disturbance. Another key parameter is the unobstructed air gap below the landing platform, which will also influence the severity of the shear layer by allowing more air to flow underneath the platform. The ideal height of the air gap will be different for each building since it is a function of the building's geometry. A 1.8 m minimum air gap is cited by the FAA in the Heliport Design Advisory Circular AC 150/5390-2D [91]. The document points to research published in FAA/RD-84/25 [19], but it is unclear how the 1.8 m criteria were derived. Regardless, there is enough justification for exploring a new threshold for AAM vehicles. The new US Federal Aviation Administration (FAA) guidelines for vertiport design has a small section on turbulence with high-level recommendations on using turbulence-mitigating design measures [92]. As technology matures and more research is conducted in this area, specific metrics and criterion can be included in future revisions of the guidelines providing design standards, which will need to be met. It is also strongly believed that aviation authorities should provide their own guidelines and regulations on turbulence and gust thresholds around vertiports instead of relying on existing building guidelines and regulations (e.g., [93]), which focus on reducing adverse wind effects that affect the quality and usability of outdoor spaces and pedestrian comfort. The modeling and measurements for the latter are very different from that required for AAM flight paths around the buildings from a probe placement and mesh refinement perspective.

Modelling building aerodynamics and the local flow fields can be performed using classical wind tunnel methods on scale buildings, or utilizing CFD similar to that presented here. There is a need to provision for the surrounding wind environment and its interaction with not only the vertiport structure but also neighboring structures which will have an impact on the local flow field [19] and can result in overspeed regions which are difficult to predict. An additional analysis, which can complement wind tunnel testing and CFD, is full scale measurements using airborne wind anemometers such as the one developed by Prudden, Fisher [94]. A swarm of such sensors are ideal for rapid simultaneous measurements that can map out the flow field accurately at full scale and later used for validation of CFD or comparison with scale experiments to account for any Reynolds number effects. Given the mobility of such systems, it can also be used to measure the perceived gust along the flight paths of UAVs.

8. Concluding Remarks

UAVs used for both delivery and human carrying systems are being introduced internationally and are intended to integrate into various civil domains. Urban and city environments provide the greatest operational challenge due to the safety considerations of operating in highly populated environments. Under even moderate winds, landing and take-off maneuvers are subjected to high levels of turbulence intensities and gusts that will impact the stability and control of these vehicles. Furthermore, the integral length scale of turbulence may be such that they are similar to the scales of UAVs; these will provide considerable control challenges in holding relatively steady flight. We are guided by existing literature on helicopter landing and take-off procedures, which is not extensive and is lacking in terms of autonomous operation. Minimization of turbulence and gusts via building or vertiport design are limited and warrant further research.

In this paper we used a CFD simulation of the ambient wind field around a nominally cuboid building in a suburban atmospheric boundary layer. Unperturbed flight paths near the building's roof were superimposed onto the simulated wind field. A possible worst-case gust for the specified wind speed and building geometry was identified when the flight path traverses the shear layer from the building's top leading edge, resulting in significant lift force variations. The analysis showed that UAVs would experience a substantial increase in angle of attack over a relatively short period of time (<1 s) as they fly through shear layer at a representative forward velocity, which can be well above typical stall angles. Due to the slow flight speeds required for landing and take-off, significant control authority of rotor systems is required to ensure safe operation due to the high disturbance effects caused by localized gusts from buildings and protruding structures. The analysis is then flowed by regulation and certification recommendations for AAM vehicles and vertiports.

CFD simulation of atmospheric flows is challenging and warrants experimental validation via collection of careful gust measurements either in a wind tunnel environment or by flying aircraft, which should be fitted with responsive anemometers capable of resolving turbulence length scales smaller than a UAV's characteristic length [94]. The resulting datasets, both computational and experimental, should be interrogated to identify two- and three-dimensional severe gusts. Subsequent work should include furthering the understanding of the transfer functions between a gust flow and the resulting aerodynamic response of the UAV, which could then be used to understand disturbances and control methods to minimize them. This paper used computational gust data to develop basic disturbance models to understand the response of a fixed wing and thrusting disk. In both instances, the effect of a gust around a cuboid building is significant and may cause significant flight perturbations that cannot be ignored. Furthermore, for larger UAV, the magnitude of corrective control required must be acknowledged and considered in the design phase when such vehicles are developed.

Author Contributions: Conceptualization, A.M.; methodology, A.M.; software, A.M. and M.M.; validation, A.M. and M.M.; formal analysis, A.M. and M.M.; investigation, All; resources, All; data curation, A.M. and M.M.; writing—original draft preparation, A.M.; writing—review and editing, All; visualization, A.M. and M.M.; project administration, A.M.; funding acquisition, A.M. All authors have read and agreed to the published version of the manuscript.

Funding: This work has been partially funded and supported by the US Air Force Office for Scientific Research (AFOSR) FA2386-22-1-4078, and Defence Science Institute (DSI) (RHD-0189).

Data Availability Statement: Not applicable.

Acknowledgments: This research was undertaken as part of the RMIT Uncrewed Aircraft Systems Research Team (ruasrt.com), within the Sir Lawrence Wackett Centre, at RMIT University. The work is part of NATO RTO AVT-282 "Unsteady Aerodynamic Response of Rigid Wings in Gust Encounters" and AVT-347 "Large-Amplitude Gust Mitigation Strategies for Rigid Wings"and input of the NATO team is appreciated. The authors are thankful for the insight provide by Mr Rex J Alexander, President of the Five-Alpha LLC and ex-Head of Aviation Infrastructure at Uber.

Conflicts of Interest: The authors declare no conflict of interest.

References

1. Stough, P.; Scanlon, C. Numerous research projects support efforts to overcome weather-related hazards. *ICAO J.* **1999**, 20–29.
2. Kohut, T. *21 Sent to Hospital after Severe Turbulence Diverts Flight AC088 to Calgary*; Global News: Toronto, ON, Canada, 2015; Available online: https://globalnews.ca/news/2427003/flight-with-20-injuries-diverted-to-calgary-international-airport/ (accessed on 10 December 2022).
3. Bedo, S. Qantas Cabin Crew Injured in Turbulence Scare. 2019. Available online: https://www.news.com.au/national/qantas-cabin-crew-injured-in-turbulence-scare/news-story/528899b3886f77242887e946a78f9c35 (accessed on 1 March 2020).
4. Souza, L. Twelve Passengers Injured Due to Severe Turbulence on an Aerolineas Argentinas Airbus A330. 2022. Available online: https://simpleflying.com/twelve-passengers-injured-severe-turbulence-aerolineas-argentinas-airbus-a330/ (accessed on 10 December 2022).
5. Pande, P. SpiceJet Passenger Injured In Turbulence Dies Five Months Later. 2022. Available online: https://simpleflying.com/spicejet-turbulence-passenger-death/ (accessed on 10 December 2022).
6. Watkins, S.; Milbank, J.; Loxton, B.; Melbourne, W. Atmospheric winds and their effects on micro air vehicles. *AIAA J.* **2006**, *44*, 2591–2600. [CrossRef]
7. Mohamed, A.; Massey, K.; Watkins, S.; Clothier, R. The Attitude Control of Fixed-Wing MAVS in Turbulent Environments. *Prog. Aerosp. Sci.* **2014**, *66*, 37–48. [CrossRef]
8. Hoblit, F.M. *Gust Loads on Aircraft: Concepts and Applications*; American Institute of Aeronautics and Astronautics: Reston, VA, USA, 1988.
9. Mohamed, A.; Watkins, S.; OL, M.; Jones, A. Flight-Relevant Gusts: Computation-Derived Guidelines for Micro Air Vehicle Ground Test Unsteady Aerodynamics. *J. Aircr.* **2020**, 1–7. [CrossRef]
10. Thorn, A. Turbulence Factor in R44 Crash that Killed Pilot and Trainee. 2022. Available online: https://australianaviation.com.au/2022/04/turbulence-factor-in-r44-crash-that-killed-pilot-and-trainee/ (accessed on 10 December 2022).
11. Severe Turbulence Caused B.C. Helicopter Crash That Seriously Injured 2, Investigation Finds. 2022. Available online: https://ca.news.yahoo.com/severe-turbulence-caused-b-c-194736055.html?guccounter=1&guce_referrer=aHR0cHM6Ly9 3d3cuZ29vZ2xlLmNvbS8&guce_referrer_sig=AQAAAJG9lcsg3uLYEePQR4Eh2qW8ph7sHzRJsRQ-E7iDNKjjW9AlwiXj1 WRyxzJs_ckdOwlVnGC2J1I_xZeZ7sgx_xdefW7GUvFAK9zMxoNmG1V6YsiR7juy2jse3VeaHKJk8y6a8j40oTrWVYIqdF_yCQ5 5auAOJlbJOrAFy8GdysH1 (accessed on 10 December 2022).
12. Severe Turbulence Led to Helicopter Collision with Terrain on Bowen Island, British Columbia. 2022. Available online: https://canadianaviationnews.wordpress.com/2022/07/29/severe-turbulence-led-to-helicopter-collision-with-terrain-on-bowen-island-british-columbia/ (accessed on 10 December 2022).
13. Smith, A.; Bell, A.; Hackett, D. *Trade-Offs in Helipad Sitting & Design*; Rowan Williams Davies & Irwin Inc.: Guelph, ON, Canada, 2017; pp. 1–10.
14. Glaser, A. Here's Why Facebook's Massive Drone Crashed in the Arizona Desert. 2016. Available online: https://www.vox.com/2016/12/18/13998900/facebooks-drone-crash-aquila-arizona-structural-failure (accessed on 8 December 2020).
15. Lye, H. Second Airbus Zephyr UAS Crashes during a Test Flight. 2019. Available online: https://www.airforce-technology.com/news/second-airbus-zephyr-uas-crashes-during-a-test-flight/ (accessed on 10 December 2022).
16. Mangan, D.; Breuninger, K. One Dead after Helicopter Crash-Lands on Roof of Midtown Manhattan Building. 2019. Available online: https://www.cnbc.com/2019/06/10/helicopter-crashes-into-building-in-midtown-manhattan.html (accessed on 3 March 2020).
17. Farell, C.; Sitheeq, M.M.; Ellis, C.R.; Voigt, R.L., Jr. *Fairview Riverside Medical Center Helistop, Minneapolis, Minnesota: Design and Operational Issues*; St. Anthony Falls Hydraulic Laboratory: Minneapolis, MN, USA, 1995.
18. Horn, J.F.; Keller, J.D.; Whitehouse, G.R.; McKillip, R.M., Jr. Analysis of Urban Airwake Effects on Heliport Operations at the Chicago Children's Memorial Hospital. In *Final Report Submitted to Illinois Department of Transportation*; Researchgate: Berlin, Germany, 2011.
19. McKinley, J.B. *Evaluating Wind Flow around Buildings on Heliport Placement*; West Palm Beach Fl Champlain Technology; Systems Control Technology Inc.: Idaho Falls, ID, USA,1984.

20. Civil Aviation Authority. *Helideck Design Considerations–Environmental Effects*; CAA Paper: Norwich, UK, 2008; Volume 3.

21. Nakayama, M.; Uchiami, Y.; Watagami, K.; Ui, K. Wind tunnel test to design helidecks on the rooftops of high-rise buildings. *J. Wind. Eng. Ind. Aerodyn.* **1991**, *38*, 459–468.

22. Perrotta, G.; Jones, A.R. Unsteady forcing on a flat-plate wing in large transverse gusts. *Exp. Fluids* **2017**, *58*, 101.

23. Andreu-Angulo, I.; Babinsky, H.; Biler, H.; Sedky, G.; Jones, A.R. Effect of transverse gust velocity profiles. *AIAA J.* **2020**, *58*, 5123–5133. [CrossRef]

24. Biler, H.; Sedky, G.; Jones, A.R.; Saritas, M.; Cetiner, O. Experimental investigation of transverse and vortex gust encounters at low Reynolds numbers. *AIAA J.* **2021**, *59*, 786–799. [CrossRef]

25. Granlund, K.; Monnier, B.; Ol, M.; Williams, D. Airfoil longitudinal gust response in separated vs. attached flows. *Phys. Fluids* **2014**, *26*, 027103. [CrossRef]

26. Biler, H.; Badrya, C.; Jones, A.R. Experimental and computational investigation of transverse gust encounters. *AIAA J.* **2019**, *57*, 4608–4622. [CrossRef]

27. Jones, A.R.; Cetiner, O.; Smith, M.J. Physics and modeling of large flow disturbances: Discrete gust encounters for modern air vehicles. *Annu. Rev. Fluid Mech.* **2022**, *54*, 469–493. [CrossRef]

28. Jones, A.R.; Cetiner, O. Overview of unsteady aerodynamic response of rigid wings in gust encounters. *AIAA J.* **2021**, *59*, 731–736. [CrossRef]

29. Eldredge, J.D.; Jones, A.R. Leading-edge vortices: Mechanics and modeling. *Annu. Rev. Fluid Mech.* **2019**, 75–104. [CrossRef]

30. Watkins, S.; Thompson, M.; Loxton, B.; Abdulrahim, M. On Low Altitude Flight through The Atmospheric Boundary Layer. *Int. J. Micro Air Veh.* **2010**, *2*, 55–67. [CrossRef]

31. Mohamed, A.; Watkins, S.; Clothier, R.; Abdulrahim, M. Influence of Turbulence on MAV Roll Perturbations. *Int. J. Micro Air Veh.* **2014**, *6*, 175–190. [CrossRef]

32. Chen, T.; Clothier, R.; Mohamed, A.; Badawy, R. An Experimental Study of Human Performance in Controlling Micro Air Vehicles in Turbulent Environment. In Proceedings of the Fourth Australasian Unmanned Systems Conference, Melbourne, Australia, 15–16 December 2014; pp. 1–8.

33. Holmes, J.D. *Wind Loading of Structures*; CRC Press: Boca Raton, FL, USA, 2001.

34. Peterka, J.; Meroney, R.; Kothari, K. Wind flow patterns about buildings. *J. Wind Eng. Ind. Aerodyn.* **1985**, *21*, 21–38. [CrossRef]

35. Leuzzi, G.; Monti, P. Particle trajectory simulation of dispersion around a building. *Atmos. Environ.* **1998**, *32*, 203–214. [CrossRef]

36. Zhang, Y.; Arya, S.; Snyder, W. A comparison of numerical and physical modeling of stable atmospheric flow and dispersion around a cubical building. *Atmos. Environ.* **1996**, *30*, 1327–1345. [CrossRef]

37. Ferreira, A.D.; Sousa, A.C.M.; Viegas, D.X. Prediction of building interference effects on pedestrian level comfort. *J. Wind Eng. Ind. Aerodyn.* **2002**, *90*, 305–319. [CrossRef]

38. Mochida, A.; Lun, I.Y.F. Prediction of wind environment and thermal comfort at pedestrian level in urban area. *J. Wind Eng. Ind. Aerodyn.* **2008**, *96*, 1498–1527. [CrossRef]

39. Lim, K.E.; Watkins, S.; Clothier, R.; Ladani, R.; Mohamed, A.; Palmer, J.L. Full-scale flow measurement on a tall building with a continuous-wave Doppler Lidar anemometer. *J. Wind Eng. Ind. Aerodyn.* **2016**, *154*, 69–75. [CrossRef]

40. Tsinober, A. *An Informal Conceptual Introduction to Turbulence*; Springer: Berlin/Heidelberg, Germany, 2009; Volume 483.

41. Baklanov, A.A.; Grisogono, B.; Bornstein, R.; Mahrt, L.; Zilitinkevich, S.S.; Taylor, P.; Larsen, S.E.; Rotach, M.W.; Fernando, H.J.S. The Nature, Theory, and Modeling of Atmospheric Planetary Boundary Layers. *Bull. Am. Meteorol. Soc.* **2011**, *92*, 123–128. [CrossRef]

42. Garratt, J.R. *The Atmospheric Boundary Layer*; Cambridge University Press: Cambridge, UK, 1994.

43. Thompson, M. The Small Scales of Turbulence in Atmospheric Winds at Heights Relevant to Mavs. Doctoral Dissertation, RMIT University, Melbourne, Australia, 2013.

44. Mohamed, A.; Carrese, R.; Fletcher, D.; Watkins, S. Scale-resolving simulation to predict the updraught regions over buildings for MAV orographic lift soaring. *J. Wind Eng. Ind. Aerodyn.* **2015**, *140*, 34–48. [CrossRef]

45. Liepmann, H. On the application of statistical concepts to the buffeting problem. *J. Aeronaut. Sci.* **1952**, *19*, 793–800. [CrossRef]

46. Dowell, E.H.; Crawley, E.F.; Curtiss, H.C., Jr.; Peters, D.A.; Scanlan, R.H.; Sisto, F. *A Modern Course in Aeroelasticity, 3rd revised and enlarged Ed ed.*; Kluwer Academic Publishers: Boston, MA, USA, 1995; pp. 327–337.

47. Von Karman, T. Progress in the statistical theory of turbulence. *Proc. Natl. Acad. Sci. USA* **1948**, *34*, 530–539. [CrossRef] [PubMed]

48. Diederich, F.W.; Drischler, J.A. *Effect of Spanwise Variations in Gust Intensity on the Lift Due to Atmospheric Turbulence*; NACA TN 3920; 1957.

49. Dryden, H.L. Turbulence investigations at the National Bureau of Standards. In Proceedings of the Fifth International Congress of Applied Mechanics, Cambridge, MA, USA, 12–16 September 1938.

50. Heisenberg, W. Zur statistischen theorie der turbulenz. *Z. Phys.* **1948**, *124*, 628–657. [CrossRef]

51. Kolmogorov, A.N. The Local Structure of Turbulence in Incompressible Viscous Fluid for Very Large Reynolds' Numbers. *Proc. R. Soc. London. Ser. A-Math. Phys. Sci.* **1991**, *434*, 9–13.

52. Liepmann, H.; Laufer, J.; Liepmann, K. *On the Spectrum of Isotropic Turbulence*; NACA TN 2473; National Advisory Committee for Aeronautics: Washington, DC, USA, 1951.

53. Standard, M. *Flying Qualities of Piloted Aircraft*; US Dept. of Defense MIL-STD-1797A; 1990.

54. Zbrozek, J. *The Relationship between the Discrete Gust and Power Spectra Presentations of Atmospheric Turbulence, with a Suggested Model of Low-Altitude Turbulence*; Royal Aeronautical Establishment, Aeronautical Research Council, Reports and Memoranda; 1961.
55. Etkin, B. *Theory of the Flight of Airplanes in Isotropic Turbulence-Review and Extension*; Advisory Group for Aeronautical Research and Development Paris (France): Neuilly-sur-Seine, France, 1961.
56. Skelton, G.B. *Investigation of the Effects of Gusts on V/STOL Craft in Transition and Hover*; Air Force Flight Dynamics Laboratory Technical Report; 1968.
57. ESDU. *Characteristics of Atmospheric Turbulence near the Ground. Part II: Single Point Data for Strong Winds (Neutral Atmosphere)*; 1985.
58. Watkins, S.; Cooper, K.R. *The Unsteady Wind Environment of Road Vehicles, Part Two: Effects on Vehicle Development and Simulation of Turbulence*; SAE Technical Paper; 2007.
59. Lissaman, P. Effects of Turbulence on Bank Upsets of Small Flight Vehicles. In Proceedings of the 47th AIAA Aerospace Sciences Meeting including the New Horizons Forum and Aerospace Exposition, Orlando, FL, USA, 5–8 January 2009.
60. Thompson, M.; Watkins, S.; White, C.; Holmes, J. Span-wise wind fluctuations in open terrain as applicable to small flying craft. *Aeronaut. J.* **2011**, *115*, 693–701. [CrossRef]
61. Wang, L.; Dai, Y.; Yang, C. Gust response analysis for helicopter rotors in the hover and forward flights. *Shock Vib.* **2017**, *2017*, 8986217. [CrossRef]
62. Diederich, F.W. The dynamic response of a large airplane to continuous random atmospheric disturbances. *J. Aeronaut. Sci.* **1956**, *23*, 917–930. [CrossRef]
63. Fadl, M.S.; Karadelis, J. CFD simulation for wind comfort and safety in urban area: A case study of Coventry University central campus. *Int. J. Archit. Eng. Constr.* **2013**, *2*, 131–143. [CrossRef]
64. Watkins, S.; Mohamed, A.; Fisher, A.; Clothier, R.; Carrese, R.; Fletcher, D. Towards Autonomous MAV Soaring in Cities: CFD Simulation, EFD Measurement and Flight Trials. *Int. J. Micro Air Veh.* **2015**, *7*, 441–448. [CrossRef]
65. Blocken, B.; Stathopoulos, T.; Carmeliet, J. CFD simulation of the atmospheric boundary layer: Wall function problems. *Atmos. Environ.* **2007**, *41*, 238–252. [CrossRef]
66. Mohamed, A.; Abdulrahim, M.; Watkins, S.; Clothier, R. Development and Flight Testing of a Turbulence Mitigation System for MAVs. *J. Field Robot.* **2016**, *35*, 639–660. [CrossRef]
67. Mohamed, A.; Watkins, S.; Clothier, R.; Abdulrahim, M.; Massey, K.; Sabatini, R. Fixed-Wing MAV Attitude Stability in Atmospheric Turbulence PART 2: Investigating Biologically-Inspired Sensors. *Prog. Aerosp. Sci.* **2014**, *71*, 1–13. [CrossRef]
68. Panta, A.; Petersen, P.; Marino, M.; Watkins, S.; Fisher, A.; Mohamed, A. Qualitative investigation of the dynamics of a leading edge control surfaces for MAV applications. In Proceedings of the International Micro Air Vehicle Conference and Flight Competition (IMAV 2017), Toulouse, France, 18–21 September 2017; pp. 1–8.
69. Medina, a.; Ol, m.V.; mancini, p.; Jones, a. Revisiting Conventional Flaps at High Deflection Rate. *AIAA J.* **2016**, *55*, 2676–2685. [CrossRef]
70. Lei, Y.; Huang, Y.; Wang, H. Effects of wind disturbance on the aerodynamic performance of a quadrotor MAV during hovering. *J. Sens.* **2021**, *2021*, 6681716. [CrossRef]
71. Nguyen, D.H.; Liu, Y.; Mori, K. Experimental study for aerodynamic performance of quadrotor helicopter. *Trans. Jpn. Soc. Aeronaut. Space Sci.* **2018**, *61*, 29–39. [CrossRef]
72. Zhou, Q.; Chen, G.; Da Ronch, A.; Li, Y. Reduced order unsteady aerodynamic model of a rigid aerofoil in gust encounters. *Aerosp. Sci. Technol.* **2017**, *63*, 203–213. [CrossRef]
73. Guo, S.; Jing, Z.; Li, H.; Lei, W.; He, Y. Gust response and body freedom flutter of a flying-wing aircraft with a passive gust alleviation device. *Aerosp. Sci. Technol.* **2017**, *70*, 277–285. [CrossRef]
74. Rajpal, D.; Gillebaart, E.; De Breuker, R. Preliminary aeroelastic design of composite wings subjected to critical gust loads. *Aerosp. Sci. Technol.* **2019**, *85*, 96–112. [CrossRef]
75. Bekemeyer, P.; Timme, S. Flexible aircraft gust encounter simulation using subspace projection model reduction. *Aerosp. Sci. Technol.* **2019**, *86*, 805–817. [CrossRef]
76. De Souza Siqueira Versiani, T.; Silvestre, F.J.; Neto, A.B.G.; Rade, D.A.; da Silva, R.G.A.; Donadon, M.V.; Bertolin, R.M.; Silva, G.C. Gust load alleviation in a flexible smart idealized wing. *Aerosp. Sci. Technol.* **2019**, *86*, 762–774. [CrossRef]
77. Bronz, M.; Gavrilovic, N.; Drouin, A.; Hattenberger, G.; Moschetta, J.-M. Flight Testing of Dynamic Soaring Part-1: Leeward Inclined Circle Trajectory. In Proceedings of the AIAA Scitech 2021 Forum, Virtual Event, 11–15 & 19–21 January 2021.
78. Gavrilovic, N.; Benard, E.; Pastor, P.; Moschetta, J.-M. Performance improvement of small unmanned aerial vehicles through gust energy harvesting. *J. Aircr.* **2018**, *55*, 741–754. [CrossRef]
79. Mohamed, A.; Taylor, G.; Watkins, S.; Windsor, S. Opportunistic soaring by birds suggests new opportunities for atmospheric energy harvesting by flying robots. *J. R. Soc. Interface* **2022**, *19*, 20220671. [CrossRef] [PubMed]
80. Poksawat, P. Control System for Fixed-Wing Unmanned Aerial Vehicles: Automatic Tuning, Gain Scheduling, and Turbulence Mitigation. Ph.D. Thesis, RMIT University, Melbourne, Australia, 2018. Available online: https://researchrepository.rmit.edu.au/esploro/outputs/doctoral/Control-system-for-fixed-wing-unmanned-aerial-vehicles-automatic-tuning-gain-scheduling-and-turbulence-mitigation/9921864086901341 (accessed on 22 December 2022).
81. Poksawat, P.; Wang, L.; Mohamed, A. Gain scheduled attitude control of fixed-wing UAV with automatic controller tuning. *IEEE Trans. Control. Syst. Technol.* **2017**, *26*, 1192–1203.

82. Poksawat, P.; Wang, L.; Mohamed, A. Automatic tuning of attitude control system for fixed-wing unmanned aerial vehicles. *IET Control. Theory Appl.* **2016**, *10*, 2233–2242.
83. Panta, A. Dynamics of leading-edge and trailing-edge control surfaces at low Reynolds number. *Nat. Sci. Rep.* 2022, *in press*.
84. Sattar, A.; Wang, L.; Mohamed, A.; Panta, A.; Fisher, A. System Identification of Fixed-wing UAV with Multi-segment Control Surfaces. In Proceedings of the 2019 Australian & New Zealand Control Conference (ANZCC), Auckland, New Zealand, 27–29 November 2019.
85. Panta, A.; Mohamed, A.; Marino, M.; Watkins, S.; Fisher, A. Unconventional control solutions for small fixed wing unmanned aircraft. *Prog. Aerosp. Sci.* **2018**, *102*, 122–135. [CrossRef]
86. Gigacz, R.; Mohamed, A.; Poksawat, P.; Watkins, S.; Panta, A. Developing a Stable UAS for Operation in Turbulent Urban Environments. In Proceedings of the International Micro Air Vehicles Conference and Flight Competition 2017, Toulouse, France, 18–21 September 2017.
87. Mohamed, A.; Clothier, R.; Watkins, S.; Sabatini, R.; Abdulrahim, M. Fixed-Wing MAV Attitude Stability in Atmospheric Turbulence PART 1: Suitability of Conventional Sensors. *Prog. Aerosp. Sci.* **2014**, *70*, 69–82.
88. Mentzoni, F.; Ertesvåg, I.S. On turbulence criteria and model requirements for numerical simulation of turbulent flows above offshore helidecks. *J. Wind. Eng. Ind. Aerodyn.* **2015**, *142*, 164–172.
89. Rowe, S.J.; Howson, D.; Turner, G. A turbulence criterion for safe helicopter operations to offshore installations. *Aeronaut. J.* **2006**, *110*, 749–758.
90. Garcia-Magariño, A.; Bardera, R.; Sor, S.; Matias-Garcia, J.C. Flow Control Devices in Cities for Urban Air Mobility. In Proceedings of the AIAA AVIATION 2020 FORUM, Virtual, 15–19 June 2020.
91. Heliport Design, in Advisory Circular. Federal Aviation Administration: Washington, DC, USA, 2023. Available online: https://www.faa.gov/documentLibrary/media/Advisory_Circular/AC_150_5390_2D_Heliports.pdf (accessed on 22 December 2022).
92. Bassey, R. Vertiport Design. In *Engineering Brief*; Federal Aviation Administration: Washington, DC, USA, 2022. Available online: https://www.faa.gov/airports/engineering/engineering_briefs/engineering_brief_105_vertiport_design (accessed on 22 December 2022).
93. City of London Corporation. *Wind Microclimate Guidlines for Developments in the City of London*; City of London Corporation: London, UK, 2019. Available online: https://www.cityoflondon.gov.uk/services/planning/microclimate-guidelines (accessed on 22 December 2022).
94. Prudden, S.; Fisher, A.; Mohamed, A.; Watkins, S. An anemometer for UAS-based atmospheric wind measurements. In Proceedings of the 17th Australian International Aerospace Congress (AIAC 2017), Melbourne, Australia, 26–28 February 2017.

Article

Strategic and Tactical Path Planning for Urban Air Mobility: Overview and Application to Real-World Use Cases

Flavia Causa *, **Armando Franzone and Giancarmine Fasano**

Department of Industrial Engineering, University of Naples "Federico II", P.le Tecchio 80, 80125 Naples, Italy
* Correspondence: flavia.causa@unina.it

Abstract: Urban air mobility requires safe and efficient airspace management, as well as effective path planning and decision-making capabilities to enable access to the urban airspace, which is predicted to be very densely populated. This paper tackles the problem of strategic and tactical path planning by presenting a framework specifically designed for accounting for several constraints and issues of the urban environment. Multi-objective and multi-constraint planner algorithms are developed to this aim, along with an innovative method for information simplification and manipulation. Navigation-aware and optimized trajectories were retrieved from the strategic approach. Tactical path planning was developed using three approaches that react differently to unpredicted conditions. The entire strategic–tactical pipeline was tested in two real-world use cases, representing common missions in urban environments, such as medical delivery and short-range air taxi. The results demonstrate the effectiveness of the proposed methodology in generating the strategic path and show the different outcomes of the proposed tactical approaches, thus highlighting their advantages and drawbacks.

Keywords: path planning; urban air mobility; tactical path planning; strategic path planning; urban airspace; urban maps; urban airspace constraints; test cases

Citation: Causa, F.; Franzone, A.; Fasano, G. Strategic and Tactical Path Planning for Urban Air Mobility: Overview and Application to Real-World Use Cases. *Drones* **2023**, *7*, 11. https://doi.org/10.3390/drones7010011

Academic Editor: Kamesh Namuduri

Received: 18 November 2022
Revised: 14 December 2022
Accepted: 22 December 2022
Published: 24 December 2022

1. Introduction

The market of unmanned aerial systems (UAS) has witnessed a rapid growth in the last decade, fostering research towards innovative algorithms and instruments to enhance their reliability, autonomy and safety. Urban air mobility (UAM) represents the next frontier of the UAS market, and a huge effort is being carried out in that direction in view of a wide variety of applications. As an example, missions such as urban taxi [1], inspection and surveillance [2] could experience a significant performance improvement (e.g., reduction of mission time) if performed by highly autonomous unmanned aerial vehicle (UAV). The development of urban air mobility will completely change our cities with a huge quantity of UAVs that require to be accommodated in the airspace. The increased UAV density also calls for new regulations [3] that are being developed to bridge the gaps between novel technologies and urban safety requirements [4]. In this framework, path planning and traffic management tasks are of paramount relevance to meet these safety requirements and enable UAV access to a traffic dense airspace.

UAV path planning, aiming to define the best route from a start to a goal point, is a widely discussed problem in the open literature, and several algorithms have been developed to specifically accommodate mission needs. In general, path planning requires accounting for UAV dynamic constraints, energy consumption (and thus wind and weather conditions [5–8]) and fixed and mobile obstacles to meet safety and effectiveness requirements [9]. When it comes to urban scenarios, navigation issues can arise for vehicles using the classical GNSS/inertial fusion scheme due to the non-nominal GNSS coverage conditions, which is experienced next to buildings and infrastructures. Indeed, these structures cause GNSS signal shadowing or deviation, leading to either signal obstruction or multipath phenomena, respectively. In these scenarios, path effectiveness, should also account for the capability to follow the planned trajectory while keeping a bounded navigation error and fulfilling the required navigation performance (RNP) [10]. Conversely, safety requirements are also linked to the need for minimizing the ground risk [11]. Several works in the open literature have addressed planning problems by accounting for ground risk mitigation [12,13], GNSS coverage fault [14], wind condition and its effect to energy consumption [6]. Nevertheless, an integrated framework which tackles all these aspects altogether is required to effectively enable safe access to the urban airspace.

The "SMARTGO" (gnsS-enabled urban air Mobility through Ai-powered environment-awaRe Techniques for strateGic and tactical path planning Operations) project [15], funded by the Italian Space Agency (ASI) and carried out by a consortium composed by the University of Naples "Federico II" (as coordinator), TopView S.r.l. and Euro.Soft S.r.l. aims to develop strategic and tactical path planning algorithms that can handle several information levels to meet urban environment requirements and constraints. Besides path planning approaches, the project also aims to gather and synthesize relevant information about the urban environment useful for path planning and at defining innovative U-space services based on the project outcomes. Data gathering includes processing satellite images with innovative algorithms, which are also based on artificial intelligence for terrain classification and ground risk estimation. Tackling both planner design and information gathering bridges the gap between data retrieval and usage and ensures information collection and representation is specifically tailored to planner's needs.

This paper briefly describes the main algorithmic outcomes of the project in terms of path planning and focuses on the application of the entire algorithmic chain to real-world test cases that have been specifically identified for the project's needs. A brief overview of the project and a preliminary version of both the strategic and tactical algorithms have already been presented in [15]. Then, a more detailed version of the tactical framework and the developed approaches to tackle with unpredicted events during flight has been described in [16]. In addition, a detailed description of the strategic path planning algorithm is reported in [17]. This work extends the previous contributions of the authors by providing the following innovative points:

1. It describes the final version of both tactical and strategical path planning algorithms.
2. It provides a systematic approach to deal with tactical planning and its multiple alternatives to overcome any type of unpredicted event.
3. It analyzes the tactical and strategic results on relevant environments by applying the entire strategic–tactical path planning flowchart and also highlighting the strategic path selection approach.

The remainder of this paper is structured as follows. Section 2 analyzes the constraints to be taken into account at both tactical and strategic level while also detailing the procedure for information retrieval. Both strategic and tactical path planning algorithms are summarized in Section 3. Test cases and their related information to be used at path planning level are reported in Section 4, and Section 5 shows the tactical and strategic path planning results. Finally, Section 6 draws conclusive remarks.

2. Environment and Vehicle Based Constraints

Risk, weather and no fly zone information, as well as 3D geometries, mobile obstacles and vehicle specifications and constraints must be taken into account to plan for a safe and effective path. *Vehicle-based constraints* include maximum airspeed and flight path angle, as well as battery capacity, maximum allowed wind velocity and navigation system performance. *Environment-based constraints* can, in general, be divided into two categories depending on whether they are connected to the airspace or not. Airspace-based constraints include (but are not limited to) no-fly-zones, maximum and minimum flight altitudes, traffic information, contingency landing site location and possible airspace structure and/or speed rules, as foreseen, for instance, in the geovectoring approach [18]. On the other hand, the other environmental constraints include general information about the scenario such as 3D geometry, weather information and estimate of ground risk.

Because the majority of the developed planning approaches are designed to deal with spatial based information, SMARTGO information gathering and organization aims at simplifying most of the aforementioned sources in multi-dimensional maps. This is done to reduce the information processing to be carried out at the path planning level. Specifically, the following maps are defined and used as planning inputs:

1. **3D map with fixed obstacles and no fly zones**. Fixed obstacle maps are obtained from publicly available representation of the environment available on open source platforms, such as CityGML [19] or OpenStreetMap.

2. **Risk map**. A 2D map that contains information about level of risk associated to each latitude and longitude coordinates. It is computed from satellite images and GIS databases. As detailed in [15], risk map retrieval first uses satellite imagery to classify terrain with a VGG16 convolutional neural network (CNN) [20] trained with the EuroSAT database. Then, information from the GIS database and building footprints is integrated to further segment terrain classification and localize critical structures, such as power plants, railway stations, subways, airports and hospitals. Level of risk going from 1 to 4, with increasing damage entity forecasted in case of vehicle fault, is extracted from segmented information of the terrain so that:

    a. Class 1 includes low-risk areas such as natural and rural ones;
    b. Class 2 includes industrial areas characterized by low people density;
    c. Class 3 includes urban environments. In this scenario a subclassification is performed to distinguish between buildings and populated areas such as squares and streets;
    d. Class 4 includes critical infrastructures (e.g., train stations and hospitals) and it is again divided in various subcategories.

    An example of risk level over a portion of Naples city (Italy) is reported in Figure 1.

3. **Landing site maps** are 2D maps containing cost information, which increases as the distance from the contingency landing area increases.

4. **Weather or wind maps** are multidimensional maps corresponding to each ground point information about wind intensity and direction in terms of azimuth and elevation. In this work, the wind dependency on altitude is not considered, which is consistent with currently available weather maps.

5. **GNSS coverage maps**. GNSS coverage maps are defined with the aim of spatially representing the information about navigation performance, thus avoiding the need to propagate navigation error covariance during the path planning process. Indeed, the navigation performance of the majority of UAVs (which are usually implementing INS/GNSS data fusion) is strictly connected to both the inertial instrument specifics and GNSS coverage. A GNSS coverage map is a 2.5D map connected to the dilution of precision (DOP) level, defining the elevation at which the DOP becomes smaller than a certain threshold. As the GNSS constellation varies as a function of time, a time-varying GNSS coverage map is expected over a selected time interval. The approach followed in the SMARTGO project samples the time interval and defines an elevation map for each sample. The so-defined elevation maps are merged in a worst-case logic to have a constant GNSS coverage map to be used during the whole mission time. Several GNSS coverage maps can be defined as a function of the selected DOP threshold. As an example, Figure 2 shows three GNSS coverage maps obtained over a portion of Naples city center with different colors. It can be noticed that the map's offset with respect to the buildings reduces as the selected DOP threshold (i.e., D_j) increases. Computing each GNSS challenging map can be very time demanding if a very large scenario is considered. However, in many cases, the need for detailed GNSS coverage maps may arise only in proximity of take-off and landing areas. The approach followed in the SMARTGO project uses this idea, thus estimating the GNSS coverage maps only in the surroundings of the start and the end point and assuming the map altitude is equal to the terrain plus an offset in the other areas.

6. **Traffic information** is provided via vehicle-to-vehicle (V2V) or infrastructure-to-vehicle (I2V) communications. This work assumes the entire flight plan of the other vehicles is fed to the ownship both in the strategic phase and during the flight. Flight plan information of the intruder is stored in 3D time-varying occupancy maps, detailed in [16]. N occupancy maps varying with time are used to prevent continuously checking for intruder possible collision, each one covering a time interval equal to Δt. The n-th occupancy map is used for checking collision in the time segment going from t_{n-1} to t_n ($t_n = t_0 + n\Delta t$, being t_0 the starting time of the mission). The representation of the intruder in each occupancy map is given by its path during the associated time interval enlarged with time and spatial margins. The nature of traffic maps allows them to be merged with the fixed obstacle maps so as to speed up the collision check operation.

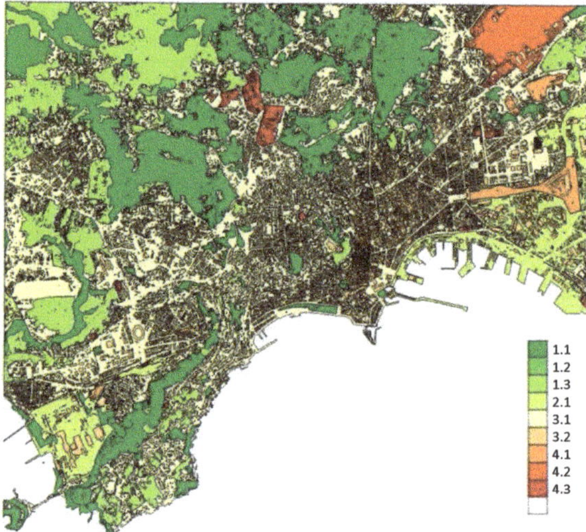

Figure 1. Risk map over a portion of the Naples city center, Italy.

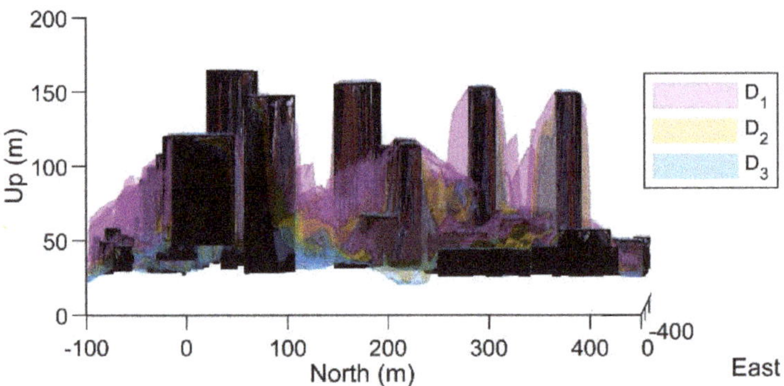

Figure 2. GNSS coverage maps as a function of the DOP threshold. $D_1 < D_2 < D_3$.

3. Path Planning Framework

Planning framework, whose flowchart is shown in Figure 3, foresees two phases, i.e., the strategic phase (detailed in Section 3.1), which is aimed at evaluating an optimized trajectory for the UAV before the flight, and a tactical phase (described in Section 3.2), which continuously checks the trajectory during the flight and takes action in the case an unpredicted event occurs that compromises the trajectory safety and effectiveness. The need for decomposing the planning approach in two phases [21] comes from guaranteeing path optimality while reducing the computational cost during flight, as tactical replanning is only demanded at finding deviations from the nominal (strategic) trajectory. To ensure the latter condition, a strategic path planner must be carried out with the largest amount of available information, following the better-informed, better-planned logic.

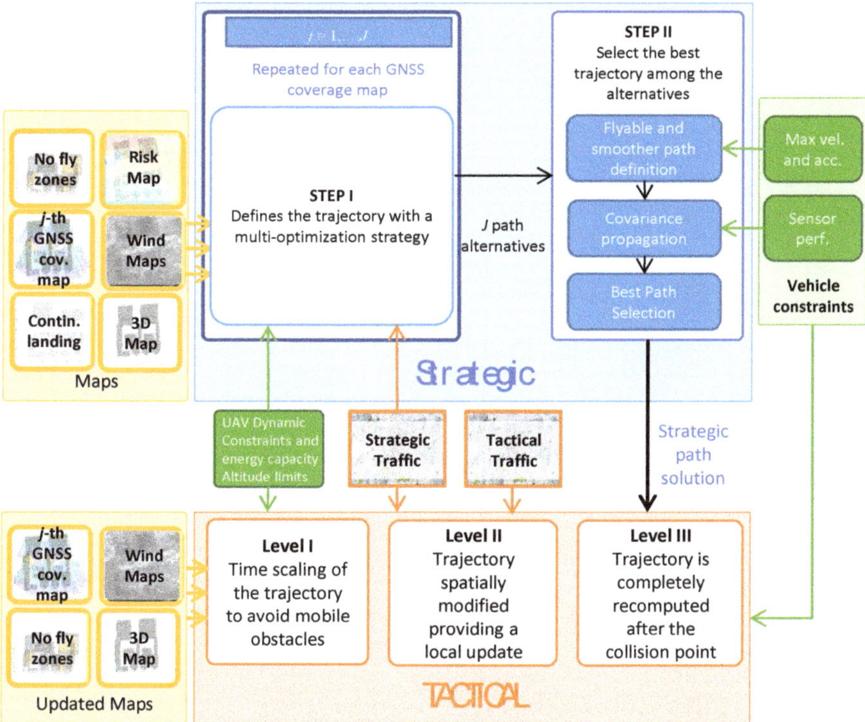

Figure 3. Strategic and tactical planning flowchart.

Both strategic and tactical solutions are conceived to deal with all the set information reported in Section 2, or a subset of them in a scalable and adaptive way. Due to the large amount of information to be dealt with and the huge dimension of the scenario where the planner is considered to operate (with mission ranges in the order of few kilometers), sampling based approaches, such as rapidly exploring random trees (RRT) [22] and its modifications, are preferred in this work to graph search method, such as A* [23] because of the high cost linked to sampling all the nodes belonging to the environment. Feasible segments to add to the solution tree are those which:

- are not intersecting with any fixed (including NFZ) or mobile obstacle;
- are compliant with the battery capacity and with the maximum velocity and flight path angle limits;
- have an altitude between the maximum and minimum flight altitude computed above the ground level;
- never lie below the GNSS coverage map used as reference;
- do not enter in areas whose wind intensity is higher than the one the UAV can tolerate.

3.1. Strategic Path planning

The strategic path planning approach developed within the SMARTGO project is characterized by two steps, which are depicted in Figure 3. The first step is based on a custom version of the batch informed three (BIT*) algorithm [24] embedded in the open motion planning library (OMPL) framework [25]. More details about STEP I algorithmic implementation are provided in [17]. Both constant (e.g., fixed obstacles and NFZ, landing site, risk and wind maps, as well as GNSS coverage volumes) and time varying information (i.e., traffic) are used as input in this step. The final path is a 4D trajectory which is constrained to the feasible trajectory conditions reported in Section 2. An optimized trajectory is obtained by minimizing:

$$f(s) = \alpha_s C_s(s) + \alpha_l C_l(s) + \alpha_r C_r(s) + \alpha_w C_w(s) \tag{1}$$

Costs C_x and weighting factor α_x can be referred to path length (s), landing site (l), risk (r) and energy (w) information. Landing site and risk costs are obtained by integrating the normalized version of the landing and risk maps along the trajectory. Energy cost is obtained by using a simplified model based on rotor theory and described by [17]. C_s represents the path length.

STEP I is repeated several (J) times, while the GNSS coverage map input changes as a function of the DOP. The second step selects the minimum cost solution among the J available ones. The trajectories are first smoothed with polynomial trajectory planning [26] and then navigation state covariance propagation is performed to verify path navigation feasibility, i.e., the fact that positioning error is always lower than a positioning error threshold (Δp_{max}). Any solutions not fulfilling this requirement is discarded and the 4D strategic (nominal) path is obtained as the one with minimum cost among the remaining alternatives.

3.2. Tactical Path Planning

Tactical planner is aimed at fast finding a path that can be followed by the UAV if safety and effectiveness of the strategic path are jeopardized during the flight. Optimality is non-accounted for at this level, and the first feasible alternative is considered as tactical solution. Therefore, all the information sources, which are only used for the aim of cost definition (i.e., landing site location and ground risk) are discarded at tactical level, and their modification cannot trigger any level of tactical replanning. Conversely, updates of GNSS maps, fixed and mobile obstacles and wind conditions may call for trajectory update. As far as wind modification is concerned, it not only alters the UAV energy consumption, but it also modifies the zones in which the UAV is allowed to fly due to the maximum admissible wind velocity. Tactical path planner assumes the ownship has a lower priority than the other UAVs, so it has to maneuver in case of conflict. It implements three different solutions, referred to as levels in Figure 3, that are specifically designed to counteract several events that could occur during the tactical phase. The three levels are characterized by an increasing level of computational complexity and their performance is summarized below. For further algorithmic details, the reader is referred to [16].

1. *Level 1* is aimed at modifying the time history of the trajectory without altering its geometry so as to keep the path optimality. Time history is modified by scaling down the UAV velocity using an ad hoc scaling function, which decelerates the vehicle before the encounter through a deterministic approach. Because spatial modification of the trajectory is not foreseen in this approach, *Level 1* can be only used for avoiding mobile obstacle collision in the case of non-frontal confliction geometries. In addition, despite the low computational time, this approach extends the mission time and can be not suitable for vehicles whose nominal path requires an energy consumption close to the battery capacity.
2. *Level 2* provides spatial modification of the trajectory in the surroundings of the location of the unfavorable event(s). The planner uses a customized version of the RRT algorithm conceived as a global replanner that only provides a local modification

of the trajectory because it is informed to return to the strategic path. The global nature of this approach avoids sequential replanning if multiple unfavorable events are experienced by the UAV, thus saving time. Due to the spatial modification of the trajectory, this solution is not only able to deal with both fixed and mobile obstacle geometries, but it can also be used to counteract wind velocity and GNSS coverage maps alteration. The heuristic nature of the RRT makes this solution non-deterministic. In addition, a higher computational time is experienced with respect to the previous approach. However, since only a local modification is provided to the strategic path, its optimality remains almost unaltered while also providing a small increase in flight time, as demonstrated in [16].

3. *Level 3* of the tactical planning provides a global modification of the trajectory starting from its last non-corrupted point. From that point, a completely new trajectory is recomputed with an algorithm still based on RRT, but not informed to return to the strategic trajectory. This solution, which completely alters the path after the unfavorable event, should be chosen when a significant modification of the flight conditions has been experienced with respect to the scenario available at the strategic level. Due to the similar algorithmic scheme, this level shares the same heuristic nature of *Level 2*, as well as the higher computational time with respect to *Level 1*.

Among the proposed solutions, the one to be chosen during tactical phase is not trivial to be identified. Although a simple geometry consideration can be made to deem whether to use *Level 1* solution or not, this is not true for the other two levels and their output must be compared to this aim. Therefore, the tactical planner is conceived to run all the levels sequentially and compare their solutions, if available, to choose the one with the minimum cost. A flowchart of the tactical planning is reported in Figure 4.

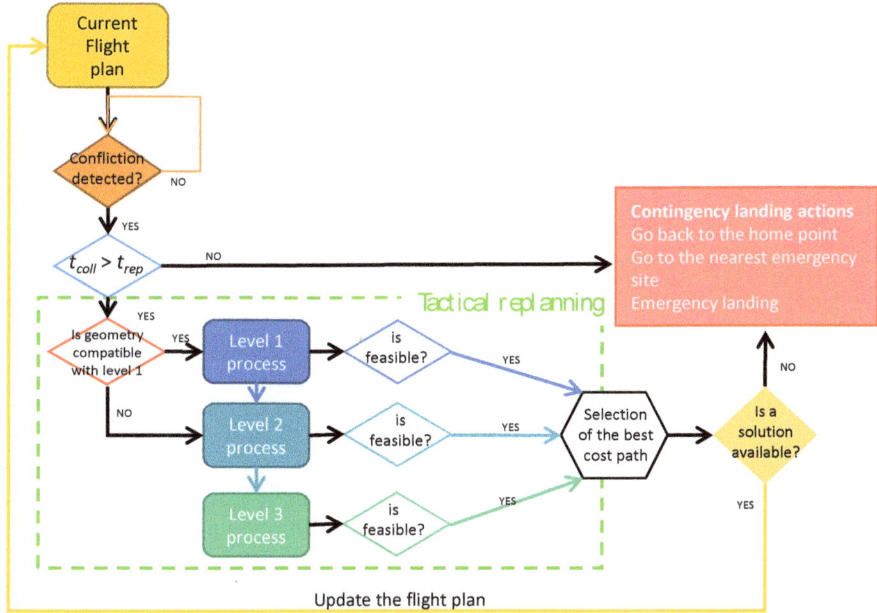

Figure 4. Tactical path planner flowchart.

Level 1, which has the lowest computational cost, is the first to be run (it runs only if compliant to the conflict geometries), then *Level 2* is queried. The level sequence runs until a timeout, and the tactical output is picked among all the available solutions at that time. During the flight, the current path is checked for any unfavorable event (i.e., contingency).

If the time to the contingency (t_{coll}) is greater than the replanning timeout (t_{rep}), the tactical replanning levels are run sequentially and the best cost path is selected to update the flight plan. In the case no available solution is found by tactical planning or $t_{coll} \leq t_{rep}$, contingency landing actions are activated.

4. Use Cases

Strategic and tactical pipeline have been applied to two test case scenarios. The first represents an urban air taxi problem, specifically designed to transport passengers from airport to business center and port. The second scenario includes the delivery of medical supplies from the mainland to an island, thereby saving time with respect to ship-based transportation. Two scenarios have been identified in the Naples area and its surroundings and are reported in Figure 5. Risk, landing site, wind and GNSS coverage maps estimated with three different thresholds, i.e., $D_1 = 2$, $D_2 = 3$, $D_3 = 4$ are reported both for air taxi and medical delivery cases in Figures 6 and 7, respectively.

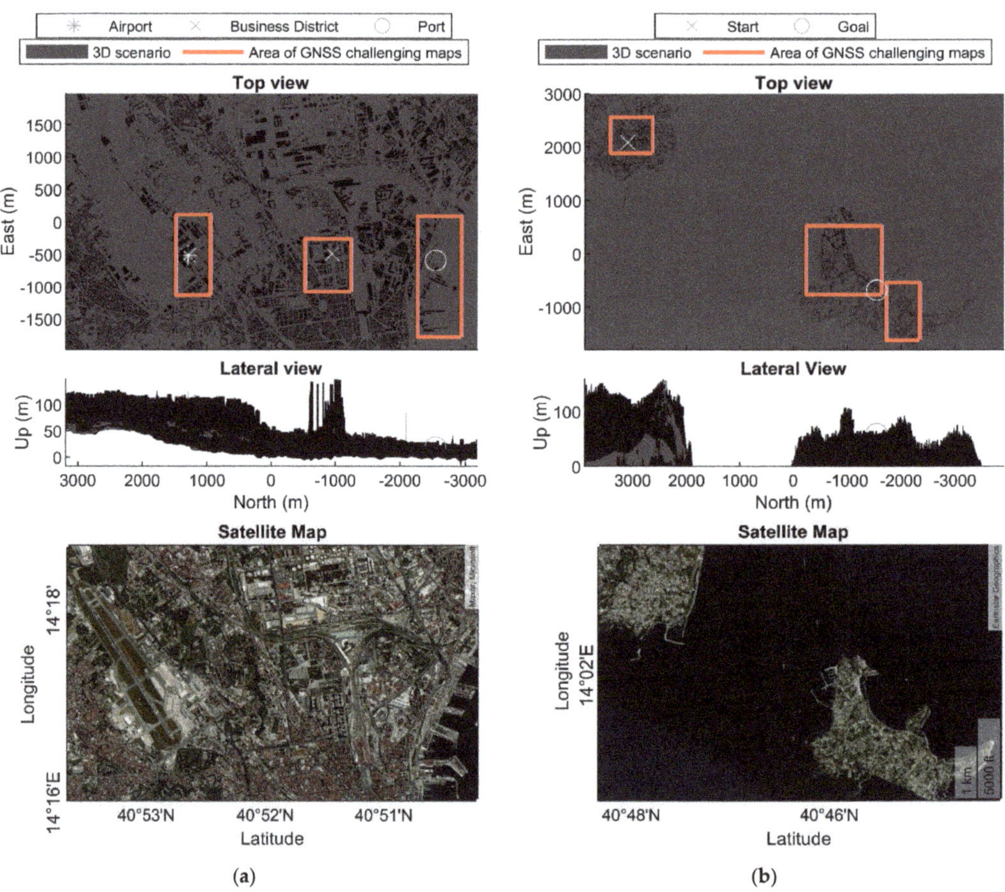

Figure 5. Test case scenarios. (**a**) Air taxi scenario. Top, lateral view and satellite map. (**b**) Medical delivery scenario. Top, lateral view and satellite map.

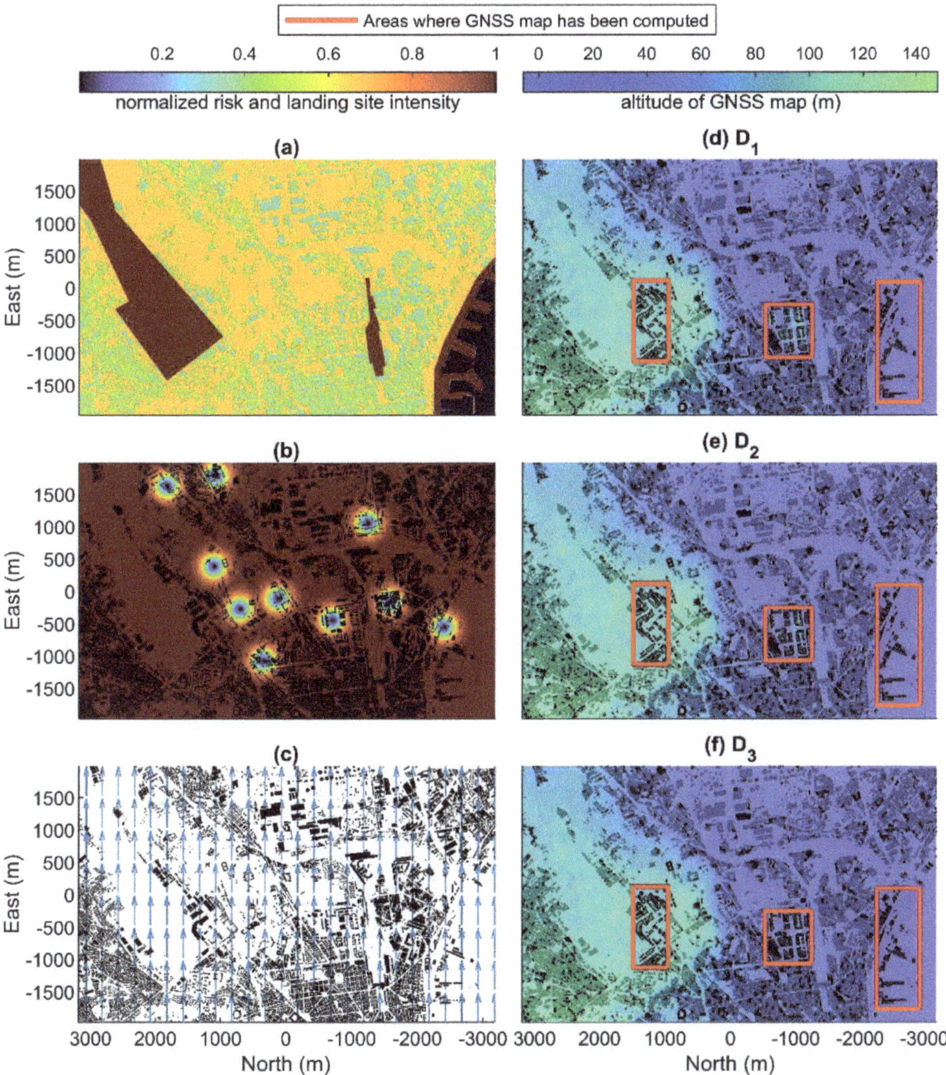

Figure 6. Air taxi scenario. (**a**) Risk and (**b**) landing site intensity maps. (**c**) Top view of wind direction, identified by blue vectors. GNSS coverage maps associated to (**d**) $D_1 = 2$, (**e**) $D_2 = 3$ and (**f**) $D_3 = 4$.

Because the two identified scenarios have a huge extension, the GNSS coverage map has been computed only in a portion of the environment which is closer to the start or the end point of the trajectory. They were obtained using a starting time of 11:30 UTC of 19th April 2022. A time interval of 20 min has been considered, with a time span of 5 min. The ground grid has 5 m spacing. A uniform wind intensity of 5 m/s has been used for both missions.

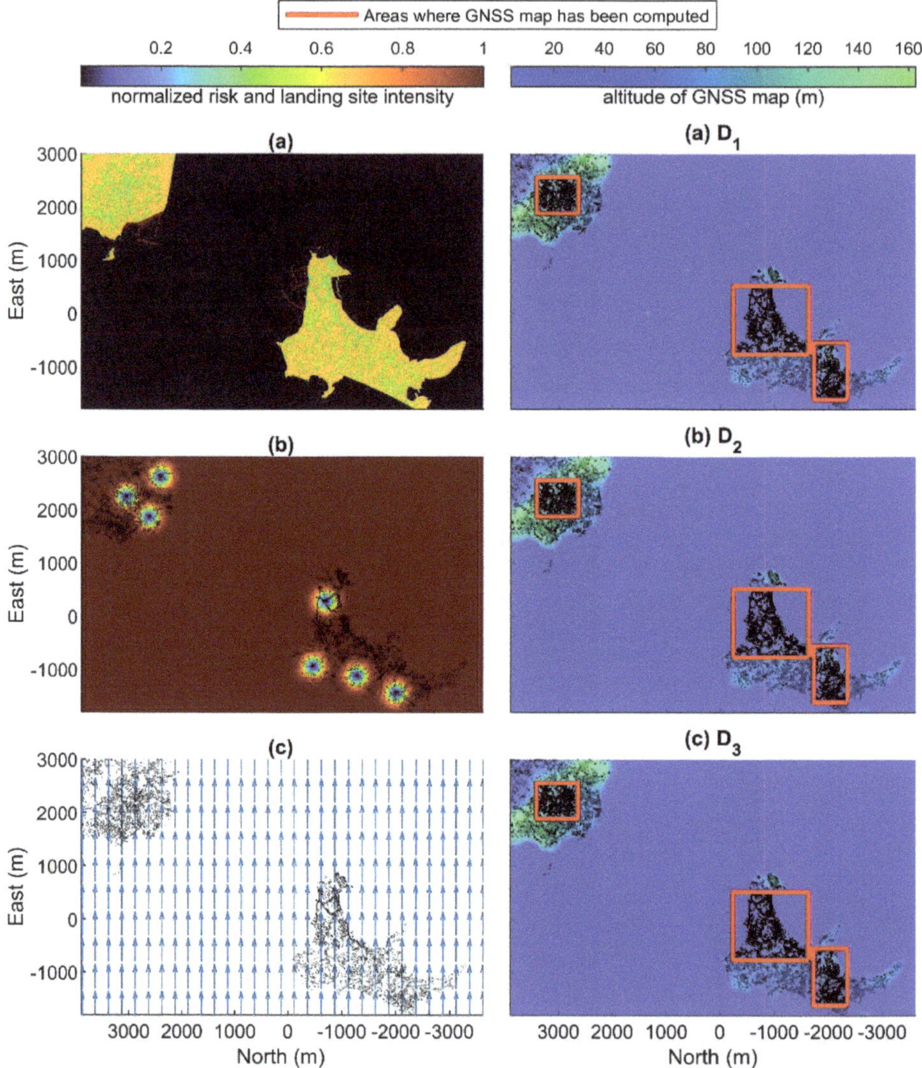

Figure 7. Medical delivery scenario. (**a**) Risk and (**b**) landing site intensity maps. (**c**) Top view of wind direction, identified by blue vectors. GNSS coverage maps associated to (**d**) $D_1 = 2$, (**e**) $D_2 = 3$ and (**f**) $D_3 = 4$.

The air taxi scenario is reported in Figure 5a. It envisages transfers from Capodichino airport to Naples business center or port. The three locations have been represented on the map with an asterisk, a cross and a circle, respectively. Both the top and lateral views have been reported in the Earth north up (ENU) coordinate frame originated at $40°51'56''$ N, $14°17'20''$ E, as well as the rectangles where GNSS coverage maps have been computed. For the sake of concreteness, the satellite map of the identified area is also reported. The lateral view highlights the high slope of the scenario. A maximum flight altitude estimated above the ground level and equal to 150 m has been assumed. Two missions have been considered, i.e.:

- **Mission 1.** From Airport ([−511, 1290, 105] ENU coordinates) to Port ([−591, −2541, 20] ENU coordinates).
- **Mission 2.** From Airport ([−511, 1290, 105] ENU coordinates) to Business district ([−485, −950, 50] ENU coordinates).

The medical delivery scenario involves the Procida island and its closest city on the mainland, i.e., Monte di Procida, which is about 4 km far. Even if several ship connections, taking about 16 min, are organized in the summer season (from June to September), very few transportations (twice a day and only on weekdays) are foreseen in the winter season, making impossible to directly deliver urgent medicines. A dedicated drone service could not only spare time but also be operated on-demand. The scenario extension (both in the lateral and top view) and its satellite view have been reported in Figure 5b. The start and arrival location includes a pharmacy in Monte di Procida (40°47′20″ N 14°3′0″ E and 100 m altitude) and the Procida local medical unit (40°45′25″ N 14°1′11″ E and 60 m altitude), which are reported with a cross and a circle in the figure, respectively. A maximum flying altitude of 150 m above the terrain level (or the sea level when the aircraft flies in the Procida channel) has been assumed.

5. Results

Strategic and tactical planning have been carried out assuming as aircraft a DJI M300 RTK [27], whose main parameters are reported in Table 1. Navigation performance of the IMU sensor has also been included, which is assumed to be the one of the medium grade IMU HG1120CA50 from Honeywell [28]. The positioning error threshold is assumed to be equal to 2 m so that any trajectory which overcomes this value at least once during the flight must be discarded. In order to trigger every tactical level to output a solution, in this work, tactical information is only limited to intruder trajectory updates. The entire path planning pipeline results will be detailed in the medical delivery scenario, which involves a single mission, in Section 5.1. On the other hand, results related to the air taxi scenario are shown in Section 5.2.

Table 1. Vehicle specifics.

Constraints		Value
Battery capacity	ξ (mAh)	11,870
Maximum airspeed	(m/s)	23
Max wind speed	(m/s)	15
Cruise speed	v_c (m/s)	10
Maximum Flight Path Angle	α (°)	15
Max Positioning error	Δp_{max} (m)	2
IMU Parameters [1]	Acc. In-run stability (mg)	0.11
	Velocity random walk (m/s/$\sqrt{h}$)	0.06

[1] Only accelerometer parameters are included since navigation error covariance propagation is run with a simplified approach.

5.1. Medical Delivery Scenario

Strategic path planning results are obtained using in STEP I $\alpha_s = \alpha_w = 1$, $\alpha_r = 4$, $\alpha_l = 2$ as weighting factors, thus privileging trajectories which reduce the ground risk and are closer to the landing sites. STEP I paths are reported in Figure 8, along with the strategic obstacles' paths. A trajectory for each GNSS challenging map is obtained, with the cost breakdown reported in Table 2. The costs are estimated over the smoothed trajectory computed in STEP II. Results of trajectory flyability test (maximum trajectory positioning error lower than Δp_{max}) are also reported in the Table. All the trajectories are compliant with navigation requirements, and D_2 (highlighted in green in Table 2), which minimizes the cost function (f) is picked as the strategic (nominal) solution. Computational time of each solution estimated on an Intel i7 pc with a 2.59 GHz processing unit has been also reported in the Table, demonstrating the planner requires less than one minute for output each trajectory.

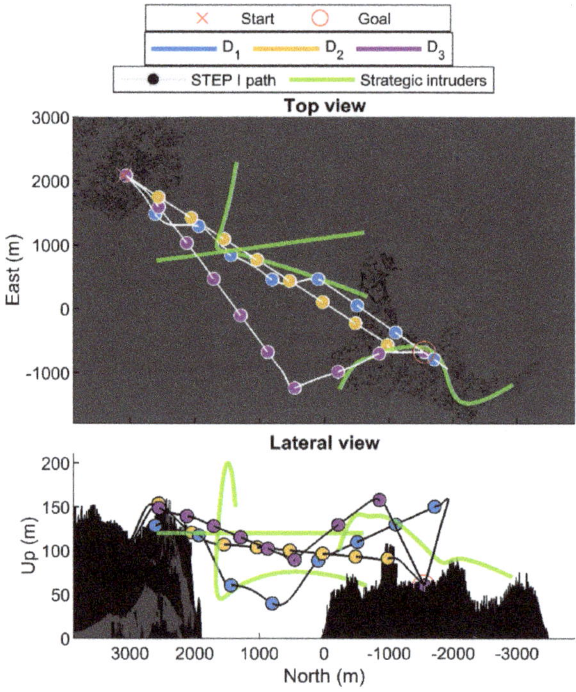

Figure 8. STEP I strategic solution and strategic mobile obstacles. Medical delivery scenarios. $\alpha_s = \alpha_w = 1$, $\alpha_r = 4$, $\alpha_l = 2$.

Table 2. Medical delivery scenario. Strategic solution costs breakdown.

GNSS Coverage Map Threshold	Cost Functions (m)					Flyable	Comput. Time (s)
	C_s	C_r	C_l	C_w	f		
D_1	6650.7	1836.8	6472.8	7018.6	33,961.9	yes	53.0
D_2	5440.8	1301.9	5098.4	5830.2	26,675.4	yes	55.3
D_3	6384.0	1481.0	5807.1	6859.1	30,781.3	yes	60.3

Tactical deconfliction accounts for unknown (tactical) obstacles that the UAV has to avoid during the flight. With the aim of testing tactical planner performance, these trajectories have been specifically designed in order to intersect the strategic path. Information about these trajectories is transferred to the UAV by the U-Space Service Provider (USSP) or via a vehicle-to-vehicle data link. This information, together with the UAVs flight plans known in the strategic phase, must be taken into account to generate a safe and collision free path. The trajectory costs obtained after tactical deconfliction are reported in Table 3, along with the maximum navigation error, the computation time and the overall flight time that (except for *Level 1* approach, which experiences a huge time delay) does not increase significantly. Using the same GNSS coverage map accounted for in the strategic path definition as a boundary allows keeping the navigation error smaller than Δp_{max}. As expected, the lowest cost solution is the one associated with *Level 2*, which is specifically designed to produce local variation from the strategic path by keeping its cost function almost unaltered. Because the *Level 1* 3D trajectory coincides with the strategic one, all the spatial based costs (risk, landing site and path length) are equal. However, this is not true for the energy cost, which is increased due to the high waiting time to avoid tactical obstacles. As far as the computation times are concerned, *Level 1* solution, based on a deterministic approach, gives the smallest contribution. On the other hand, about 5 s are required to solve the deconflictions for spatial based solutions. These values are compatible with typical values of tactical replanning cut off time which is of the order of 10 s.

Table 3. Medical delivery scenario. Tactical solution costs breakdown.

GNSS Map Thr.	Tactical Level	Total Time (s)	Max Nav Err. (m)	Comp. Time (s)	Cost Functions (m)				
					C_s	C_r	C_l	C_w	f
	Strategic	543.5	1.19		5440.8	1301.9	5098.4	5830.2	26,675.4
D_2	1	1174.6	1.25	1.7	5440.8	1301.9	5098.4	6300.2	27,145.3
	2	546.0	1.20	4.2	5457.5	1292.3	5110.4	5838.5	26,686.1
	3	584.0	1.19	5.2	5958.5	1747.0	5484.7	6244.4	30,160.3

Tactical results are reported in Figure 9, either for spatial based solution (i.e., associated to *Level 2* and *3*) and time scaling results (*Level 1*), which are depicted in Figure 9a,b, respectively. Figure 9a shows both the lateral and the top view of the *Level 2* and *3* trajectories by also reporting the information of strategic and tactical intruders (top view) and the GNSS coverage map associated to the nominal trajectory, i.e., whose threshold is D_2 (lateral view). The *Level 3* solution has a larger deviation from the strategic path than *Level 2*, as expected. This deviation from the optimal path produces an increase of the trajectory cost. Figure 9b compares the velocity history of the strategic path with respect to the tactical one, noting the huge delay produced by the time scaling approach. Indeed, the ownship is slowed down twice and the avoidance of the second intruder produces a huge velocity reduction (near to zero) and a very long waiting time to avoid collision.

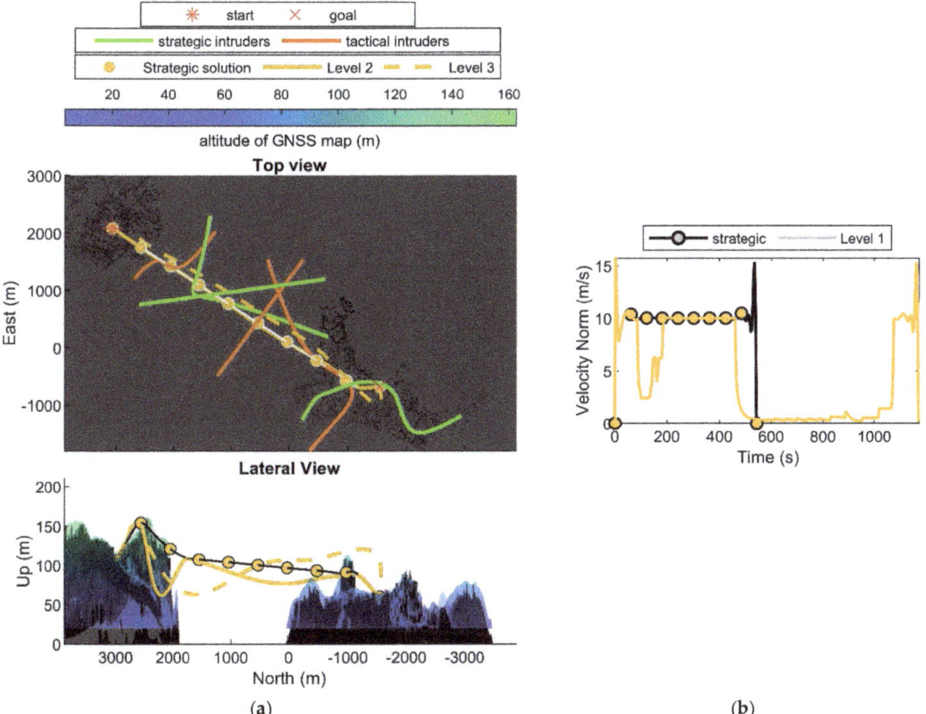

Figure 9. Tactical solution—Medical delivery scenario. (**a**) *Level 2* and *3* trajectories. (**b**) *Level 1* velocity norm history.

5.2. Air Taxi Scenario

Air taxi strategic paths have been obtained using the same weights of the previous section. The strategic costs breakdown and computational time for the two missions are reported in Table 4, where the nominal solution (i.e., the one having the minimum cost) for each mission has been highlighted in green. A lower computational time (about 20 s) is required in this case to obtain the solution, which depends on the different scenario geometry. Mission 2 does not have the D_1 solution, because the point located in the business center falls below its associated GNSS coverage map. The selected strategic path is the one associated with D_2. When mission 1 is accounted for, the lowest cost trajectory is the one associated to D_1, even if a shorter length is obtained using D_2. This is due to the large landing site weighting factor, which tries to push the trajectory far from the shortest length one in order to make it pass over landing site locations. Results for Mission 1 and 2 are reported for the first step of the strategic planning algorithm in Figure 10a,b, respectively, along with the paths of the strategic intruders.

Table 4. Air taxi scenario. Strategic solution costs breakdown.

Miss. No	GNSS Map Threshold	Cost Functions (m)					Flyable	Computation Time (s)
		C_s	C_r	C_l	C_w	f		
	D_1	4168.0	2.4905	2.8956	4189.7	24,110.9	yes	26.3
1	D_2	4098.6	2449.0	3347.8	4118.3	24,708.4	yes	22.0
	D_3	4110.2	2580.3	3125.7	4131.2	24,813.9	yes	25.4
	D_1							
2	D_2	2358.5	1449.9	2042.7	2349.2	14,592.7	yes	21.9
	D_3	2502.2	1510.7	1925.3	2490.8	14,886.4	yes	23.6

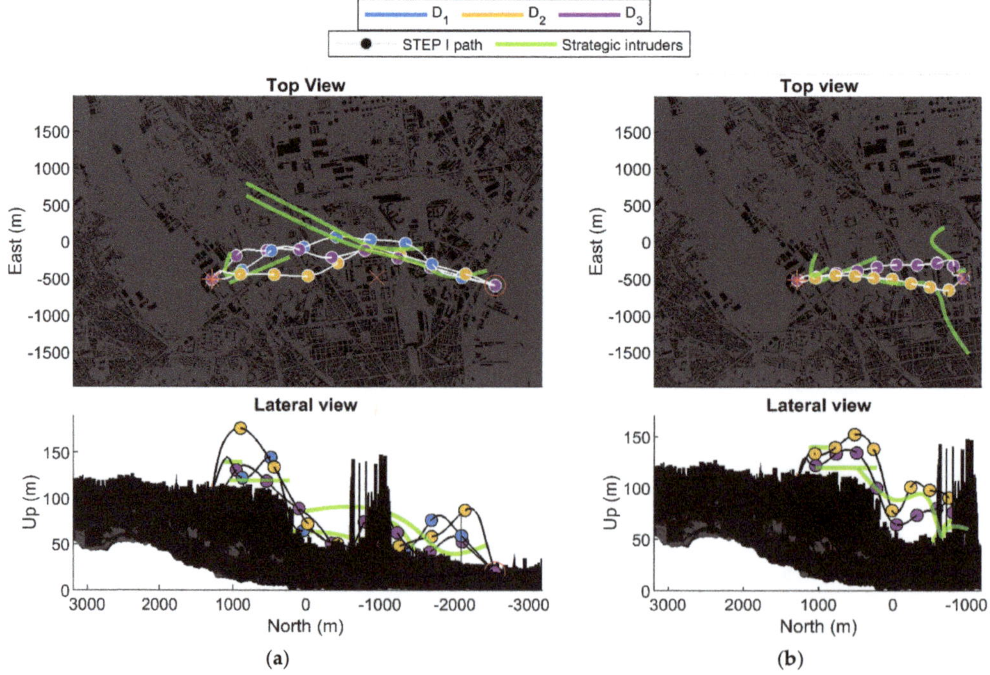

Figure 10. STEP I strategic solution and strategic mobile obstacles. Air taxing scenario (a) Mission 1 and (b) Mission 2. $\alpha_s = \alpha_w = 1$, $\alpha_r = 4$, $\alpha_l = 2$.

The trajectory updates during the tactical phase are shown in Figures 11 and 12 for mission 1 and 2, respectively. A zoomed portion of the scenario, which encloses the trajectory in each mission, has been reported to better visualize the tactical variation of the path. For each figure, the *Level 1* solution in terms of velocity history is reported in subfigure b, whereas the *Level 2* and *3* trajectory deviation from the strategic path are shown in subfigure a. As in the previous section, the strategic (nominal) trajectory is also reported, as well as the trajectories of both the strategic and tactical intruders (in top view) and the GNSS coverage map associated with the strategic solution (in lateral view).

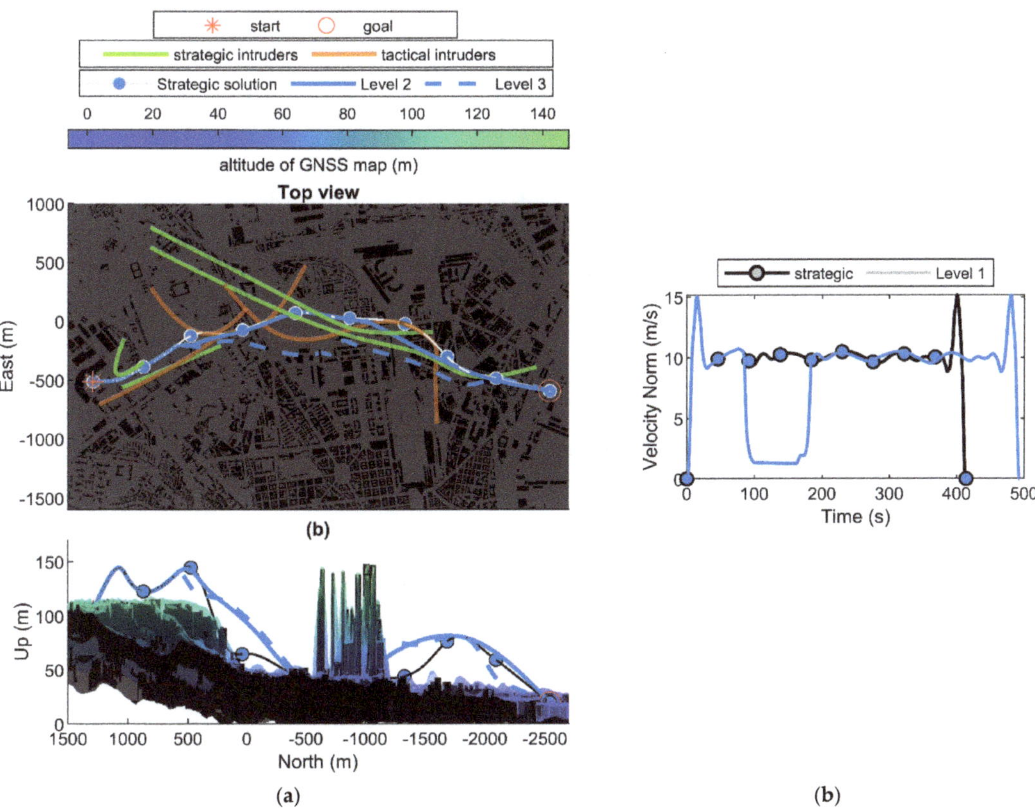

Figure 11. Tactical solution—Air taxi scenario, mission 1. (**a**) *Level 2* and *3* trajectories. (**b**) *Level 1* velocity norm history.

The tactical paths' cost breakdown is reported in Table 5, which also states the computational cost and the maximum navigation error. The latter, as in the previous case, slightly differs from the strategic one, thus not exceeding the maximum limits. Computation time is very low for the *Level 1* (below the second) and is at a maximum 6 s when *Level 2* and *3* are considered. Figure 12 again shows that the path obtained with *Level 2* locally deviates from the strategic trajectory by providing less modification, also in terms of path cost. Conversely, when the *Level 3* solution is used, path cost increases because the path is completely rebuilt without any knowledge of the ground information and costs. This could sometimes lead to a reduction of the trajectory length and duration (as in Mission 1). However, in all the cases, an increase of overall cost is provided.

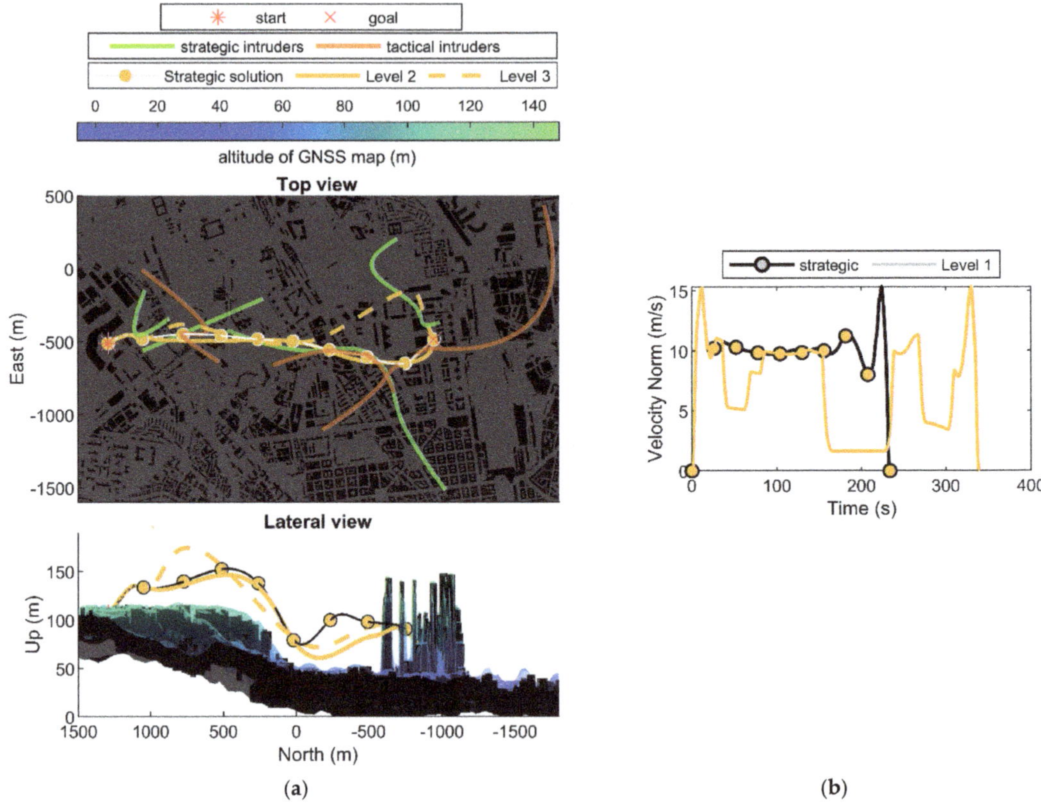

Figure 12. Tactical solution—Air taxi scenario, mission 2. (**a**) *Level 2* and *3* trajectories. (**b**) *Level 1* velocity norm history.

Table 5. Air taxi scenario. Tactical solution costs breakdown.

Miss. No.	GNSS Map Thresh	Tactical Level	Total Time (s)	Max Nav Err. (m)	Comp. Time (s)	Cost Functions (m)				
						C_s	C_r	C_l	C_w	f
		Strategic	414.1	1.19		4168.0	2.4905	2.8956	4189.7	24,110.9
1	D_1	1	492.4	1.21	0.2	4168.0	2.4905	2.8956	4223.2	24,142.5
		2	414.6	1.19	2.0	4156.9	2504.5	2990.0	4191.1	23,246.0
		3	407.3	1.19	6.2	4115.3	2519.4	3006.3	4114.9	24,320.3
		Strategic	233.6	1.35		2358.5	1449.9	2042.7	2349.2	14,592.7
2	D_2	1	339.7	1.36	0.4	2358.5	1449.9	2042.7	2367.2	14,611.4
		2	235.7	1.36	2.7	2365.6	1465.7	2123.9	2369.6	14,845.7
		3	252.1	1.41	2.0	2558.5	1556.3	2029.6	2546.7	15,389.6

6. Conclusions

Strategic and tactical planning algorithms to tackle UAV flight in urban environments have been presented and tested in this work, with the aim to provide an adaptive and scalable framework for urban operations. Indeed, the developed planning algorithms can deal with multiple sources of information by using the whole set of data or a subset of them. The design of the strategic path can be tailored to the user's needs by acting on the weighting cost factor, which spatially deviates the solution path towards the highest priority requirement. In addition, tactical modification to the trajectory allows reacting to unfavorable conditions, still ensuring the safety and effectiveness of the path. The entire algorithmic chain has been tested on two scenarios that involve air taxi within a very complex and obstacle-dense urban environment, and medical delivery from mainland to island. Results demonstrate the effectiveness of the proposed algorithms in yielding optimized and time-saving trajectories, thus highlighting the advantage of using unmanned aircraft to perform such operations. The promising results of the current work fulfill the SMARTGO ambition by creating an approach that can be used as a milestone for future urban air mobility planning algorithm design. Future efforts are aimed at further developing the conceived architecture and assessing its performance in very high traffic density, including flight rules and/or structured airspace. As an example, the tactical planner computational burden can be further reduced in order to better comply with dense, rapidly evolving scenarios, thus requiring better software engineering, which is foreseen as further algorithm improvement.

Author Contributions: Conceptualization, F.C. and G.F.; methodology, F.C., A.F. and G.F.; software, F.C. and A.F.; validation, F.C. and A.F.; formal analysis, F.C.; investigation, F.C. and A.F.; resources, F.C.; data curation, F.C.; writing—original draft preparation, F.C.; writing—review and editing, A.F. and G.F.; visualization, F.C.; supervision, G.F. and F.C.; project administration, G.F.; funding acquisition, G.F. All authors have read and agreed to the published version of the manuscript.

Funding: This work has been carried out in the framework of the project SMARTGO, funded by the Italian Space Agency (ASI).

Institutional Review Board Statement: Not applicable.

Informed Consent Statement: Not applicable.

Data Availability Statement: Data sharing is not applicable to this article.

Acknowledgments: The authors want to thank Valerio Pisacane from Euro.Soft S.r.l. and Alberto Mennella and Annamaria Tortora from TopView S.r.l. for their contribution to risk map derivation and use cases scenario definition, respectively.

Conflicts of Interest: The authors declare no conflict of interest.

References

1. Ong, S. Electric air taxi flies over Singapore—[News]. *IEEE Spectr.* **2019**, *56*, 7–8. [CrossRef]
2. Hayajneh, M.; Al Mahasneh, A. Guidance, Navigation and Control System for Multi-Robot Network in Monitoring and Inspection Operations. *Drones* **2022**, *6*, 332. [CrossRef]
3. Kopardekar, P.H. Safely Enabling UAS Operations in Low-Altitude Airspace, NASA UTM. Available online: http://utm.arc.nasa.gov/docs/pk-final-utm2015.pdf (accessed on 30 November 2015).
4. SESAR 3 Joint Undertaking. *Multiannual Work Programme 2022-2031*; SESAR Joint Undertaking: Brussels, Belgium, 2022.
5. Jayaweera, H.M.P.C.; Hanoun, S. Path Planning of Unmanned Aerial Vehicles (UAVs) in Windy Environments. *Drones* **2022**, *6*, 101. [CrossRef]
6. Xue, M.; Wei, M. Small UAV Flight Planning in Urban Environments. In *AIAA Aviation 2020 Forum*; AIAA AVIATION Forum; American Institute of Aeronautics and Astronautics: Reston, VA, USA, 2020.
7. Hong, D.; Lee, S.; Cho, Y.H.; Baek, D.; Kim, J.; Chang, N. Energy-Efficient Online Path Planning of Multiple Drones Using Reinforcement Learning. *IEEE Trans. Veh. Technol.* **2021**, *70*, 9725–9740. [CrossRef]
8. Lou, J.; Yuksek, B.; Inalhan, G.; Tsourdos, A. An RRT* Based Method for Dynamic Mission Balancing for Urban Air Mobility Under Uncertain Operational Conditions. In Proceedings of the 2021 IEEE/AIAA 40th Digital Avionics Systems Conference (DASC), San Antonio, TX, USA, 3–7 October 2021; pp. 1–10.
9. Blasi, L.; D'Amato, E.; Mattei, M.; Notaro, I. UAV Path Planning in 3D Constrained Environments Based on Layered Essential Visibility Graphs. *IEEE Trans. Aerosp. Electron. Syst.* **2022**, 1–30. [CrossRef]
10. Watanabe, Y.; Veillard, A.; Chanel, C. Navigation and Guidance Strategy Planning for UAV Urban Operation. In *AIAA Infotech @ Aerospace*; AIAA SciTech Forum; American Institute of Aeronautics and Astronautics: Reston, VA, USA, 2016.
11. la Cour-Harbo, A. Quantifying Risk of Ground Impact Fatalities for Small Unmanned Aircraft. *J. Intell. Robot. Syst.* **2019**, *93*, 367–384. [CrossRef]
12. Sláma, J.; Váňa, P.; Faigl, J. Risk-aware Trajectory Planning in Urban Environments with Safe Emergency Landing Guarantee. In Proceedings of the 2021 IEEE 17th International Conference on Automation Science and Engineering (CASE), Lyon, France, 23–27 August 2021; pp. 1606–1612.

13. Primatesta, S.; Guglieri, G.; Rizzo, A. A Risk-Aware Path Planning Strategy for UAVs in Urban Environments. *J. Intell. Robot. Syst.* **2019**, *95*, 629–643. [CrossRef]
14. Delamer, J.-A.; Watanabe, Y.; Chanel, C.P.C. Safe path planning for UAV urban operation under GNSS signal occlusion risk. *Rob. Auton. Syst.* **2021**, *142*, 103800. [CrossRef]
15. Fasano, G.; Causa, F.; Franzone, A.; Piccolo, C.; Cricelli, L.; Mennella, A.; Pisacane, V. Path planning for aerial mobility in urban scenarios: The SMARTGO project. In Proceedings of the 2022 IEEE International Workshop on Metrology for AeroSpace, Pisa, Italy, 27–29 June 2022.
16. Causa, F.; Franzone, A.; Fasano, G. Comparison and integration of tactical path planning approaches for Urban Air Mobility. In Proceedings of the 2022 IEEE/AIAA 41st Digital Avionics Systems Conference (DASC), Portsmouth, VA, USA, 18–22 September 2022; pp. 1–10.
17. Causa, F.; Fasano, G. Multi-objective modular strategic planning framework for Urban Air Mobility. *Submitt. IEEE Trans. Aerosp. Electron. Syst.* **2023**, in press.
18. Hoekstra, J.M.; Ellerbroek, J.; Sunil, E. Geovectoring: Reducing Traffic Complexity to Increase the Capacity of UAV airspace. In Proceedings of the International Conference for Research in Air Transportation (ICRAT), Barcelona, Spain, 25–29 June 2018.
19. Yao, Z.; Nagel, C.; Kunde, F.; Hudra, G.; Willkomm, P.; Donaubauer, A.; Adolphi, T.; Kolbe, T.H. 3DCityDB—A 3D geodatabase solution for the management, analysis, and visualization of semantic 3D city models based on CityGML. *Open Geospatial Data Softw. Stand.* **2018**, *3*, 5. [CrossRef]
20. Simonyan, K.; Zisserman, A. Very Deep Convolutional Networks for Large-Scale Image Recognition. In Proceedings of the International Conference on Learning and Representations, San Diego, CA, USA, 7–9 May 2015; pp. 1–14.
21. Chakrabarty, A.; Stepanyan, V.; Krishnakumar, K.; Ippolito, C. Real-time path planning for multi-copters flying in UTM-TCL4. In Proceedings of the AIAA Scitech 2019 Forum, San Diego, CA, USA, 7–11 January 2019; ISBN 9781624105784.
22. LaValle, S.M. *Rapidly-Exploring Random Trees: A New Tool for Path Planning*; Iowa State University: Ames, IA, USA, 1998.
23. Guruji, A.K.; Agarwal, H.; Parsediya, D.K. Time-efficient A* Algorithm for Robot Path Planning. *Procedia Technol.* **2016**, *23*, 144–149. [CrossRef]
24. Gammell, J.D.; Srinivasa, S.S.; Barfoot, T.D. Batch Informed Trees (BIT*): Sampling-based optimal planning via the heuristically guided search of implicit random geometric graphs. In Proceedings of the 2015 IEEE International Conference on Robotics and Automation (ICRA), Seattle, WA, USA, 26–30 May 2015; pp. 3067–3074.
25. Sucan, I.A.; Moll, M.; Kavraki, L.E. The Open Motion Planning Library. *IEEE Robot. Autom. Mag.* **2012**, *19*, 72–82. [CrossRef]
26. Richter, C.; Bry, A.; Roy, N. Polynomial Trajectory Planning for Aggressive Quadrotor Flight in Dense Indoor Environments. In Proceedings of the Robotics Research: The 16th International Symposium ISRR; Inaba, M., Corke, P., Eds.; Springer International Publishing: Cham, Switzerland, 2016; pp. 649–666.
27. DJI. Matrice 300 RTK. Available online: https://www.dji.com/it/matrice-300/specs (accessed on 21 December 2022).
28. Honeywell. HG1120 MEMS Inertial Measurement Unit. Available online: https://aerospace.honeywell.com/en/~{}/media/aerospace/files/brochures/n61-1524-000-004-hg1120-mems-inertial-measurement-unit-bro.pdf (accessed on 10 June 2019).

 *drones*

Article

LightMAN: A Lightweight Microchained Fabric for Assurance- and Resilience-Oriented Urban Air Mobility Networks

Ronghua Xu [1], Sixiao Wei [2], Yu Chen [1,*], Genshe Chen [2] and Khanh Pham [3]

[1] Department of Electrical and Computer Engineering, Binghamton University, Binghamton, NY 13902, USA
[2] Intelligent Fusion Tech, Inc., Germantown, MD 20876, USA
[3] The U.S. Air Force Research Lab, Space Vehicles Directorate, Albuquerque, NM 87110, USA
* Correspondence: ychen@binghamton.edu

Abstract: Rapid advancements in the fifth generation (5G) communication technology and mobile edge computing (MEC) paradigm have led to the proliferation of unmanned aerial vehicles (UAV) in urban air mobility (UAM) networks, which provide intelligent services for diversified smart city scenarios. Meanwhile, the widely deployed Internet of drones (IoD) in smart cities has also brought up new concerns regarding performance, security, and privacy. The centralized framework adopted by conventional UAM networks is not adequate to handle high mobility and dynamicity. Moreover, it is necessary to ensure device authentication, data integrity, and privacy preservation in UAM networks. Thanks to its characteristics of decentralization, traceability, and unalterability, blockchain is recognized as a promising technology to enhance security and privacy for UAM networks. In this paper, we introduce LightMAN, a lightweight microchained fabric for data assurance and resilience-oriented UAM networks. LightMAN is tailored for small-scale permissioned UAV networks, in which a microchain acts as a lightweight distributed ledger for security guarantees. Thus, participants are enabled to authenticate drones and verify the genuineness of data that are sent to/from drones without relying on a third-party agency. In addition, a hybrid on-chain and off-chain storage strategy is adopted that not only improves performance (e.g., latency and throughput) but also ensures privacy preservation for sensitive information in UAM networks. A proof-of-concept prototype is implemented and tested on a micro-air–vehicle link (MAVLink) simulator. The experimental evaluation validates the feasibility and effectiveness of the proposed LightMAN solution.

Keywords: unmanned aerial vehicle (UAV); lightweight blockchain; drone security; assurance; authentication; resilience

 **check for updates**

Citation: Xu, R.; Wei, S.; Chen, Y.; Chen, G.; Pham, K. LightMAN: A Lightweight Microchained Fabric for Assurance- and Resilience-Oriented Urban Air Mobility Networks. *Drones* **2022**, *6*, 421. https://doi.org/10.3390/drones6120421

Academic Editors: Ivana Semanjski, Antonio Pratelli, Massimiliano Pieraccini, Silvio Semanjski, Massimiliano Petri and Sidharta Gautama

Received: 31 October 2022
Accepted: 13 December 2022
Published: 16 December 2022

Publisher's Note: MDPI stays neutral with regard to jurisdictional claims in published maps and institutional affiliations.

1. Introduction

Thanks to rapid advancements in artificial intelligence (AI), big data, information fusion, and Internet of Things (IoT) technologies, it has become realistic for the concept of smart cities to provide seamless, intelligent, and safe services for communities [1,2]. As a class of robotic vehicles in the IoT, unmanned aerial vehicles (UAV), commonly known as drones, are widely adopted in smart city scenarios for sensing data, carrying payloads, and performing specific missions guided either by remote control centers or in autonomous ways [3]. Thanks to fifth-generation (5G) communication networks and mobile edge computing (MEC) technology, UAVs demonstrate higher mobility than other robotic vehicles, and they can provide on-the-fly communication capabilities in a remote area where terrestrial infrastructure is under-developed or disaster-struck areas where physical or technology has infrastructure been destroyed [4]. Moreover, drones equipped with different types of sensors, such as environmental sensors or cameras, can form UAV networks to guarantee better quality-of-service (QoS) or quality-of-experience (QoE) for users who demand a large number of network-based intelligent services in smart cities, such as video surveillance [5], disaster management, smart transportation, medical suppliers, and public safety [6,7].

With an ever-increasing presence of UAVs in urban air mobility (UAM) networks, the highly connected internet of drones (IoD) also raises new concerns on performance, security, and privacy. On an architectural level, conventional UAV-enabled applications rely on a centralized framework, which is prone to a single point of failure (SPF). As centralized servers coordinate flying drones and perform decision-making tasks, the entire UAV system may be paralyzed if control centers experience malfunctions or are under attacks such as denial of service (DoS) attacks. In addition, complete centralized frameworks that swarm a large number of distributed drones are prone to performance bottlenecks (PBN). As a result, increasing end-to-end network latency degrades QoS or QoE in real-time applications. Moreover, the dynamicity of UAV networks including resource-constrained drones also meets security and privacy challenges within a distributed network environment. Security threats that can severely affect UAV networks can be categorized as firmware attacks (e.g., false code injection, firmware modification, malware infection, etc.) and network attacks (e.g., spoofing, jamming, command injection, network isolation, etc.) [8]. Owing to encrypted data transmission between drones and unauthorized access to data stored on servers, privacy breaches lead to revealing sensitive information such as location, flying path, or other identity-related data.

Thanks to multiple attractive features, such as decentralization, immutability, transparency, and traceability, blockchain has demonstrated great potential to revolutionize centralized UAV systems. By utilizing a cryptographic consensus mechanism and peer-to-peer (P2P) networking infrastructure for message propagation and data transmission, blockchain allows all participants to maintain a transparent and immutable public distributed ledger. The decentralization provided by blockchain is promising for the mitigation of the impact of SPF and PBN by reducing the overhead of the central server in UAV networks. In addition, encryption algorithms, consensus protocols, and tamper-proof distributed ledgers of blockchain enhance the privacy and security of UAV networks. As a result, blockchain provides a "trust-free" network to guarantee the integrity, accountability, and traceability of UAV data. Furthermore, smart contracts (SC) introduce programmability into a blockchain to support a variety of customized business logic rather than classic P2P cryptocurrency transactions [9]. Therefore, blockchain is promising to enhance governance, regulation, and assurance in UAM networks with the help of decentralized security services, such as identification authentication [10], access control [11], and data validation [12].

The shift from centralized UAV networks to decentralized blockchain-assisted UAV systems improves the efficiency of system operations and ensures security and privacy guarantees. Existing blockchain-based UAV solutions mainly consider blockchain as a trusted network and immutable storage to improve the efficiency of communications [13,14], incentive mechanisms [15], security of access authentication [16,17], and data sharing processes [18,19]. However, directly adopting conventional blockchains to build decentralized UAV networks still meets tremendous challenges in IoD scenarios. The current solutions based on permissionless blockchains (e.g., Bitcoin [20] or Ethereum [21]) demand high computation resources in proof-of-work (PoW) mining processes such that they are not affordable to resource-constrained drones. While using permissionless blockchains such as Hyperledger [22] can achieve low energy consumption and high throughout, they are highly limited in terms of scalability and communication complexity.

To address the aforementioned limitations of integrating blockchain into UAV networks, this paper proposes LightMAN, a lightweight microchained fabric for data assurance and operation resilience-oriented UAM networks. Unlike existing works [6,8,18,19] that rely on computation-intensive PoW blockchains, LightMAN adopts microchain [23], a lightweight-designed blockchain, to achieve efficiency and security guarantees for a small-scale permissioned UAV network. As drone information and flight logs are securely and accurately stored on the immutable distributed ledger of the microchain, participants within a UAM network can verify the authenticity of drones and verify tamper-proof data sent to/from drones without relying on a third-party agency. Compared with blockchain-based UAV networks that either directly save raw data on the distributed ledger [18] or outsource raw data to a cloud server [19], our LightMAN allows encrypted data to be stored on a distributed data storage (DDS), while the microchain only records references of data as checkpoints. Such a hybrid on-chain and off-chain storage strategy not only improves performance (e.g., latency and throughput) but also ensures privacy preservation for sensitive information in UAM networks.

In brief, the key contributions of this paper are highlighted as follows:

(1) A complete LightMAN system architecture is presented along with details of key components and functionalities;

(2) A machine learning-based anomaly detection (MLAD) method to monitor the UAM networks in real time is proposed. To generate the source data (MAVLink message) for creating the cyber-resiliency scenario, we implemented a software-in-the-loop (SITL) simulator and associated demonstration package (pymavlink) in a python environment to emulate the message communications among UAVs;

(3) A lightweight blockchain called microchain is leveraged to guarantee security and privacy requirements in UAV data access and sharing scenarios; and

(4) A proof-of-concept prototype is implemented and tested on a small-scale physical network. The experimental results show that the proposed LightMAN only incurs less than two seconds of latency while committing transactions on the distributed ledger and no more than 18% overhead during access authentication.

The remainder of the paper is organized as follows: Section 2 provides background knowledge of UAV and blockchain technologies and reviews existing state-of-the-art blockchain-based UAV systems. Section 3 introduces the rationale and system architecture of LightMAN. Section 4 presents the prototype implementation, experimental setup, and performance evaluation. Finally, Section 5 summarizes this paper with a brief discussion on current limitations and future directions.

2. Background and Related Work

This section describes the fundamentals of the UAV concept, explains blockchain technology, and introduces the state-of-the-art decentralized solutions to secure UAM networks.

2.1. Unmanned Aerial Vehicles

Unmanned aerial vehicles (UAVs), simply called drones, are specific robotic IoTs, which have electronic components, mechanical power modules, and onboard operating systems to execute complicated tasks. According to their flying mechanisms, UAVs can be categorized as multi-rotor-wing drones, fixed-wing drones, and hybrid fixed/rotary-wing drones [24]. Regarding the range and altitude that a done can be remotely operated at, UAV platforms can be classified into two types: low-altitude platforms (LAPs) and high-altitude platforms (HAPs). Original UAVs were mainly used for battlefields, with advancements in hardware, software, and networking infrastructure, but there has been increasing usage of UAVs in civilian and commercial applications.

Owing to their unmanned nature and requirements for remote wireless communication, modern UAV-aided systems are vulnerable to different attacks [25]. Thus, the continued use of UAVs increases the need for cyber-awareness including UAVs in the airspace, the development of the automatic dependent surveillance broadcast (ADS-B), and the risk of cyber intrusion. The Federal Aviation Administration (FAA) mandates the national adoption of ADS-B, which uses "plaintext" to broadcast messages in avionics networks. Such an unencrypted ADS-B manner introduces serious privacy and security vulnerabilities, such as message spoofing for false aircraft position reports. As a result, current radar-based air traffic service (ATS) providers seek to preserve privacy and corporate operations of flight plans, position, and state data. Moreover, the privacy of aircraft track histories is mandatory and only accessible to authorized entities within UAM networks. In addition, it is necessary to ensure confidentiality, availability, and integrity for urban aircraft data accessing and sharing data during UAM operations.

2.2. Blockchain Technology

From the system architecture aspect, a typical blockchain system consists of three essential components: a distributed ledger, a consensus protocol, and smart contracts [26]. Essentially, distributed ledger technology (DLT) is a type of distributed database that is shared, replicated, and maintained by all participants under a P2P networking environment. Each participant maintains a local view of the distributed ledger in the context of a distributed computing environment, and a well-established consensus allows all participants to securely reach an agreement on a global view of the distributed ledger under consideration of failures (Byzantines or crash faults). Given different consensus algorithms and network models, distributed consensus protocols are categorized into Nakamoto consensus protocols [20] or Byzantine fault-tolerant (BFT) consensus protocols [27]. From a topology aspect, blockchain can be classified into three types: public (permissionless) blockchains, private (permissioned) blockchains, and consortium blockchains [28].

By using cryptographic and security mechanisms, a *smart contract* (SC) combines protocols with user interfaces to formalize and secure the relationships over computer networks [29]. Essentially, SCs are programmable applications containing predefined instructions and data stored at a unique address on the blockchain. Through exposing the public functions or application binary interfaces (ABIs), an SC acts as the trusted autonomous agent between parties to perform predefined business logic functions or contract agreements under specific conditions. Owing to the secure execution of predefined operational logic, unique addresses and public, exposed ABIs, using a SC provides an ideal decentralized app (Dapp) backbone to support upper-level IoT applications.

2.3. Blockchain-Based UAV Networks

There have been many studies in the past that have explored blockchain and smart contracts to enable decentralized UAV networks. In general, existing blockchain-based UAV networks can be categorized into three branches: securing UAV communications, maintaining data integrity and improving identity authentication.

2.3.1. UAV Communication

By utilizing the blockchain concept in the development of drone networks, a blockchain-empowered drone network called BeDrone allows drones in service to act as the miners of the blockchain [15]. Each drone can acquire computing and storage resources from nearby edge service providers to carry on the blockchain processes, such as mining blocks and storing ledgers. BeDrone uses game theory to design incentive mechanisms for resource allocation, acquisition, and trading among participants. However, details of the underlying blockchain framework are not discussed.

To ensure ultra-reliability and security for intelligent transport during drone-catching in multi-access edge computing (MEC) networks, a neural-blockchain-based transport model (NBTM) [13] was proposed by forming a distributed decision neural network for multiple blockchains. NBTM uses neural networks to formulate policies and rules as the drone-caching model for reliable communication and content sharing. A hierarchical blockchain model consisting of three blockchains and a master blockchain provides security mechanisms for content sharing and data delivery. The simulation results demonstrate that the proposed NBTM can enhance the reliability of UAV networks with a lower failure rate. However, the performance of using multi-blockchains is not mentioned.

To build agile and resilient UAV networks for the collaborative application of large-scale drone groups, a software-defined UAV network called SUV [30] was proposed by combining software-defined networking (SDN) and blockchain technology to achieve a decentralized, efficient and flexible network infrastructure. By decoupling the control panel and the data panel of a UAV network, SDN allows SUV to optimally manage all drones and simplify functions of data forwarding. Blockchain facilitates the decentralization of the SDN control panel and ensures the credibility of the SDN controller identity and behavior in an open networking environment. The proposed SUV is promising for the provision of flexibility, survivability, security, and programmability for 5G-oriented UAV networks [30]. However, its implementation and performance evaluation are not described.

Similar to the works [13,30] that focusesd on improving security in UAV communications, a lightweight blockchain based on a proof-of-traffic (PoT) consensus algorithm was proposed to provide secure routing for swarm UAVs [14]. PoT leverages the traffic status of swarm UAVs to construct a consensus rather than the computation resources used by PoW. The evaluation shows that PoT can reduce the burden of energy consumption and computational resource allocation for swarm UAV networking. However, the performance of PoT consensus is not discussed, such as transaction latency and throughput.

2.3.2. UAV Data Integrity

Some early works used blockchain as tamper-proof storage to protect the UAVs' data integrity during sharing and operating processes. To secure drone communications and preserve data integrity, a blockchain-based drone system called DroneChain [19] was proposed using a PoW blockchain and a cloud server. The collected data of each drone are associated with its device ID and are saved into a cloud server, while a hash of each data record is stored in the blockchain. DroneChain allows for data assurance, provenance, and resistance against tampering. Moreover, the distributed nature of DroneChain also improves the availability and resilience of data validation for potential failures and attacks. However, using a centralized cloud server for UAV raw data storage is prone to privacy violations and SPF in data querying and sharing.

To address issues of DroneChain that adopts the traditional cloud server and PoW blockchain in UAV networks, a secure data dissemination model based on a consortium blockchain was proposed for IoD [18]. All users and drones are divided into multiple clusters, and one master controller (MC) within a cluster can work as a normal node in a public Ethereum blockchain network. A forger node selection algorithm on the basis of utility function using game theory periodically selects one forger node for block generation. The experimental results evaluate the performance of the data dissemination model, such as the computation time of block creation and validation. However, details of blockchain design and data storage are not mentioned.

2.3.3. UAV Authentication

By storing identification and access control information in the distributed ledger, blockchain can provide decentralized authentication services for UAV networks. To solve issues of authentication of drones during flights, a secure authentication model with low latency for IoD in smart cities was proposed by using a drone-based delegated proof-of-stake (DDPOS) blockchain atop zone-based network architecture [16]. Similar to [18], a drone controller in each zone of a smart city is responsible for the management and authentication mechanism for drones, and it also handles all operations related to the blockchain. Compared to the original PoS algorithm, a customized DDPOS algorithm can mitigate mining centralization and the flaws of real-life voting in the UAV network. The experimental results show the efficiency of the proposed solution under a simulated environment, such as low package loss rate, high throughput, and end-to-end delay.

To address the challenges of centralized authentication approaches in cross-domain operations, a blockchain-based cross-domain authentication scheme for an intelligent 5G-enabled IoD was proposed [17]. The proposed solution uses a local private blockchain based on Hyperledger fabric to support drone registration and identity management. As multiple signatures based on threshold sharing are used to build an identity federation for collaborative domains, a smart contract contains access control policies, and multi-signatures aims to secure mutual authentication between terminals across different domains.

3. Design Rationale and System Architecture

UAM offers the potential to create a faster, cleaner, safer, and more integrated transportation systems. However, recent events have shown that modern UAVs are vulnerable to attack and subversion through faulty or sometimes malicious devices that are present on UAM communication networks, which increases the need for cyber awareness to include UAVs in the airspace and the risk of cyber intrusion. Aiming at a secure-by-design, intelligent and decentralized network architecture for assurance and resilience-oriented UAM networks, LightMAN leverages deep learning (DL) and microchains to enable efficient, secure, and privacy-preserving data access and sharing among participants in UAV networks. Figure 1 demonstrates the LightMAN architecture that consists of two sub-frameworks: (i) the UAM network and (ii) the microchain fabric.

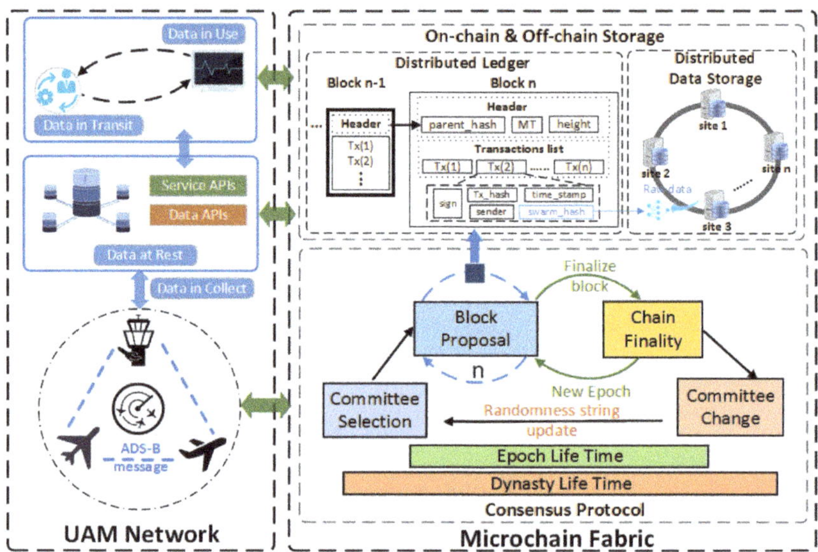

Figure 1. System Architecture of LightMAN.

A UAM network encompasses air traffic operations for manned and unmanned aircraft systems in a metropolitan area. The left part of Figure 1 shows a UAV application that provides on-demand, automated transportation services. Each drone uses its onboard sensors to enroll and capture raw mission data, such as ADS-B messages or MAVLink messages, and these data can be digitized and converted to key features, such as aircraft identification and trajectories. The operation centers (ground stations) can collect data for flight planning and monitoring. In addition, raw data can be transferred to an avionic data center that provides long-term storage services (data at rest) for high-level information fusion and analysis. Finally, a cloud server performs high-level computing extensive and big-data-oriented tasks such as multi-airborne collaborative planning and decision-making reasoning. Based on a thorough analysis of shared avionics data, intelligent avionic services (data in transit) incorporates AI technologies to optimize UAV services and protect against never-before-seen attacks. Information visualization (data in use) provides context-based human–machine interactions for authorized users to learn dynamic mission priorities and resource availability [31].

The microchain fabric acts as a security and trust networking infrastructure to provide decentralized security and privacy-preserving guarantees for UAM data. Microchain leverages a permissioned UAV network management and assumes that the system administrator is a trustworthy oracle to maintain registered identity profiles of UAM. Thus, each drone or user uses their unique ID to identify authentication and access control procedures. In addition, cryptographic primitives such as public key infrastructure (PKI) and encryption algorithms can guarantee the confidentiality and integrity of drone data (e.g., ADS–B) in communication. Moreover, microchain integrates a lightweight consensus protocol with a hybrid on-chain and off-chain storage to ensure UAV data and flight logs are stored securely and distributively without relying on any centralized server.

3.1. Deep Learning (DL)-Powered UAM Security

To better detect anomalous behaviors (e.g., aircraft route anomalies) to constantly collecti high-resolution cyber-attack information across avionics flight data, we have designed and developed DL-based cybersecurity monitoring techniques against cyber threats for UAM situation awareness (SAW). The developed LightMAN with cognitive-based decision support is not intended to replace human interaction and decision-making; rather, it is meant to support the operator to combine data, identify potential threats rapidly for a pre-planned mission, and provide timely recommended actions.

Learning directly from high-dimensional sensory inputs is one of the long-standing challenges. Our objective is to develop machine learning (ML)-based anomaly detection (MLAD) and reinforcement learning (RL) artificial agents that can achieve a good level of performance and generality on diagnostics and prognostics. Similar to a human operator, the goal for the agents is to learn strategies that lead to the greatest long-term rewards. Formally, MLAD can be described as a Markov decision process (MDP), which consists of s set of states, S, plus a distribution of starting states, $P(s_0)$; a set of actions, A; transition dynamics, $T(s_{t+1} \mid s_t, a_t)$, that map a state-action pair at time t to the distribution of states at time t + 1; a reward function, $R(s_t, a_t, s_{t+1})$; and a discount factor, $\delta \in [0, 1]$, where smaller values place more emphasis on immediate rewards. It is assumed that an agent interacts with an environment, S, in a sequence of actions, actions, observations, and rewards. At each time-step, the agent selects an action, $a_t \in A, A = 1, \ldots, K$, which is passed to the environment and modifies its internal state and the corresponding reward [32]. In general, S may be stochastic. The system's internal state is not observable to the agent most of the time, instead, it observes various target features of interest from the environment, such as the signal features. It receives a reward R representing the change in overall system performance.

Based on the MLAD-RL strategy, we developed an automated monitoring mechanism for system-level source analytics. The monitoring data are defined as a set of metrics (e.g., route latitude/longitude, transmission delay, traffic buffer queue length, etc.) on each UAM edge and associated applications and processes. Given a large number of features, LightMAN uses feature extraction and reduction techniques in collected log data to select a set of the most critical features and implement deep learning-based detection schemes for identifying anomalous statuses. The general steps of the proposed anomaly monitoring technique are as follows: (i) *Data Collection*: The relevant sensory data collected across the system are assembled into a set of feature matrices. We define the feature as an individually measurable variable of the node being monitored (e.g., data frames, MAVLink messages, command and control (C2) mission logs, controller area network (CAN) buses, etc.); (ii) *Feature Extraction*: To effectively deal with high-dimensional data, we implement feature extraction techniques via named entity recognition (NER) [33] and the vector space model (VSM), which can reduce data dimensionality and improve analysis by removing inherent data dependencyl (iii) *Deep Learning-Based Detection*: LightMAN applies DL techniques (e.g., L-CNN, RNN/LSTM, etc.) to characterize the dynamic state of the monitored system. With the trained model in place, the operator can conduct the detection and classification of potential attacks.

As shown in Figure 2, the detection process consists of two main steps: the training process and the detecting process. In the training process, the collected log data are converted to a uniform data format for the learning process. We then train the classifier model for both normal and abnormal system states. In the online monitoring process, LightMAN monitoring tools collect real-time flight data, and the processed traffic data are sent to the learned classifier for anomaly detection. The effectiveness of the monitoring schemes is characterized by the true positive rate, false positive rate, monitoring time, overhead, etc.

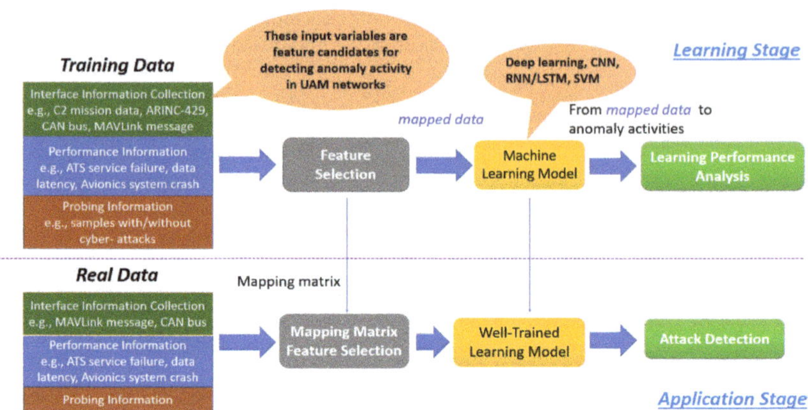

Figure 2. ML/DL Learning Process for UAM Monitoring.

3.2. Microchain Fabric for UAM Data Sharing

As the right part of Figure 1 shows, a microchain fabric consists of two sub-systems: (i) a lightweight consensus protocol that relies on a randomly selected consensus committee to achieve a low latency when committing transactions on the distributed ledger; (ii) a hybrid on-chain and off-chain storage strategy that improves efficiency and privacy-preservation. For details regarding the consensus protocol in the microchain, interested readers can refer to our earlier work [23,34]. The core functionalities and workflows are briefly described as follows:

- The lifetime of a committee is defined as a *dynasty*, and all nodes within the network use a random committee election mechanism to construct a new committee at the beginning of a new dynasty. The new committee members rely on their neighboring peers, which use a node discovery protocol to reach out to each other. Finally, all committee members maintain a fully connected consensus network, and non-committee nodes periodically synchronize states of the current dynasty. Until the current dynasty's lifetime is ending, committee members utilize an epoch randomness generation protocol to cooperatively propose a global random seed for the next committee election.

- Given a synchronous network environment, operations of consensus processes are coordinated in sequential rounds called *epochs*. The block proposal leverages an efficient proof-of-credit (PoC) algorithm, which allows the consensus committee to continuously publish blocks containing transactions and extend the main chain length. The block proposal process continues running multiple rounds until the end of an epoch. Then, a voting-based chain finality protocol allows committee members to make an agreement on a checkpointing block. As a result, temporary fork chains are pruned, and these committed blocks are finalized on the unique main chain.

- The organization of on-chain and off-chain storage is illustrated by the upper right part of Figure 1. As the basic unit of on-chain data recorded on the distributed ledger, a block contains header information (e.g., previous block hash and block height) and orderly transactions. The distributed data storage (DDS), which is built on a swarm [35] network, is used as off-chain storage. The UAV data and flight logs that require heterogeneous formats and various sizes are saved on the DDS, and they can be easily addressed by their swarm hash. In an optimal manner, each transaction only contains a swarm hash as a reference pointing to its raw data on the DDS. Compared with raw data, a swarm hash has a small and fixed length (32 or 64 bytes); therefore, all transactions have almost the same data size. It is promising to improve efficiency in transaction propagation without directly padding raw data into transactions.

4. Experimental Results and Evaluation

In this section, experimental configuration based on a proof-of-concept prototype implementation is described. Following that, we evaluate the performance of running LightMAN based on numerical results, which especially focus on microchain operations. Finally, a comparative evaluation among previous work highlights the main contributions of LightMAN in terms of lightweight blockchain design, performance improvement, security, and privacy properties.

4.1. Prototype Implementation

A proof-of-concept prototype of LightMAN was implemented and tested in a physical network environment. The microchain was implemented in Python with Flask [36] as a web-service framework. All security primitives such as digital signature, encryption algorithms, and hash functions were developed by using standard python library cryptography [37]. MAVLink [38] implemented a Software-In-The-Loop (SITL) simulator consisting of Pymavlink, ArduPilot, MAVProxy and QGroundControl. As a package of Python MAVLink libraries, Pymavlink was used to implement drone communication protocol and analyze flight logs. ArduPilot [39] is an open-source autopilot software that was used to simulate many drone types on a local server without any special hardware support. MAVProxy acted as the ground control station for ArduPilot, and QGroundControl provided the graphical user interface (GUI) for ArduPilot. We combined the SITL simulator and Pymavlink package to emulate UAM scenarios and collect MAVLink messages as UAV data.

Table 1 describes devices used for the experimental setup. Each validator of microchain was deployed on a Raspberry Pi (RPi) while a SITL simulator was deployed on the Redbarn HPC. The microchain test network contained 16 RPis. Regarding a test Swarm network, 6 service sites were deployed on six separate desktops that each had an Intel Core 2 Duo CPU E8400 @ 3 GHz and 4 GB of RAM. All devices were connected through a local area network (LAN).

Table 1. Configuration of Experimental Devices.

Device	Redbarn HPC	Raspberry Pi 4 Model B
CPU	3.4 GHz, Core i7-2600K (8 cores)	1.5 GHz, Quad core Cortex-A72 (ARM v8)
Memory	16 GB DDR3	4 GB SDRAM
Storage	500 GB HHD	64 GB (microSD card)
OS	Ubuntu 18.04	Raspbian GNU/Linux (Jessie)

4.2. MAVLink Message Data Acquisition

To better perform the machine learning-based anomaly detection (MLAD) within LightMAN among UAM networks, we leveraged the MAVLink Protocol, which stands for micro-air–vehicle link, and its related messages as our starting point for the security analysis of UAM networks. It is an open-source protocol, and it is supported by many closed-source projects for drones to send way-points, control commands, and telemetry data [40]. Usually, it contains two types of messages: state messages and command messages. State messages refer to these messages sent from the unmanned system to the ground station and contain information about the state of the system, such as its ID, location, velocity, and altitude. Command messages are usually sent from the ground station to the unmanned system to execute some actions by autopilot. Those messages are transmitted through WiFi, Ethernet, or other serial telemetry channels. We also utilized a SITL simulator (ArduPilot) [40] to emulate the MAVLink message communication. Specifically, we ran the ArduPilot directly on a local server without any special hardware. While running, the sensor data came from a flight dynamics model in a flight simulator.

Figure 3 presents an example of obtained MAVLink message source data. We recorded and saved this key information for MLAD training. For instance, *GPS_RAW_INT* refers to the absolute geolocation of GPS, latitude, longitude, and altitude. *AHRS* refers to the attitude and heading reference system (AHRS), which consists of sensors on three axes that provide attitude information for aircraft, including roll, pitch, and yaw. *EKF_STATUS_REPORT* indicates that an extended Kalman filter (EKF) algorithm was used to estimate vehicle position, velocity, and angular orientation based on rate gyroscopes, accelerometer, compass, GPS, airspeed, and barometric pressure measurements.

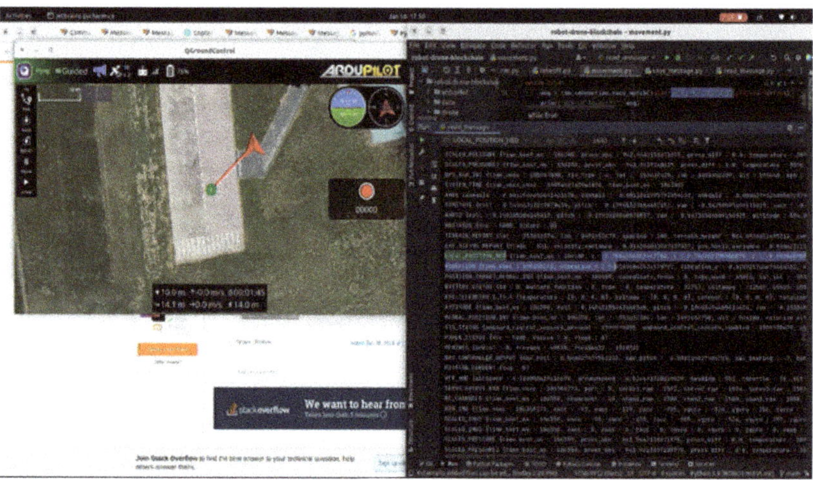

Figure 3. Software-In-The-Loop Simulation for Data Acquisition.

4.3. Performance Evaluation

During the identity authentication stage, the system administrator or data owners can launch a transaction to the microchain, which encapsulates a capability access token assigned to an entity. Then, any user can query such a token from microchain participants and verify it during the access validation process. We designed a capability-based access control (CapAC) scenario [11] in which one HPC simulates a service owner to record CapAC tokens into the microchain, and another RPi simulates a service provider to query CapAC tokens from the microchain for the access control process. We conducted 100 Monte Carlo test runs and used the average of results for evaluation.

4.3.1. End-to-End Latency of Authorizing Access Tokens

Figure 4 demonstrates how committee size K represented by the number of validators and access authorization transaction throughput Th_S measured by the transactions per second (tps) affects the end-to-end latency incurred by committing a transaction on a microchain network. As the microchain executes an efficient consensus protocol within a small consensus committee, it brings a lower total latency, which has marginal impacts for an increasing committee size K. As a trade-off, a small consensus committee containing resource-constrained RPi devices as validators has limited capability to process large volumes of transactions. Thus, the end-to-end latency is almost dominated by Th_S, as Figure 4 shows. We assume that each node within LightMAN waits no less than 5 s to collect UAV data and then launch a transaction. Thus, the network latency of committee transactions on microchain can satisfy real-time requirements of access authorization.

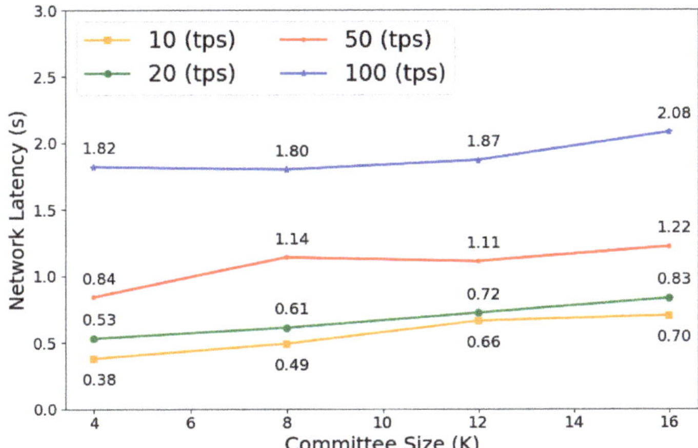

Figure 4. End-to-end latency of committing CapAC tokens on Microchain: committee size vs. tps.

4.3.2. Processing Time and Throughput in Access Authentication

For comparing our LightMAN's performance metrics with conventional centralized frameworks in access authentication, we designed basic scenarios as a benchmark, which did not cooperate with any access control strategy for UAV data access requests. To evaluate the processing time and throughput of access authentication operations, we used an HPC to simulate a cloud-based UAV server, which provided drone data query services given basic and LightMAN scenarios. Then, we let an RPi send multiple access requests to a UAV server and wait until all responses are correctly received.

Figure 5 shows average delays that evaluate how long a CapAC access request can be successfully handled by the UAV data server as increasing Th_S from 20 tps to 1000 tps. Regarding the fixed bandwidth of the test network, the capacity of UAV servers dominates the performance of handling access requests. Thus, the delays of access authentication are almost linear scale to Th_S given basic and LightMAN scenarios. However, LightMAN still demonstrates efficiency in the decentralized access authentication process that queries CapAC tokens from microchain and verifies access control policies, and it only incurred limited extra overheads (no more than 18%) compared with basic scenarios.

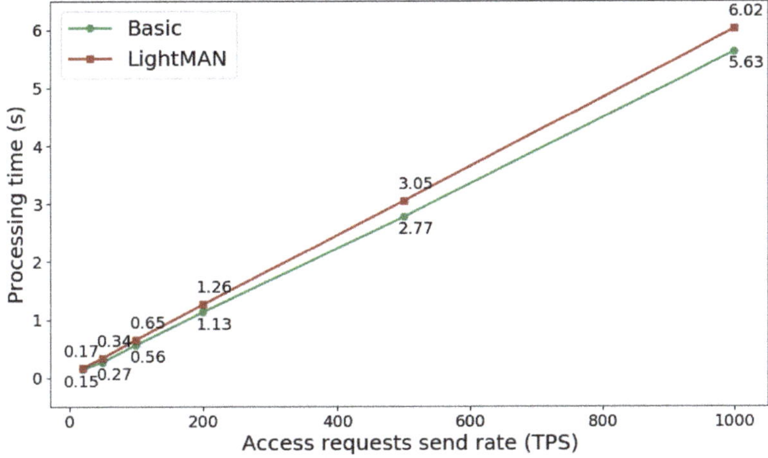

Figure 5. Processing Time of querying CapAC tokens and validating access rights.

To evaluate the data processing capability, we calculated throughput as $\frac{Th_S}{T_D}$, where T_D is the time latency of completing Th_S data tasks. A higher throughput indicates better system performance. Figure 6 presents the transaction throughput of handling access authentication requests, given that Th_S varies from 20 tps to 1000 tps. Each access request in LightMAN mode demands more computation resources on CapAC token validation; therefore, LightMAN demonstrates a lower transaction throughput than the basic mode even if the access request send rate Th_S is the same. Owing to system capacities, such as the network bandwidth and computation power of service providers, the transaction throughput of LightMAN and the basic mode become saturated under conditions where $Th_S \geq 500$ tps. Compared with the baseline, our solution can provide security and privacy features without significantly reducing system performance.

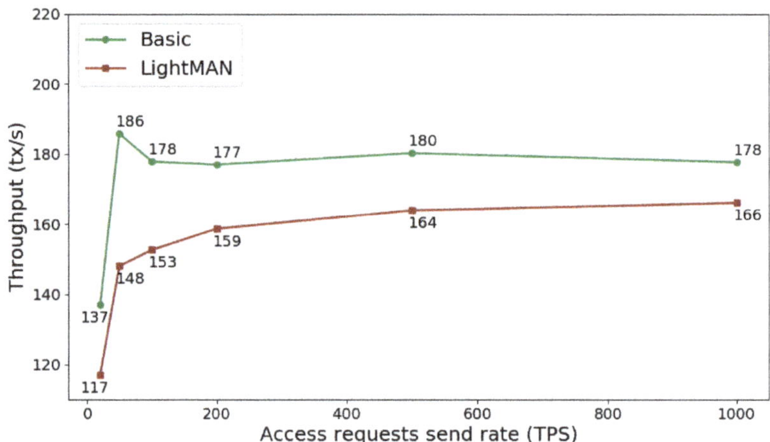

Figure 6. Throughput of querying CapAC tokens and validating access rights.

4.3.3. Computation Cost by Preserving Data Privacy

We assumed that MAVLink message data streams of a drone were encrypted and then recorded into DDS for each 60 s duration. As a result, each data file was about 1 MB, and we used these sample data files to evaluate computation overheads incurred by sharing UAV data via DDB along with data encryption and decryption procedures. Figure 7 shows the processing time of accessing data from Swarm and data encryption algorithms given different host platforms. Regarding DDS operations such as uploading files onto and downloading files from a private Swarm network, delays are almost the same on both platforms. Unlike downloading data, which simply query data from a DDS service site, uploading data onto DDS takes a longer time than is used to synchronize data units across distributed service sites within a Swarm network. Owing to constrained computation resources, RPi takes a longer process time to encrypt and decrypt data than the desktop does, even if sample data files have the same size. Compared with a 60 s cycle time of recording a drone's data, encrypting a data file and then uploading it onto DDS only brings marginal delays on both platforms (2.4 s on desktop and 3.2 s on RPi). Given data-in-use scenarios that frequently download files from a DDS service node and then decrypt them, the encryption algorithm incurs more computation overheads than Swarm operations. Given a data query request rate $Th_S = 500$ tps that takes an average of 3.05 s on access authentication, accessing UAV data incurs an extra 19% (0.57/3.05) of delays on desktop and 59% (1.79/3.05) of delays on RPi. As a trade-off, using encrypted data to protect private information is inevitable at the cost of a longer processing time.

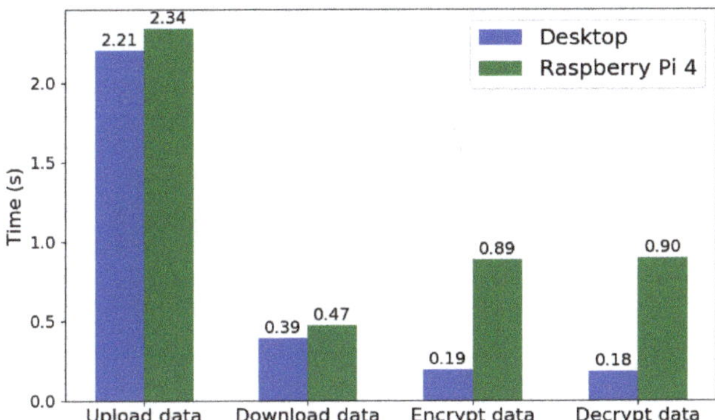

Figure 7. Processing time of data operations: accessing DDS and symmetric encryption.

4.4. Comparative Evaluation

Table 2 presents the comparison between our LightMAN and previous blockchain-based solutions for UAV networks. The symbol $\sqrt{}$ indicates that the scheme guarantees the security properties or implements some prototypes to evaluate the system performance or other specifications. The symbol $\times$ indicates the opposite case. Existing blockchain-based solutions that are developed to secure UAV communications [13–15] lack details on underlying blockchain frameworks, and most of them assumed that the cryptocurrency-oriented blockchain designs can be adopted in the UAV communication systems. Being fully aware of the specific performance requirements and resource constraints, we demonstrate a complete system architecture consisting of ML-based UAM monitoring and a lightweight microchain. Compared with solutions that adopt conventional PoW and BFT consensus protocols [17,19], LightMAN focuses on a lightweight blockchain design for IoD, which leverages a novel PoC+VCF consensus protocol to reduce computation and communication overheads on IoT systems. We especially evaluate blockchain performance (e.g., network latency, transaction throughput, and computation overheads) by applying a microchain-enabled security mechanism to access authentication and data sharing process scenarios, which are not considered or sufficiently discussed in related work [16,18].

In terms of the optimization for UAV data storage, a DDS is adopted atop the Swarm network as the off-chain storage to store raw UAV data. Therefore, LightMAN is promising for the enhancement of the system robustness (availability and recoverability) for data-sharing applications compared with existing solutions that rely on centralized storage [19]. Furthermore, LightMAN stores encrypted sensitive information on the DDS while only recording references of raw data on the transparent distributed ledger. As a result, blockchain transactions only contain references of small size rather than large volumes of UAV data. Such a hybrid on-chain and off-chain data storage structure not only reduces communication and storage overheads but also ensures privacy preservation in the data-sharing process by exposing hash-style references as proofs.

Table 2. Comparison among existing solutions.

	Consensus	Storage	Performance	Security	Privacy
BeDrone [15]	$\times$	$\times$	$\times$	$\sqrt{}$	$\times$
NBTM [13]	$\times$	$\times$	$\times$	$\sqrt{}$	$\times$
SwarmUAV [14]	PoT	$\times$	$\sqrt{}$	$\sqrt{}$	$\times$
DroneChain [19]	PoW	Centralized	$\sqrt{}$	$\sqrt{}$	$\times$
SecureIoD [18]	PoS	$\times$	$\times$	$\sqrt{}$	$\times$
ZoneIoD [16]	DDPoS	$\times$	$\sqrt{}$	$\sqrt{}$	$\times$
5G-IoD [17]	BFT	$\times$	$\sqrt{}$	$\sqrt{}$	$\times$
LightMAN	PoC+VCF	Decentralized	$\sqrt{}$	$\sqrt{}$	$\sqrt{}$

5. Conclusions and Future Work

This paper presents LightMAN, which combines DL-powered UAM security and a lightweight microchained fabric to support assurance and resilience-oriented UAM networks. The DL-based cybersecurity monitoring techniques can prevent cyber threats and provide cognitive-based decision support for UAM. A lightweight microchain works as a secure-by-design network infrastructure to enable decentralized security solutions for UAV access authentication and data sharing. The experimental results based on a prototype implementation demonstrate the effectiveness and efficiency of our LightMAN. However, there are open questions that need to be addressed before applying LightMAN to real-world UAM scenarios. We leave these limitations to our future work:

(1) Although the microchain is promising for providing a lightweight blockchain for a small-scale UAV network such as a drone cluster, it is not suitable for a large-scale UAM system demanding scalability and dynamicity in multidomain coordination. A hierarchical integrated federated ledger infrastructure (HIFL) [41] is promising for the improvement of scalability, dynamicity, and security for multi-domain IoD applications. Thus, our ongoing efforts include validating LightMAN in a real-world UAV network and investigating the integration of microchain and HIFL to support secure inter-chain transactions in a large-scale UAM system.

(2) There are still unanswered questions regarding an incentive mechanism that motivates users and drones to devote their resources (e.g., computation, storage, and networking) to participant consensus processes and gain extra profits. In our future work, we will use game theory to model incentive strategies and evaluate the effectiveness, security, and robustness of LightMAN in IoD scenarios.

(3) The third important milestone is an in-field validation of LightMAN in the context of practical applications. Once all the functional blocks and integrated systems are successfully tested in the lab environment, a small-scale drone network will be created with drones that are designed by the team. The completely customized drones will allow us to mount the LightMAN system on top of multiple application-determined sensing blocks, such as smart surveillance cameras or motion sensors. Specifically, to better validate the effectiveness of LightMAN, we plan to test our implementation with a hardware-in-the-loop (HITL) design in a hierarchical practical environment. We will deploy our validator devices on the hardware drones and establish a small-scale decentralized platform. Each drone will function as an individual node with communication protocols (e.g., MAVLink, TCP/IP) within LightMAN. Some typical communication-related anomalies (e.g., GPS spoofing, and channel access attacks) will be crafted to perform a practical injection attack onto the device sensors. In the future study, we will also build multiple clients and servers onboard to stream the shared data (e.g., MAVLink messages) and process the UAM monitoring among UAVs in real time.

Author Contributions: Conceptualization, R.X., S.W., Y.C., G.C. and K.P.; methodology, R.X., S.W. and Y.C.; software, R.X. and S.W.; validation, R.X., S.W. and Y.C.; formal analysis, R.X., S.W. and Y.C.; funding acquisition, Y.C.; investigation, R.X., S.W. and Y.C.; resources, R.X., S.W. and G.C.; data curation, R.X. and S.W.; writing—original draft preparation, R.X., S.W. and Y.C.; writing—review and editing, R.X., S.W. and Y.C.; visualization, R.X. and S.W.; supervision, Y.C.; project administration, Y.C. All authors have read and agreed to the published version of the manuscript.

Funding: This research was partially funded by the United State National Science Foundation (NSF) under the grant CNS-2141468.

Institutional Review Board Statement: Not applicable.

Informed Consent Statement: Not applicable.

Data Availability Statement: Not applicable.

Acknowledgments: The authors want to thank Erik Blasch for his guidance and suggestions during the writing of this manuscript. The views and conclusions contained herein are those of the authors and should not be interpreted as necessarily representing the official policies or endorsements, either expressed or implied, of the Air Force Research Laboratory or the U.S. government.

Conflicts of Interest: The authors declare no conflict of interest.

Abbreviations

The following abbreviations are used in this manuscript:

ABI	Application Binary Interfaces
AC	Access Control
ADS-B	Automatic Dependent Surveillance Broadcast
AI	Artificial Intelligence
ATS	Air Traffic Service
CAN	Controller Area Network
CapAC	Capability-based Access Control
DApp	Decentralized App
DDS	Distributed Data Storage
DDoS	Distributed Denial-of-Service
DL	Deep Learning
DLT	Distributed Ledger Technology
IoD	Internet of Drones
IoT	Internet of Things
MC	Master Controller
MEC	Multi-Access Edge Computing
ML	Machine Learning
PBN	Performance Bottleneck
PoC	Proof-of-Credit
PoT	Proof-of-Traffic
PoW	Proof-of-Work
PKI	Public Key Infrastructure
QoE	Quality-of-Experience
QoS	Quality-of-Service
RL	Reinforcement Learning
SAW	Situational Awareness
SC	Smart Contract
SDN	Software-defined Networking
SITL	Software-In-The-Loop
SPF	Single Point of Failure
UAM	Urban Air Mobility
UAV	Unmanned Aerial Vehicle

References

1. Xu, R.; Nikouei, S.Y.; Nagothu, D.; Fitwi, A.; Chen, Y. Blendsps: A blockchain-enabled decentralized smart public safety system. *Smart Cities* **2020**, *3*, 928–951. [CrossRef]
2. Xu, R.; Lin, X.; Dong, Q.; Chen, Y. Constructing trustworthy and safe communities on a blockchain-enabled social credits system. In Proceedings of the 15th EAI International Conference on Mobile and Ubiquitous Systems: Computing, Networking and Services, New York, NY, USA, 5–7 November 2018; pp. 449–453.
3. Alladi, T.; Chamola, V.; Sahu, N.; Guizani, M. Applications of blockchain in unmanned aerial vehicles: A review. *Veh. Commun.* **2020**, *23*, 100249. [CrossRef]
4. Hassija, V.; Chamola, V.; Agrawal, A.; Goyal, A.; Luong, N.C.; Niyato, D.; Yu, F.R.; Guizani, M. Fast, reliable, and secure drone communication: A comprehensive survey. *IEEE Commun. Surv. Tutor.* **2021**, *23*, 2802–2832. [CrossRef]
5. Chen, N.; Chen, Y.; Blasch, E.; Ling, H.; You, Y.; Ye, X. Enabling smart urban surveillance at the edge. In Proceedings of the 2017 IEEE International Conference on Smart Cloud (SmartCloud), New York, NY, USA, 3–5 November 2017; IEEE: Piscataway, NJ, USA, 2017; pp. 109–119.
6. Han, T.; Ribeiro, I.D.L.; Magaia, N.; Preto, J.; Segundo, A.H.F.N.; de Macêdo, A.R.L.; Muhammad, K.; de Albuquerque, V.H.C. Emerging drone trends for blockchain-based 5G networks: Open issues and future perspectives. *IEEE Netw.* **2021**, *35*, 38–43. [CrossRef]

7.   Aloqaily, M.; Bouachir, O.; Boukerche, A.; Al Ridhawi, I. Design guidelines for blockchain-assisted 5G-UAV networks. *IEEE Netw.* **2021**, *35*, 64–71. [CrossRef]

8.   Blasch, E.; Xu, R.; Chen, Y.; Chen, G.; Shen, D. Blockchain methods for trusted avionics systems. In Proceedings of the 2019 IEEE National Aerospace and Electronics Conference (NAECON), Dayton, OH, USA, 15–19 July 2019; IEEE: Piscataway, NJ, USA, 2019; pp. 192–199.

9.   Xu, R.; Zhai, Z.; Chen, Y.; Lum, J.K. BIT: A blockchain integrated time banking system for community exchange economy. In Proceedings of the 2020 IEEE International Smart Cities Conference (ISC2), Piscataway, NJ, USA, 28 September–1 October 2020; IEEE: Piscataway, NJ, USA, 2020; pp. 1–8.

10.   Xu, R.; Chen, Y.; Blasch, E.; Chen, G. Exploration of blockchain-enabled decentralized capability-based access control strategy for space situation awareness. *Opt. Eng.* **2019**, *58*, 041609. [CrossRef]

11.   Xu, R.; Chen, Y.; Blasch, E.; Chen, G. Blendcac: A smart contract enabled decentralized capability-based access control mechanism for the iot. *Computers* **2018**, *7*, 39. [CrossRef]

12.   Nikouei, S.Y.; Xu, R.; Nagothu, D.; Chen, Y.; Aved, A.; Blasch, E. Real-time index authentication for event-oriented surveillance video query using blockchain. In Proceedings of the 2018 IEEE International Smart Cities Conference (ISC2), Kansas City, MO, USA, 16–19 September 2018; IEEE: Piscataway, NJ, USA, 2018; pp. 1–8.

13.   Sharma, V.; You, I.; Jayakody, D.N.K.; Reina, D.G.; Choo, K.K.R. Neural-blockchain-based ultrareliable caching for edge-enabled UAV networks. *IEEE Trans. Ind. Inform.* **2019**, *15*, 5723–5736. [CrossRef]

14.   Wang, J.; Liu, Y.; Niu, S.; Song, H. Lightweight blockchain assisted secure routing of swarm UAS networking. *Comput. Commun.* **2021**, *165*, 131–140. [CrossRef]

15.   Chang, Z.; Guo, W.; Guo, X.; Chen, T.; Min, G.; Abualnaja, K.M.; Mumtaz, S. Blockchain-empowered drone networks: Architecture, features, and future. *IEEE Netw.* **2021**, *35*, 86–93. [CrossRef]

16.   Yazdinejad, A.; Parizi, R.M.; Dehghantanha, A.; Karimipour, H.; Srivastava, G.; Aledhari, M. Enabling drones in the internet of things with decentralized blockchain-based security. *IEEE Internet Things J.* **2020**, *8*, 6406–6415. [CrossRef]

17.   Feng, C.; Liu, B.; Guo, Z.; Yu, K.; Qin, Z.; Choo, K.K.R. Blockchain-based cross-domain authentication for intelligent 5G-enabled internet of drones. *IEEE Internet Things J.* **2021**, *9*, 6224–6238. [CrossRef]

18.   Aggarwal, S.; Shojafar, M.; Kumar, N.; Conti, M. A new secure data dissemination model in internet of drones. In Proceedings of the 2019 IEEE International Conference on Communications (ICC 2019), Shangai, China, 20–24 May 2019; IEEE: Piscataway, NJ, USA, 2019; pp. 1–6.

19.   Liang, X.; Zhao, J.; Shetty, S.; Li, D. Towards data assurance and resilience in IoT using blockchain. In Proceedings of the 2017 IEEE Military Communications Conference (MILCOM 2017), Baltimore, MD, USA, 23–25 October 2017; IEEE: Piscataway, NJ, USA, 2017; pp. 261–266.

20.   Nakamoto, S. Bitcoin: A Peer-to-Peer Electronic Cash System. Decentralized Business Review. 2008. Available online: https://bitcoin.org/bitcoin.pdf (accessed on 20 October 2022).

21.   Welcome to Ethereum. Available online: https://ethereum.org/en/ (accessed on 30 August 2022).

22.   Hyperledger Fabric. Available online: https://hyperledger-fabric.readthedocs.io/en/latest/ (accessed on 30 July 2022).

23.   Xu, R.; Chen, Y.; Blasch, E. Microchain: A Light Hierarchical Consensus Protocol for IoT Systems. In *Blockchain Applications in IoT Ecosystem*; Springer: Cham, Switzerland, 2021; pp. 129–149.

24.   Fotouhi, A.; Qiang, H.; Ding, M.; Hassan, M.; Giordano, L.G.; Garcia-Rodriguez, A.; Yuan, J. Survey on UAV cellular communications: Practical aspects, standardization advancements, regulation, and security challenges. *IEEE Commun. Surv. Tutor.* **2019**, *21*, 3417–3442. [CrossRef]

25.   Blasch, E.; Sabatini, R.; Roy, A.; Kramer, K.A.; Andrew, G.; Schmidt, G.T.; Insaurralde, C.C.; Fasano, G. Cyber awareness trends in avionics. In Proceedings of the 2019 IEEE/AIAA 38th Digital Avionics Systems Conference (DASC), San Diego, CA, USA, 8–12 September 2019; pp. 1–8.

26.   Xu, R.; Nagothu, D.; Chen, Y. Decentralized video input authentication as an edge service for smart cities. *IEEE Consum. Electron. Mag.* **2021**, *10*, 76–82. [CrossRef]

27.   Lamport, L.; Shostak, R.; Pease, M. The Byzantine generals problem. *ACM Trans. Program. Lang. Syst.* **1982**, *4*, 382–401. [CrossRef]

28.   Ferrag, M.A.; Derdour, M.; Mukherjee, M.; Derhab, A.; Maglaras, L.; Janicke, H. Blockchain technologies for the internet of things: Research issues and challenges. *IEEE Internet Things J.* **2018**, *6*, 2188–2204. [CrossRef]

29.   Szabo, N. Formalizing and securing relationships on public networks. *First Monday* **1997**, *2*. [CrossRef]

30.   Hu, N.; Tian, Z.; Sun, Y.; Yin, L.; Zhao, B.; Du, X.; Guizani, N. Building agile and resilient UAV networks based on SDN and blockchain. *IEEE Netw.* **2021**, *35*, 57–63. [CrossRef]

31.   Blasch, E.; Raz, A.K.; Sabatini, R.; Insaurralde, C.C. Information Fusion as an Autonomy enabler for UAS Traffic Management (UTM). In Proceedings of the AIAA Scitech Forum 2021, Virtual Event, 11–15 January 2021; pp. 1–12.

32.   Mnih, V.; Kavukcuoglu, K.; Silver, D.; Graves, A.; Antonoglou, I.; Wierstra, D.; Riedmiller, M.A. Playing Atari with Deep Reinforcement Learning. *arXiv* **2013**, arXiv:1312.5602.

33.   Li, J.; Sun, A.; Han, J.; Li, C. A Survey on Deep Learning for Named Entity Recognition. *IEEE Trans. Knowl. Data Eng.* **2020**, *34*, 50–70. [CrossRef]

34.   Xu, R.; Chen, Y. μDFL: A Secure Microchained Decentralized Federated Learning Fabric atop IoT Networks. *IEEE Trans. Netw. Serv. Manag.* **2022**, *19*, 2677–2688. [CrossRef]

35. Swarm. Available online: https://ethersphere.github.io/swarm-home/ (accessed on 30 September 2022).
36. Flask: A Pyhon Microframework. Available online: https://flask.palletsprojects.com/ (accessed on 30 September 2022).
37. Pyca/Cryptography Documentation. Available online: https://cryptography.io/ (accessed on 30 September 2022).
38. MAVLink Developer Guide. Available online: https://mavlink.io/en/ (accessed on 30 September 2022).
39. ArduPilot Project. Available online: https://github.com/ArduPilot/ardupilot (accessed on 30 September 2022).
40. Taylor, M.; Chen, H.; Qin, F.; Stewart, C. Avis: In-Situ Model Checking for Unmanned Aerial Vehicles. In Proceedings of the 2021 51st Annual IEEE/IFIP International Conference on Dependable Systems and Networks (DSN), Taipei, Taiwan, 21–24 June 2021.
41. Xu, R.; Chen, Y.; Li, X.; Blasch, E. A Secure Dynamic Edge Resource Federation Architecture for Cross-Domain IoT Systems. In Proceedings of the 2022 International Conference on Computer Communications and Networks (ICCCN), Waikiki Beach, Honolulu, HI, USA, 25–27 July 2022; IEEE: Piscataway, NJ, USA, 2022; pp. 1–8.

Article

Study of Urban Logistics Drone Path Planning Model Incorporating Service Benefit and Risk Cost

Quan Shao [1,*,†], Jiaming Li [1,†], Ruoheng Li [1], Jiangao Zhang [1] and Xiaobo Gao [1,2]

1   The College of Civil Aviation, Nanjing University of Aeronautics and Astronautics, Nanjing 211106, China
2   Civil Aviation Electronic Technology Co., Ltd., Chengdu 610043, China
*   Correspondence: shaoquan@nuaa.edu.cn
†   These authors contributed equally to this work.

Abstract: The application of drones provides a powerful solution for "the last-mile" logistics services, while the large-scale implementation of logistics drone services will threaten the safety of buildings, pedestrians, vehicles, and other elements in the urban environment. The balance of risk cost and service benefit is accordingly crucial to managing logistics drones. In this study, we proposed a cost-benefit assessment model for quantifying risk cost and service benefit in the urban environment. In addition, a global heuristic path search rule was developed to solve the path planning problem based on risk mitigation and customer service. The cost-benefit assessment model quantifies the risk cost from three environmental elements (buildings, pedestrians, and vehicles) threatened by drone operations based on the collision probability, and the service benefit based on the characteristics of logistics service customers. To explore the effectiveness of the model in this paper, we simulate and analyse the effects of different risk combinations, unknown risk zones, and risk-benefit preferences on the path planning results. The results show that compared with the traditional shortest-distance method, the drone path planning method proposed in this paper can accurately capture the distribution of risks and customers in the urban environment. It is highly reusable in ensuring service benefits while reducing risk costs and generating a cost-effective path for logistics drones. We also compare the algorithm in this paper with the A* algorithm and verify that our algorithm improves the solution quality in complex environments.

Keywords: urban logistics; drones; path planning; risk cost; service benefit; optimization algorithms

1. Introduction

The daily parcel of e-commerce enterprises has attracted huge attention due to their rapidly growing volume. In 2021, the global parcel shipping volume exceeded 159 billion parcels, which is expected to reach 256 billion in 2027 at a compound annual growth rate of 8.5 per cent [1]. Meanwhile, the variability of customer demand characteristics, such as different service locations and service times, has led to the need for logistic service providers to invest large-scale capacity and resources in "the last mile" transportation of parcels [2]. Thus, more and more companies are trying to find innovative and autonomous delivery methods for "the last mile" transport, such as drone logistics, to improve the quality of logistics. With the development of technology, drones' airworthiness and cargo-carrying capacity have improved significantly. Electric-powered logistics drones are not restricted by road networks and can reduce environmental costs and increase service flexibility [3]. The contactless services provided by drone logistics are also widely recognised due to the coronavirus outbreak [4]. Overall, the above advantages make drone logistics a powerful solution to solving the problems of traditional logistics [5]. Internationally renowned logistics companies such as Amazon, DHL Express, and Jingdong Logistics have begun developing drone logistics versions [6]. Statistics from BusinessWire also show that the global business value of drone package delivery has grown from USD 0.68 billion in 2020 to approximately USD 1 billion in 2021 and is expected to be USD 4.4 billion in 2025 [7].

Citation: Shao, Q.; Li, J.; Li, R.; Zhang, J.; Gao, X. Study of Urban Logistics Drone Path Planning Model Incorporating Service Benefit and Risk Cost. *Drones* **2022**, *6*, 418. https://doi.org/10.3390/drones6120418

Academic Editors: Ivana Semanjski, Antonio Pratelli, Massimiliano Pieraccini, Silvio Semanjski, Massimiliano Petri and Sidharta Gautama

Received: 25 November 2022
Accepted: 13 December 2022
Published: 15 December 2022

Publisher's Note: MDPI stays neutral with regard to jurisdictional claims in published maps and institutional affiliations.

However, the accident risks associated with the large-scale application of logistics drones must be effectively assessed and managed. The drone would not only threaten the safety of people and vehicles on the ground in urban environments [8], but also may collide with high-rise buildings [9]. To ensure the safety of other aircraft, people, and property after a drone crash, aviation organisations, including the Federal Aviation Authority (FAA), require a risk mitigation assessment in the pre-flight state [10]. Hence, the study of path risk assessment and mitigation methods is a critical technical prerequisite for logistics drone applications.

Throughout the existing research, the vehicle path problem is a classical mathematical model for studying urban last-mile logistics. It is based on the travelling salesman problem (TSP) [11], which ensures the minimum transportation time or cost by planning the service sequence of customers. The existing research evolved on the basis of this problem model. Murray and Chu [12] proposed a collaborative path-planning model for trucks and drones considering drone service range and load capacity constraints. In this work, they reported two new variants of the traditional TSP problem, the flying sidekick travelling salesman problem (FSTSP) and the parallel drone scheduling travelling salesman problem (PDSTSP). Yurek and Ozmutlu [13], Freitas and Penna [14], and Mbiadou Saleu et al. [15] also presented various algorithms for these problems. The above-simplified approaches assumed that the order of customer service at different locations remains consistent with the drone's service path, while ignoring the problem of safety risks inevitably involved in the actual operation. Inspired by this factor, existing studies have started to consider the risk assessment of drone operations. These mainly include the risk of collision in flight and the impact on the ground.

Falling drones would threaten the safety of pedestrians and vehicles on the ground. Mitici and Blom [16] proposed a mathematical model for collision probability estimation, which provides a research solution for the collision risk assessment of drones. Bertrand et al. [17] studied the probability of drone operations threatening road traffic, defined the range of ground a falling drone could affect, and developed a collision probability model to identify high-risk areas in the road network. Koh et al. [18] and Clothier et al. [19] studied the extent of injury to pedestrians struck by drones and proposed weight limits for drones based on the associated injury scales and criteria. Drone aerial collision risks mainly originate from buildings, no-fly zones, unstable weather, and other drones [20]. To assess the risk of aerial collisions, existing studies have established various collision models, mainly including the REICH model, the EVENT model, and the position probability model based on the concept of position error. The REICH model [21] lays the foundation of flight safety interval assessment and is mainly applied to assess the risk of collision between two aircraft in parallel flight paths. It finds that the collision probability and relative velocity in each direction determine the flight collision risk. The EVENT model proposed by Brooker [22] combines radar and controller operations to analyse lateral and longitudinal separation, which can calculate the probability of collision risk in each direction. The probabilistic model based on position error focuses on collecting and processing information about the positioning error and trajectory deviation of the drone, in order to predict the probability of the flight trajectory conflicting with the risk area [23].

Based on the conflict risk assessment research, most research on drone path risk mitigation aims to find no-conflict paths. One intuitive approach is geometry-based. The closest proximity point approach is used to solve the potential conflict warning problem by measuring the position between two drones, thus avoiding collision risk and ensuring the safety of the planned path [24,25]. As an improvement to the geometric method, Fan et al. [26] and Tang et al. [27] introduced artificial potential fields (APF) and simulated the environment by designing virtual attractive and repulsive potential fields for autonomous guidance of the drone to avoid obstacles. Driven by efficiency, many researchers have tried to use heuristic search to find the optimal no-conflict path. For example, a node-based optimal algorithm is a special form of dynamic programming. When a map or graph is already constructed, they first define a cost function, and then search each node and arc to find a path with minimum cost. It mainly includes the A* algorithm [28], Lifelong Planning A* (LPA) [29], Theta* [30], Lazy Theta* [31], D*-Lite [32], Harmony Search [33], etc. Evolutionary algorithm, which contains genetic algorithm [34], memetic algorithm [35], particle swarm optimisation [36], ant colony optimisation [37], and shuffled frog leaping algorithm [38]. The evolutionary algorithm starts by selecting randomly feasible solutions as the first generation. Then, taking the environment, drone capacity, goal, and other constraints into consideration, the planner evaluates the fitness of each individual. In the next step, a set of individuals is selected as parents for the next generations according to their fitness. The last step is a mutation and crossover step and stops the process when a pre-set value is achieved. The best fitness individual is decoded as the optimal path. Recent studies have treated drones as intelligent agents for stochastic dynamic threats in urban environments and used reinforcement learning to guide drones to avoid collisions [39–41].

Nevertheless, it is still an open problem for drone logistics to plan effective service paths in complex urban environments and ensure service completion based on reducing the threat to pedestrians, vehicles, buildings, etc. Many works focus on only considering obstacles in the environment during the finding phase of collision-free paths, while little attention has been paid to the fact that the risk cost from the threat is simultaneous with the service benefit of providing services to customers. To address the shortcomings in the above studies, we propose an urban environment model considering the coupling effect of customer service requirements and complex risks and develop a path point search strategy for improving the exploration of feasible paths in the environment. We summarise the main contributions of this paper as follows.

(1) We studied the complex risk factors of drone operation in urban environments and established a risk quantification model, which considers three primary risk sources in urban environments, including pedestrians, vehicles, and buildings.
(2) We established a logistics service benefit quantification model and proposed a multi-drone path planning method that integrates risk cost and service benefit, with the goal of guiding drones to find a path with the highest service benefit and lowest risk cost under the constraints of flight performance indicators, such as energy consumption and step length.
(3) We proposed a path point search strategy to solve a dynamic path planning problem driven by customer demand and risk. The strategy ensures that drones can adjust local paths in dynamic environments through regular global searches.

The rest of this paper is organised as follows: Section 2 analyses the critical elements affected by drones in the urban environment and illustrates the concept of path planning that combines customer needs and risks. The proposed methodology is described in Section 3, followed by simulation validations and case studies in Section 4. The summary of our work is in Section 5.

2. Problem Definition

Drones operate at low altitudes below 400 feet above the ground in cities. Once there is a collision, they can cause threats to buildings and other non-cooperative drones in the air. On the other hand, they can threaten pedestrians and vehicles on the ground when a crash occurs [42,43]. We conclude the primary environmental elements threatened by drone operations into four categories as follows.

(1) Drone impacts pedestrians, causing fatalities;
(2) Drone impacts vehicles, causing traffic accidents;
(3) Drone impacts high-rise buildings, resulting in property loss;
(4) Drone collision with other non-cooperative drones.

In this work, we ignore other risk factors, such as noise and privacy impacts on the public, due to their insignificance [44]. Pedestrians, vehicles, buildings, and logistics service customers are randomly dispersed in the city. Therefore, the core problem in logistics drone path planning is quantifying risk cost and service benefit for different locations. The urban environment for drone flight is divided into discrete 2D grids, and each grid's risk cost and service benefit are derived from environmental risk elements and customer demand. The cost-benefit value within each grid is used to guide drones to serve more customers and avoid high-risk areas in the complex urban environment.

The technology framework of the proposed work is presented in Figure 1. There are five steps to quantify risk cost and service benefit in the environment. First, the threat of drone operations to pedestrians, vehicles, and buildings in the city is analysed. Then, we develop three risk cost assessment models to quantify the various types of risk costs from the above elements under threat. Thirdly, we develop a service benefit assessment model based on the characteristics of logistics customers. Fourth, we synthesise the integrated risk cost and service benefit into cost-benefit values. Fifth, we construct the cost-benefit map. The urban environment is gridded, and the cost-benefit value calculation method of the flight path is established. Based on the cost-benefit map, we propose a drone path planning model with energy limitation constraints and a search algorithm with heuristic factors. To explore the effectiveness of the model in this paper, we next simulate and analyse the effects of different risk combinations, unknown risk zones, and risk-benefit preferences on the path planning results. We also compare the algorithm of this paper with the A* algorithm to verify the solving ability of this paper's algorithm. Finally, the reusability of the method in this paper is demonstrated by statistical analysis.

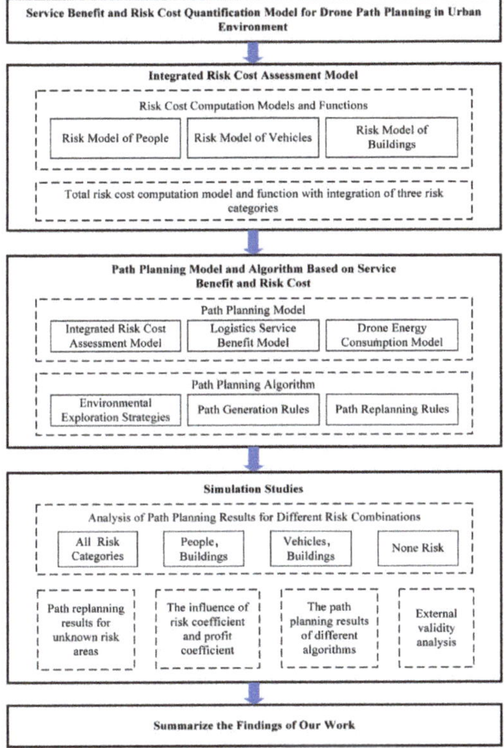

Figure 1. The technology framework of the proposed work.

3. Materials and Methods

3.1. Risk Assessment Model

This section presents a quantitative model of the integrated risk cost. We analyse the risks derived from the operation of drones in an urban environment. Figure 2 depicts the impact of considering risk cost mitigation on path planning results. The environment is divided into equal-sized grids, and the risk cost within each grid is calculated according to the integrated risk model. The colours in the map show the distribution of risk cost, with red representing areas of high risk cost and blue representing areas of low risk cost. The dashed line indicates the path with the shortest distance from the starting point to the end, and the solid line is the path that considers risk cost mitigation, which chooses the area with lower risk cost to pass, and the final path has a much lower risk cost than the shortest path.

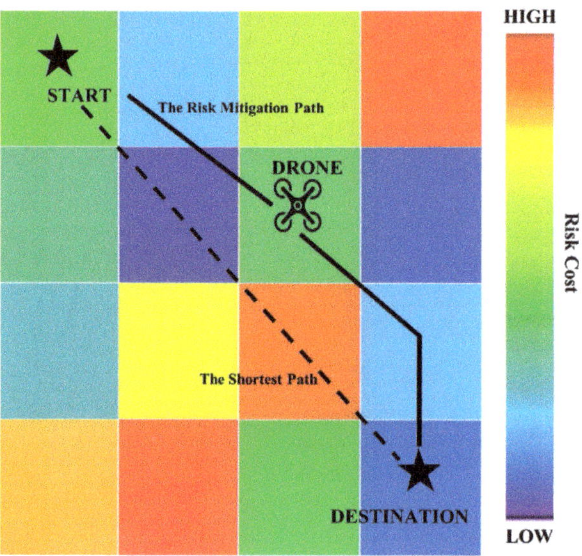

Figure 2. Risk cost mitigation affects path planning results.

3.1.1. Quantifying the Risk Cost Associated with Pedestrians

The risk cost of a drone striking a pedestrian is modelled according to the three components of collision [45,46]: (a) a drone crash, (b) a drone striking pedestrians, and (c) resulting in the death of pedestrians, as shown in Figure 3.

Figure 3. Drones and pedestrians.

We quantify the risk cost due to drones affecting pedestrians by potential fatalities as in Equation (1):

$$Cost_1 = Cost_p = P_{crash}S_d\rho_p P_d \tag{1}$$

where $Cost_1$ is the quantified value of pedestrian risk cost, P_{crash} is the probability of a drone crash, S_d is the exposed area of the ground impacted by the falling drone, ρ_p is the population density, and P_d is the fatality rate associated with the kinetic energy of the drone. The speed of the drone hitting the ground is shown in Equation (2):

$$v_d = \int_0^t (g - f)dt = \int_0^t \left(g - \frac{f_d S_d \rho_A v_{d-real}^2}{2m} \right) dt = \sqrt{\frac{2mg}{f_d S_d \rho_A} \left(1 - e^{-\frac{H f_d S_d \rho_A}{m}} \right)} \quad (2)$$

where $g = 9.8 m/s^2$, f is the resistance acceleration, f_d is the drag coefficient, ρ_A is the air density, v_{d-real} is the actual airspeed of the falling drone, m is the mass of the drone, H is the height of the drone falling point. The energy generated by the descending drone is shown in Equation (3)

$$E_{fd} = \frac{1}{2}mv_d^2 \quad (3)$$

Considering that the buffering effect of buildings and trees can mitigate the injury of falling drones to pedestrians, a sheltering factor S_f, $S_f \in (0,1]$ is introduced to consider this sheltering effect when calculating risk costs. A higher value implies a better sheltering effect and a lower probability of death. By combining the sheltering element into the kinetic energy equation, the lethality P_d of a falling drone can be obtained as shown in Equation (4):

$$P_d = \left(1 + \sqrt{\frac{\mu}{v}} \left(\frac{v}{E_{fd}} \right)^{\frac{1}{4S_f}} \right)^{-1} \quad (4)$$

where μ is the energy that might cause a 50% fatality with $S_f = 0.5$, v is the impact energy threshold required to cause fatality as S_f approaches zero. The values of S_f for different environments are shown in Table 1.

Table 1. Sheltering coefficients [47].

S_f	Type of Shelters
0	None
0.25	Trees
0.50	Low-rise buildings
0.75	High-rise buildings
1	Industrial buildings

3.1.2. Quantifying the Risk Cost Associated with Vehicles

Vehicles are another key element in the urban environment that can shelter falling drones; different from buildings and trees, the sheltering effect of vehicles mainly occurs while driving. Similar to the modelling of drone strikes on pedestrians, falling drones cause road traffic accidents in three components [17]: (a) a drone crash, (b) a drone striking vehicles, (c) resulting in traffic accidents, and (d) causing human fatalities, as shown in Figure 4.

Quantify the risk cost due to drones affecting vehicles by potential fatalities, as shown in Equation (5)

$$Cost_2 = Cost_V = P_{crash} P_V N_V \quad (5)$$

where P_V is the probability of a falling drone hitting a vehicle, proportional to the traffic density, and N_V is the average number of fatalities caused by a crash. The probability of a drone hitting vehicles on the ground is defined as the ratio of the total area occupied by vehicles to the entire scope of the road, as shown in Equation (6)

$$P_V = \frac{\overline{S_V} \rho_V}{D_{road}} \quad (6)$$

where $\overline{S_V}$ is the average projected area of the vehicle, ρ_V is the traffic density, and D_{road} is the road width.

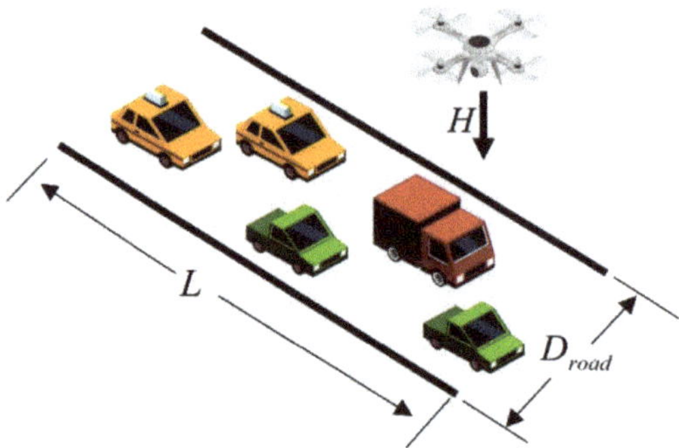

Figure 4. Drones and vehicles.

3.1.3. Pedestrian Density and Vehicle Density

The density distribution of pedestrians and vehicles in the urban environment can directly affect the risk cost of drone operations. Their density distribution is highly correlated with attractive facilities [48]. To quantitatively assess this correlation, gravity models are used to calculate pedestrian and vehicle density [49]. Inspired by gravity models, the pedestrian density in urban environments is shown in Equation (7).

$$\rho_P = e^{(1-r^2)}\rho_P^0 \tag{7}$$

where ρ_P^0 is the average pedestrian density, r is the distance from the centre of gravity. If there is an increase in r, it leads to a decrease in ρ_P, as shown in Figure 5.

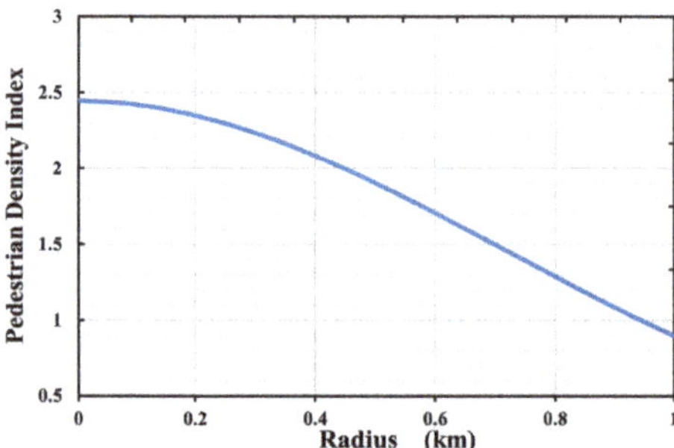

Figure 5. Gravity model for pedestrian distribution.

Similarly, the road traffic density distribution is shown in Equation (8):

$$\rho_V = e^{(1-r^2)}\rho_V^0 \tag{8}$$

where ρ_V^0 is the average traffic density.

3.1.4. Quantifying the Risk Cost Associated with Buildings

As shown in Figure 6, the operation of drones in urban airspace inevitably involves potential conflicts with buildings, and this potential conflict incurs risk costs [50]. Considering the overlapping locations of logistics customers and buildings, buildings cannot simply be set up as no-fly grids.

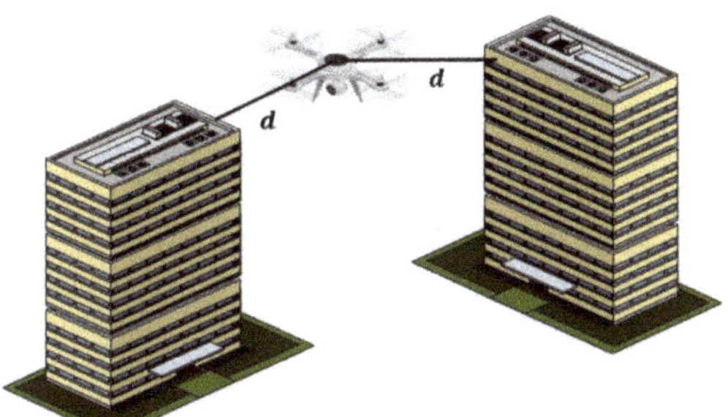

Figure 6. Drones and buildings.

The flight risk decreases as the distance between the drone and the building increases. However, due to the different sizes and shapes of buildings in the city, the influence range of buildings on drones is also different. For a normal distribution, setting different variances can reflect the different influence ranges of buildings, which is simple compared to other distributions or describing the shape and dimensions of the buildings. To simplify the model calculations, the distribution of the risk cost due to the influence of the building is assumed to be a normal distribution with different variances [51].

For n independent buildings in the map, given the central location $B_i = (X_i, Y_i)$ of the i-th building, $Cost_{Bu}(x, y)$ denotes the risk cost of the point (x, y) when considering the impact of the building B_i, as shown in Equation (9)

$$Cost_3 = Cost_{Bu}(x, y) = \frac{1}{\sqrt{2\pi}\sigma} e^{-\frac{d^2}{2\sigma^2}} \tag{9}$$

where $d = \sqrt{(x - X_i)^2 + (y - Y_i)^2}$, $i \in \{1, 2, \cdots, n\}$ indicates the Euclidean distance between the drone and the centre of the building.

3.1.5. Comprehensive Risk Model

In the previous section, risk cost quantification models were constructed for three elements: pedestrian, vehicle, and building. The different calculation methods would obtain different values of risk cost magnitude, which cannot be measured by the same standard. Therefore, the risk costs of the three elements need to be standardised to describe the total risk cost in the urban environment.

The risk costs of all three elements can be calculated through a particular distribution, and then each type of risk cost contained in a raster would be divided by the maximum risk value generated by the risk source separately. It is guaranteed that all risk cost values for each type are in the range of $(0, 1]$.

The weights of the three risks may vary with the difference in their importance or preference, and the contribution of each risk may also vary with the cost [52]. For example, aviation regulators emphasise the risk of pedestrian fatalities caused by drones. The risk cost of pedestrians will be weighted much more than the other two factors. Traversing the areas with high pedestrian density will result in higher costs, so the planned paths will be more inclined to avoid these areas.

For point (x, y), its cumulative risk value needs to consider a pedestrian risk zones, b vehicle risk zones, and c building risk zones. The total risk cost of the point (x, y) is calculated as shown in Equation (10),

$$R_{total}(x, y) = \alpha_i \sum_{i=1}^{a} \frac{C_1^i}{C_{1-max}^i} + \alpha_j \sum_{j=1}^{b} \frac{C_2^j}{C_{2-max}^j} + \alpha_k \sum_{k=1}^{c} \frac{C_3^k}{C_{3-max}^k} \tag{10}$$

where α_i, α_j, α_k are the weighting factors, $\alpha_i + \alpha_j + \alpha_k = 1$.

The cumulative risk $R_{total}(x, y)$ of the path C is shown in Equation (11)

$$\int_{(x,y)\in C} R_{total}(x,y) \tag{11}$$

3.2. Drone Path Planning Model Based on Risk Cost and Service Benefit

The flight path of drones performing logistics services needs to mitigate the path risk cost based on ensuring service completion. Therefore, the objectives of logistics drone path planning include risk mitigation and customer service. The integrated risk cost quantification model established in the previous section can be used for risk mitigation. Customers' locations in cities often overlap with risk factors, such as buildings, crowds, and roads. This would cause drone service paths to pass through risk areas, so it is necessary to balance the path risk cost and service benefit. Figure 7 depicts the impact of considering risk cost mitigation and service benefits on the path planning results. The solid white arrows indicate the shortest path, the white dashed arrow is the path considering risk cost mitigation, and the white dotted line indicates the path that balances risk cost and service benefit, where the path is changed to fulfil customer needs based on the most risk cost-effective path.

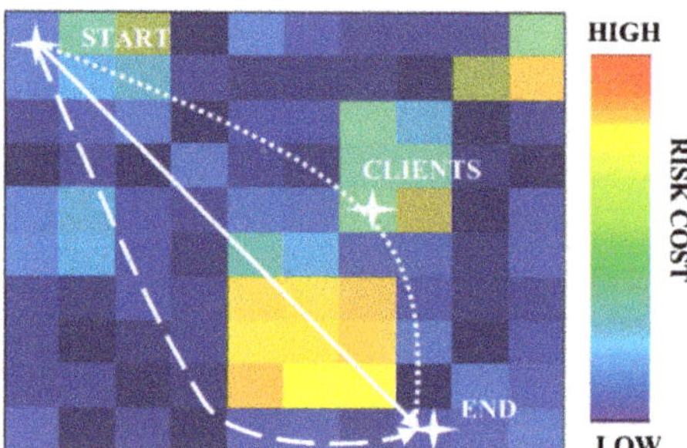

Figure 7. The path planning based on risk and customer.

In this section, our primary work is to establish a multi-drone path planning method to guide drones to find a path with the highest service benefit and lowest risk cost under the constraints of flight performance indicators such as energy consumption and step length. Furthermore, a global search strategy is proposed to solve the above paths.

3.2.1. Service Benefits Modelling

Assume that each customer has an initial requirement $C^0_{demand-j}$ that needs to be handled by drone. We also assume that the drone can only serve a certain distance from the customer's location. Therefore, for a customer j, the range that can be served is denoted as $s(p_j, R)$, where p_j is the location of the customer j and R is the radius of the acceptable service range. Service starts when the drones enter the service range of the customer j. Each drone has a constant service speed τ. The remaining demand $C_{demand-j}$ of the customer served by k drones simultaneously over time Δt is shown in Equation (12)

$$C^t_{demand-j} = C^{t+\Delta t}_{demand-j} - \tau k \Delta t \tag{12}$$

Assuming a nonlinear relationship between customer residual demand $C_{demand-j}$ and service revenue $C_b(C_{demand-j})$, this paper uses a sigmoid-like function to improve performance, as shown in Equation (13)

$$C_b(C_{demand-j}) = 1 - \exp\left[-\frac{(C_{demand-j})^\chi}{C_{demand-j} + \psi}\right] \tag{13}$$

where χ and ψ are control parameters. For each customer, the service revenue $C_b(C_{demand-j})$ decreases rapidly with its remaining demand $C_{demand-j}$. It is guaranteed that serving the customer with the highest remaining demand generates the greatest revenue, thus increasing global customer service completion.

3.2.2. Energy Consumption Modelling of Drones

Assuming that the lifting and lowering process of the drone is ignored and only straight-line flight is considered, the energy consumed for moving a distance d at a constant speed v is shown in Equation (14),

$$E_v = P(w)\frac{d}{v}, \ \Delta E_v = P(w)\Delta t \tag{14}$$

where $P(w)$ is the power of the drone moving at a constant speed v. For the n-rotor drone, its power is shown in Equation (15),

$$P(w) = (W+w)^{\frac{3}{2}}\sqrt{\frac{g^3}{2\rho_A \varsigma n}} \tag{15}$$

where W is the self-weight of the drone, w is the weight of the load carried by the drone, ρ_A is the fluid density of air, ς is the area of the rotating blades, and g is the acceleration of gravity. The total power of the drone is shown in Equation (16),

$$E_{total} = \eta C V_n \tag{16}$$

where η is the energy conversion efficiency, C is the capacity of the cell, and V_n is the nominal voltage of the n cells.

The drone departs with an empty load. As the drone services the customer, the drone's load increases while the customer's remaining demand decreases. After the current customer is served, the drone maintains the current load until it starts serving the next customer.

Assuming that the demand is proportional to the load and the scale factor is ε, then for a drone i serving x customers at the same time, the load varies with time, as shown in Equation (17),

$$w_i^{t+\Delta t} = w_i^t + \tau x \varepsilon \Delta t \tag{17}$$

3.2.3. Global Path Planning Model

Based on the risk cost and service benefit quantification model, We introduce a cost-benefit matrix to measure the benefits and costs between any two points on the map. The map is represented as an $N \times N$ grid, and the cost-benefit matrix TC_{mn} between any points p_m and p_n is shown in Equation (18)

$$TC_{mn} = d_{p_m,p_n} + \frac{M_{benefit}}{1 + \sum_{n \in s(p_j,R)} C_b(C_{demand-j})} + M_{risk}\int_{(x,y)\in C} R_{total}(x,y) \tag{18}$$

where $p_m, m \in \{1,2,\cdots,N^2\}$ is the current position of the drone, $p_n, n \in \{1,2,\cdots,N^2\}$ is the next position of the drone. d_{p_m,p_n} is the Euclidean distance between p_m and p_n. $\sum_{n \in s(p_j,R)} C_b(C_{demand-j})$ is the benefit generated by the demand of all customers that can be served at point p_n. $M_{benefit}$ and M_{risk} are the coefficients of service benefit and risk cost, which affect the path planning strategy. In practice, $M_{benefit}$ and M_{risk} can be adjusted according to preference. For example, if the tolerance for risk cost is poor, then M_{risk} can be set to a higher value to amplify the impact of risk cost.

The goal of the present work is to plan a service path with minimum total cost. The total cost includes the risk cost and the inverse of the service benefit. The objective function is shown in Equation (19)

$$\min : TC(P) = \sum_{e_{ir}\in P} TC(e_{ir}), i > 0, r = i+1 \tag{19}$$

where P is the flight path consisting of edge e, $TC(P)$ is the total cost of the path P, and $TC(e_{ir})$ is the cost of the edge e_{ir}.

According to the drone energy consumption model, the power available for flight is limited. Therefore, the logistics drone must complete the service and reach the endpoint as soon as possible before consuming the planned available power. The constraint is defined as

$$\forall l_{ir} \geq l_{\min}, e_{ir} \in P, r = i + 1, i, r > 0 \tag{20}$$

$$E_{consume} = \sum_{e_{ir} \in P} E_{consume}^{ir} \leq E_{plan} \tag{21}$$

$$\frac{(x_i - x_{i-1}, y_i - y_{i-1})^T (x_{i+1} - x_i, y_{i+1} - y_i)}{\|(x_i - x_{i-1}, y_i - y_{i-1})\| \cdot \|(x_{i+1} - x_i, y_{i+1} - y_i)\|} \geq \cos \beta_{\max} \tag{22}$$

Equation (20) represents the shortest distance constraint for an edge between two adjacent nodes in the drone path, $l_{\min}$ is the minimum distance of the edge, and l_{ir} is the length of the edge e_{ir}. Equation (21) represents that the total energy consumption of the drone must not exceed the available power, $E_{consume}$ is the total energy consumption of the path P, $E_{consume}^{ir}$ is the energy consumption of each side e_{ir} in the path P, and E_{plan} is the total available power. Equation (22) represents the constraint on the maximum turning angle of the drone, (x_i, y_i), (x_{i-1}, y_{i-1}), (x_{i+1}, y_{i+1}) are the coordinates of three consecutive path points, and $\beta_{\max}$ is the maximum acceptable turning angle.

3.2.4. Path Planning Algorithm

To solve the least-cost flow problem for large scale in this study, heuristic methods (e.g., A* algorithm) have better performance in terms of computational time to solve the path planning problem. The standard A* algorithm generally uses the Manhattan or Euclidean distance to select the following move location. However, in the cost-benefit environment established in this paper, the cost of each raster is different and unevenly distributed, so considering only the distance cannot reflect the actual cost of the path. As the complexity of the environment increases, the traditional A* algorithm has difficulty finding a suitable path and deadlocks. Therefore, the following path search rule is proposed to improve the environment's exploration, and the rule's effectiveness is verified in the experimental stage.

(1) Environmental exploration strategy

In this work, a heuristic factor is set according to the Boltzmann distribution to ensure a complete exploration of the environment. The drone is currently at the path point p_i, $i \in \{1, 2, \cdots, N^2\}$, and the probability of the point $p_r, r \in \{1, 2, \cdots, N^2\}$ being selected as the next path point is calculated based on the value TC_{ir}, as shown in Equation (23)

$$p(i, r) = \frac{\exp\left[\frac{T}{TC_{ir}}\right]}{\sum\limits_{k \in R, k \neq i} \exp\left[\frac{T}{TC_{ik}}\right]} \tag{23}$$

where T is the temperature parameter that controls the degree of environment exploration, R is the set of all N^2 points in the map. At the beginning of exploration, since the drone knows little information about the environment, a smaller T value is set to ensure that the drone can explore the environment quickly in the early stages. As the exploration time increases and the drone has enough information about the environment, the value of T is increased to ensure that the algorithm can reach convergence within a specific time.

(2) Original global path generation rules

A sequence of points forms a drone path. The calculation of the cost-benefit value TC_{mn} for the drone moving between two points in the map is established in Equation (18). The path point exploration rule based on the cost-benefit value TC_{mn} is established in Equation (23). Based on this, our global path planning is divided into two steps. Based on this, our global path planning is divided into two steps. Firstly, based on the cost-benefit value in the environment at the planning start time t_0, a series of paths satisfying the constraints are iteratively generated according to the global search method (as shown in Algorithm 1), and the path with the optimal cost-benefit value is selected as the original global path. The second step performs local replanning on the basis of the original global path (as shown in Algorithm 2). The generation of the original global path is described as follows.

(1) For the i-th drone (UAV_i), for each *episode* repeat (2)–(6).

(2) Initialise $Path_i$ to an empty list $Path_i[]$. The initial position P_0 of the drone is the first point $Path_i[1]$ in $Path_i$.

(3) For each step in each *episode*, repeat (4)–(5).

(4) For the current location point p_s, select the next point p_{s+1} according to the Boltzmann exploration strategy.

(5) Add p_{s+1} to $Path_i[]$, as the $s + 1$-th path point $Path_i[s + 1]$. Return to (3) until the target point is reached or the power is exhausted.

(6) Finish this *episode*, *episode* $+ 1$, and return to (2).

(7) Until *episode* $= MAX$, the learning process ends and the current optimal *Path* is output.

The process of global path planning is defined in Algorithm 1.

Algorithm 1 Original global path generation

1	**For** i **in** $UAVnum$ **do**
2	**For** $episode \leftarrow 0\, toMAX$ **do**
3	$Path_i \leftarrow Path_i[\,]$
4	$Path_i[1] \leftarrow P_0$
5	**For** $p_s \leftarrow P_0\ to\ Target$ **do**
6	$p_{s+1} \leftarrow Boltzmann$
7	$Path_i[s + 1] \leftarrow p_{s+1}$
8	$s = s + 1$
9	**End for**
10	$episode = episode + 1$
11	**End for**
12	**End for**

---

(3) Drones Movement and local path replanning rules

Based on the original global path defined in Algorithm 1, we need to further establish the rules that the drone moves according to the original path and simulate the actual operation of the drone on the original path. During the flight of drones, new risk areas may appear on the map as time changes, causing the subsequent part of the original global path to cross high-risk cost areas, and then the original path needs to be locally replanned. The reason for the above situation is that global path planning is carried out at time t_0, and some risk zones in the environment do not exist at this time but appear at time $t = t_0 + \Delta t$ (e.g., the temporary gathering of pedestrians due to time-predictable activities). This risk zone needs to be addressed by local path replanning rules during the actual flight of the UAV based on the original global path. This does not require real-time path planning, only further pre-planning for new risk zones that are known to occur during flight. The process of local path replanning by the drones to avoid the newly generated risk zone is defined in Algorithm 2.

For the ith drone (UAV_i) *Scan* is executed after moving one step along the original path. After *Scan* is executed, there are two scenarios. The first scenario is the discovery of new obstacles (including other drones), and the cost-benefit matrix will be recalculated for replanning the subsequent paths. The second scenario is that the surrounding environment remains unchanged, and the path also keeps the same. For each time interval Δt, the step length of the drone movement is fixed as *Step*. If the distance between the current position and the subsequent path point is less than *Step*, the drone will move directly to the subsequent path point.

Algorithm 2 Drone Movement and local path replanning

---

13   **For** i **in** $UAVnum$ **do**
14     $Path_i \leftarrow$ **Pathpoint Generation**( )
15   **End for**
16   **For** i **in** $UAVnum$ **do**
17   **If** $pos_i ==$ Target **then**
18   Stop(i)
19   **else**
20     $Obstacle \leftarrow Scan$
21   **If** $Obstacle$ **then**
22      $Path_i \leftarrow$ **Pathpoint Generation**( )
23   **End if**
24   **If** $pos_i == Path_i[1]$ **then**
25      $Path_i \leftarrow Path_i[2 \ldots end]$
26   **End if**
27     $pos_i ==$ Move($Step, Path_i[1]$)

28   **For** j **with** UAV_i **in** (P_j, R) **do**
29     $C_{demand-j} \leftarrow C_{demand-j} - \tau$
30     $E_v \leftarrow E_v - P(w_i)\Delta t$

31       $w_i \leftarrow w_i + \varepsilon\tau$
32   **End for**
33   **End if**
34   **End for**

---

4. Results

In order to validate the path planning model coupling risk cost and service benefit, we perform simulations and analyses in a constructed urban environment containing pedestrian risk zones, vehicle risk zones, building risk zones, and logistics customers.

First, the urban environment model is constructed based on the modelling of risk areas and customer demands above. Then, we apply the proposed path search algorithm to search the logistics service path with the lowest risk cost and the highest service benefit.

Based on the above, the effect of risk combinations and the dynamic addition in risk areas are investigated to verify the reliability of the model and algorithm, which can mitigate the three risk costs while ensuring the response to the dynamic environment. Next, sensitivity analysis is conducted for the risk and benefit coefficients to study the balance of risk cost and service benefit in path planning. To evaluate the effectiveness of the algorithm proposed in this paper, the three most critical metrics in logistics path planning, namely, service completion, average path length, and average risk, are considered to compare with the A* algorithm. Finally, simulations and statistical analyses were performed to evaluate the effectiveness of the proposed path planning model for balancing risk cost and service benefit when extended to other urban environments.

4.1. Path Planning for Multiple Drones

The urban environment model proposed in this paper includes pedestrian risk, vehicle risk, and building risk, and it is verified that drones can ensure the completion of customer service while reducing the cost of path risk. In this section, the required parameters for simulation experiments are shown in Table 2 [46,53,54], and the optimisation effect of the model in this study compared with the traditional method is analysed.

Table 2. Simulation parameters.

Parameters	Value	Parameters	Value
$l_{\min}/m$	2	$\rho_V/vehicle \cdot m^{-1}$	7.12×10^3
P_{crash}	6.04×10^{-5}	N_V	0.25
$\rho_A/kg \cdot m^{-3}$	1.23	$\alpha_i, \alpha_j, \alpha_k$	0.50, 0.25, 0.25
f_d	0.30	χ, ψ	2, 8
S_d/m^2	0.02	R/m	200
$v/kg \cdot m^2 \cdot s^{-2}$	232	$M_{risk}, M_{benefit}$	20, 1
$\mu/kg \cdot m^2 \cdot s^{-2}$	106	$C/A \cdot s$	1/144
$\rho_p/people \cdot m^{-2}$	0.007	$V_n/kg \cdot m^2 \cdot A^{-1} \cdot s^{-3}$	22.80
$\overline{S_V}/m^2$	9.68	W/kg	20
D_{road}/m	20	n	6
η	0.70	ε	0.50

The flight area with a range of 1000×1000 m is divided into 50×50 grids. In the environment model, we assume that the drone starting points are represented by black circles; the endpoint is represented by a black cross; the building risk zones are randomly generated variance σ; the crowd risk zones and the road vehicle risk zones are randomly generated risk radius r; the customer zones to be served are assigned random initial demand $d_j \in (0, 10]$.

Considering that the size of the drones is much smaller than the size of the grids, in this paper, we use the integral method to obtain the path risk cost, and the calculation result is not affected by the size of the grids and drones. Therefore, the drone is considered a prime point to simplify the calculation. The drone path planning is guided based on the risk cost distribution consisting of pedestrians, buildings, and vehicles in the environment and the service benefit distribution determined by the customer's location and acceptable service range. The path planning is performed in MATLAB using the algorithm described above. The initial environment modelling and path planning results are shown in Figure 8. The paths of the three drones departing from different locations are represented by three colours. Path group 1 represents the result of path planning considering the balance of service benefit and risk cost, where Drone 1 serves Customers 1, 2, and 3 according to the solid red path, Drone 2 serves Customers 4 and 5 according to the solid blue path, Drone 3 serves Customers 4, 5, and 6 according to the solid green path. Path group 2 is the path only considering customer service without risk. Path group 3 is the path only considering risk without customer service. The colours in the map show the distribution of risk cost, with red representing areas of high risk cost and blue representing areas of low risk cost. The contour lines represent the distribution of risk cost due to building risk and pedestrian risk, and road 1 and road 2 represent the vehicle risk cost distributed along the road. The specific path parameters are shown in Table 3.

As shown in Figure 8 and Table 3, the result of path planning without considering the risk model (path group 2; no risk considered) traverses the high-risk area to ensure the shortest path to complete the customer service and reach the endpoint, resulting in increased risk cost. The path without considering customer service (path group 3; no customer considered) ignores customers overlapping with the location of high-risk areas for ensuring the shortest length and lowest risk cost path to reach the endpoint. The customers overlapping with high risk cost areas are completely ignored, leading to a decrease in service completion. The model of this paper, which considers both risk avoidance and customer service completion as the driving force, can balance the risk cost and service benefit. Risk cost is reduced by 81.25% compared with path group 2, and service completion is improved by 57.00% compared with path group 3.

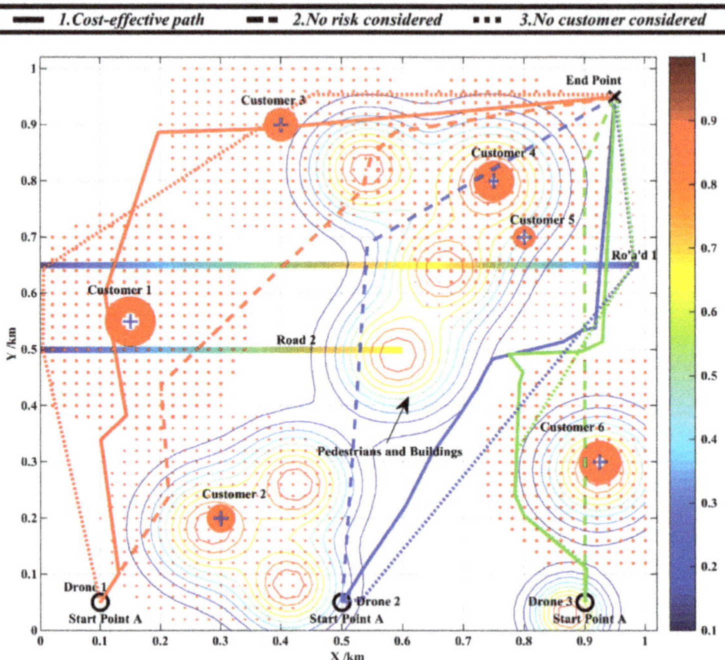

Figure 8. Environment modelling and path planning.

Table 3. Comparison of path planning results.

Path Group	Average Path Length/km	Completion Ratio	Average Risk Cost
Cost-effective path	1.24	1.00	6.20
No risk considered path	1.11	1.00	33.07
No customer considered path	1.19	0.43	3.01

4.2. Path Planning with Different Risk Combinations

According to the result in the previous subsection, it can be seen that the model obviously mitigates the risk cost in path planning. However, the comprehensive risk model proposed considers three types of risks: pedestrian, vehicle, and building. The path planning results also are affected to an extent by the difference in risk models.

Therefore, further quantitative analysis is required to study the effects of different risk combinations on drone path planning and risk costs in urban environments. This section simulates and studies path planning in the above flight area with four risk combinations: (a) Group A considers three risks, (b) Group B considers pedestrians and buildings, (c) Group C considers buildings and vehicles, (d) Group D considers pedestrians and vehicles, and (e) Group E does not consider risks.

Figure 9 presents the effect of different risk combinations on path planning. Path A has a total risk cost of 6.20. Path E is the worst because it does not mitigate any risks, with 433.23% higher total risk cost than Path A. Path B and Path C have similar results, with Path C being 7.99% higher than Path B due to dense pedestrian areas being more relevant to buildings. The risk cost of Path D increases by 53.99% relative to Path A. Due to the gravity model, the distribution of pedestrians and vehicles is associated with buildings, and disregarding building risks leads to a subsequent small increase in pedestrian and vehicle risks, but this increase is significantly lower than Path B and Path C, where the corresponding risks are not considered.

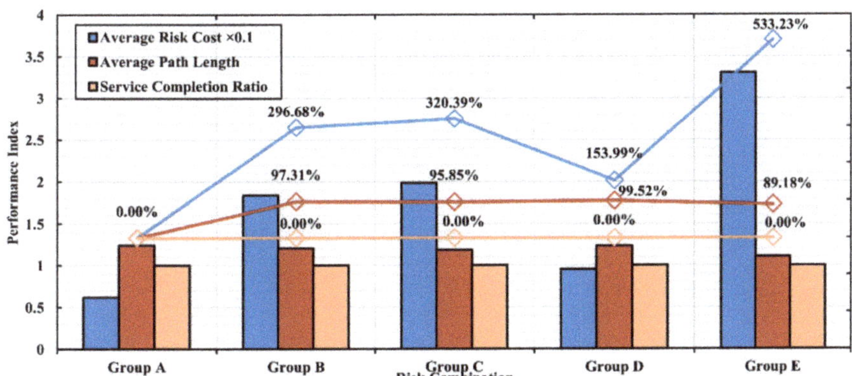

Figure 9. Impact of different risk combinations in the environment on path planning.

The average path length is affected by the combination of risks, and Path E has the shortest length without considering risks. Considering all three risk types, the model proposed in this paper only increases the path length by 12.00% over Path E.

For the increase in path length, on the one hand, the 12.00% increase in path length is minimal compared to the 433.23% increase in risk. On the other hand, we add the constraint of drone energy consumption to the path planning model. Although the length of Path A increases, it still completes all customer service requirements and reaches the target point within the energy consumption constraint, indicating that the increase in path length is negligible.

The results show the path planning under different risk combinations to further understand the differences in the various types of risk costs of the path planning results while considering the risk combinations.

This paper investigates each type of risk cost (pedestrian risk cost R1, vehicle risk cost R2, and building risk cost R3) in the above five risk combinations. The results are shown in Table 4. Path C was planned without considering the risks associated with pedestrians in the environment. The drone path enters dense pedestrian areas, resulting in a pedestrian risk cost R1 of 15.48, which is higher than the case of Path A and Path B, where pedestrian risk is considered. On the contrary, the risk combination considered in Path B includes pedestrian risk, thus avoiding the area with high pedestrian risk costs. However, vehicle risk is not considered, resulting in a higher vehicle risk cost of 14.47. The exclusion of building risk in Path D leads to an increase in building risk by 452.63%.

Table 4. Split comparison of path risk costs.

Risk Cost	Group A	Group B	Group C	Group D	Group E
Total Risk Cost	6.20	18.40	19.87	9.55	33.07
Pedestrian Risk Cost R1	3.35	3.56	15.48	5.64	16.62
Vehicle Risk Cost R2	2.66	14.47	3.11	2.86	15.11
Building Risk Cost R3	0.19	0.37	1.28	1.05	1.33

The gravity model leads to an overlap of the three risk types, which is similarly demonstrated in the variation of the three types of risk cost. Path B ignores vehicle risk, while pedestrian risk and building risk increase respectively by 6.27% and 94.74%; Path C ignores pedestrian risk, but vehicle risk and building risk increase respectively by 16.92% and 573.68%; Path D ignores building risk, and pedestrian and vehicle risk increase respectively by 68.35% and 7.52%. Although Path D ignores the building risk, the drone path does not intrude into the high building risk zones due to the presence of pedestrian risk, so the building risk is reduced by 17.97% compared to Path B. The relevance of the variation in different risk cost types also proves the importance of studying the integrated risk assessment model in this paper.

Path E presents that all three types of risk values are the highest among the five paths due to the correlation of various risk areas in the urban environment, such as the dense distribution of pedestrians and vehicles around the buildings. Therefore, the path planning results without considering any risk, the cost of all three risk categories is higher than the value of the corresponding risk category in any other combination.

For the mitigation effect of each type of risk, comparing Path E with Path A, it is shown that Path E in construction risk is 1.33 and Path A is 0.19, decreasing the risk by about 85.61%. Path E in vehicle risk is 15.11, and Path A is 2.66, decreasing the risk by approximately 82.40%. Path E for pedestrian risk is 16.62, and Path A is 3.35, decreasing the risk by about 79.85%. The total risk is reduced by approximately 81.25%. As a result, the model in this paper has a good mitigation effect on all three types of risks, and the proportion of the three risk reductions is kept at about 80.00%.

We can conclude that more risk sources in path planning can effectively mitigate the total path risk cost. This is because capturing more comprehensive risk sources is beneficial for avoiding more high-risk areas. It also further demonstrates the importance of our analysis and modelling for various types of elements threatened by drones in cities, which can guarantee the effectiveness of capturing risk costs in path planning.

4.3. Temporary Response Effect of the New Risk Area

During the process of drone logistics transportation in urban environments, the obstacles and risk areas in the environment can basically be examined and commanded to go around in the global path pre-planning stage before starting the mission due to the more comprehensive network coverage. However, due to the complexity of the urban environment, it is still challenging to avoid unknown obstacles in advance, such as flocks of birds, which require commanding the drone to change its route to avoid them.

In the path planning algorithm of this study, the drone scans the global environment at each step. Once there are new risk areas that affect the original flight path of the drone, the subsequent path is replanned to ensure that the drone adapts to the dynamic urban low-altitude environment. This section focuses on analysing the effect of the avoidance strategy proposed by the algorithm.

As shown in Figure 10 and Table 5, when a new risk area appears at the location of the point (0.2, 0.9), drone 1 moves one step according to the original path and finds that the subsequent original path passes through the new high risk cost area, so a local path replanning is performed to avoid the new risk area. The red dashed line in Figure 10 represents the locally replanned path of drone 1, and the solid red line indicates the original path. The solid blue line indicates the path of drone 2, the solid green line indicates the path of drone 3. As the new risk zone does not affect the original paths of drone 2 and drone 3, the paths of these two drones do not change. For analysing the impact on drone 1, which was replanned to avoid the new risk zone, we further compared and analysed the path parameters. The path length of drone 1 increased from 1.62 of the original path to 1.69, and the growth rate was 4.32%. The path risk cost was affected by the new risk zone, which increased from 5.20 to 5.22, with a growth rate of 0.35%. The service completion was always 100%, indicating that the path length increase was negligible. It is clear that the avoidance strategy proposed by the algorithm allows the drone to change the original path before entering the new risk zone. It could ensure that the risk cost from the new risk zone is mitigated and the increase in path length is minimal.

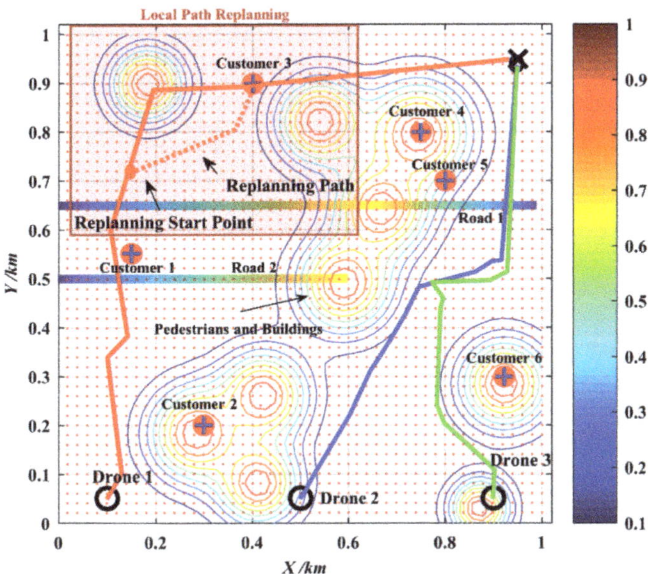

Figure 10. Path replanning due to new risk zones.

Table 5. Results of path replanning.

Drone 1	Path Length/km	Completion Ratio	Path Risk Cost
Original Path	1.62	1	5.20
Replanning Path	1.69	1	5.22

In order to further study new risk zones, this paper investigates the effect of the number of new risk zones on the path planning results. As shown in Figure 11, the length of path 1 increased by about 3.99% on average with the addition of each new risk area, the service completion always remained at 1, and the path risk cost increased by about 0.30% on average.

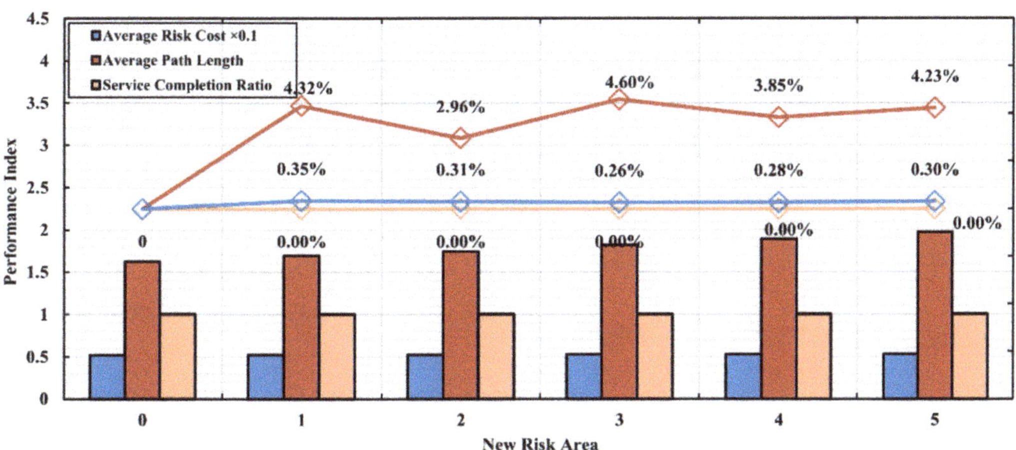

Figure 11. Path replanning due to new risk zones.

In summary, the temporary response of the algorithm to new risk zones can reduce new risk costs on the basis of service completion. The avoidance strategy is influenced by the time when the risk zone is discovered. The above discussed that a new risk zone is discovered before the drone enters that risk zone. Because the avoidance strategy requires the drone to scan and judge whether there is a new risk zone once in each step, it can guide the drone to update the next path point in time to avoid the risk zone. In the case that the drone has already entered the new risk zone when it is found, it is obvious that the drone will change the next path point according to the strategy, thus leaving the risk zone with the shortest distance. This case does not need to be discussed.

4.4. Sensitivity Analysis on Coefficients

After the above analysis, it can be seen that the model in this paper has a good effect on mitigating the path risk cost based on the assurance of service completion. Trade-off effects of service benefits and risk costs will be discussed in this part. In the model, the parameter $M_{benefit}$ determines the priority for the service, thus affecting service completion. When M_{risk} has a fixed value, a larger $M_{benefit}$ makes the drone more inclined to satisfy more customers, and the drone will bear more risk costs and path lengths due to the overlap of customer locations and risk areas. On the contrary, a smaller $M_{benefit}$ means that the drone will ignore some customers but reach the destination directly with a shorter path and lower risk cost. As shown in Figure 12, service completion and average risk will increase with the increase of $M_{benefit}$.

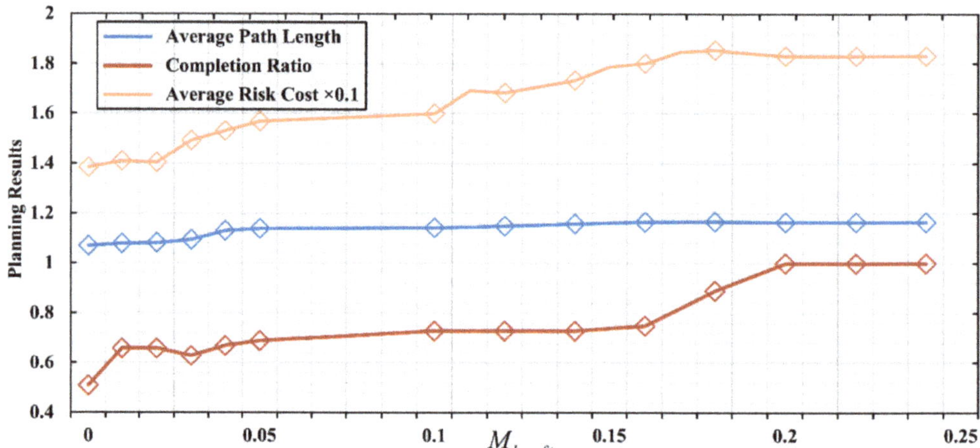

Figure 12. Sensitivity analysis of the parameter $M_{benefit}$ ($M_{risk} = 5$).

Since $M_{benefit}$ is the coefficient of customer service benefit, its change had the most significant impact on service completion among the three indicators, which was 0.51 when $M_{benefit} = 0$ and increased to 1 when $M_{benefit} = 2$ with a growth ratio of 49%; while the average path length increased from 1.07 to 1.165 with a growth ratio of 8.87%; the average path risk increased from 1.385 to 1.832 with a growth ratio of 32.27%.

Average path length and average path risk increased much less than service completion. Due to the increase in $M_{benefit}$, drones tend to complete more services, resulting in the drones needing to detour farther to reach the customer service area. The path risk also increased due to the overlap of customers and risk areas. However, since the minimisation objective in this paper's model includes path length and risk cost, this constraint ensures that the path length and risk cost remain stable when customer service completion increases rapidly. It can be found that our model achieves a flexible balance of service benefit with risk and path cost by adjusting $M_{benefit}$.

Similar to parameter $M_{benefit}$, M_{risk} controls the drone's tolerance for risk. When M_{risk} increases, drones are more inclined to avoid the risk zone to reduce path risk, which leads to a rapid decrease in the average risk. The average path risk decreases by 79.09%, with a 52.90% decrease from $M_{risk} = 0$ to $M_{risk} = 10$ and a 26.19% decrease from $M_{risk} = 11$ to $M_{risk} = 20$. Customer service completion remained at 100% when $M_{risk} \leq 10$. Due to the overlap between customer location and risk area, when $M_{risk} \geq 11$, the coefficient of risk cost was much higher than customer service benefit, drones tended to avoid risk instead of serving customers in the risk area, leading to a decrease in customer service completion rate, which decreased by 25% when $M_{risk} = 20$.

With the increase of M_{risk}, drones tend to move away from the risk area, leading to an increase in path length. Due to customer demand, the drone still needs to enter the customer area while avoiding the risk area, so the path length grows faster with an increase of 16.37% when $M_{risk} \leq 10$. In the stage of $M_{risk} = 11$ to $M_{risk} = 20$, the influence of the customer is significantly weaker than the risk area, which can be proved by the 25% drop in demand completion analysed above. A sufficiently large M_{risk} value made the drone less likely to extend the detour distance, which can be demonstrated by the average risk value decreasing by 26.19% from $M_{risk} = 11$ to $M_{risk} = 20$, which is about 50% less than $M_{risk} = 0$ to $M_{risk} = 10$. The reasons mentioned above eventually led to a significant slowdown in the growth of drone path length, which increased by only 0.5% from $M_{risk} = 11$ to $M_{risk} = 20$. The results for the parameter M_{risk} are shown in Figure 13.

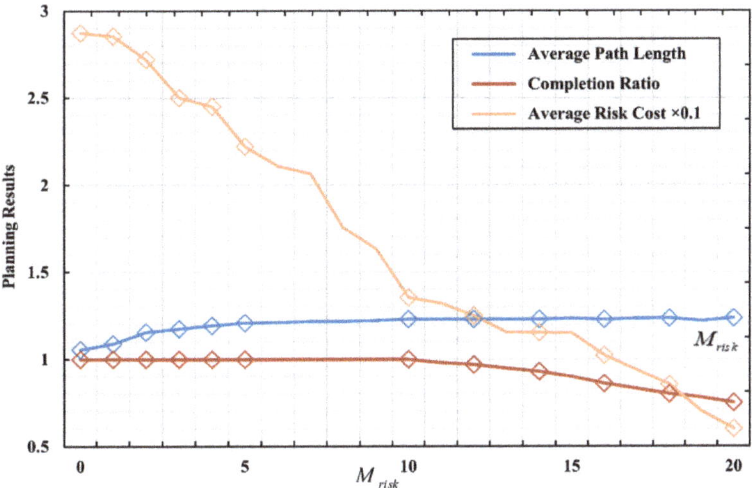

Figure 13. Sensitivity analysis of the parameter M_{risk} ($M_{benefit} = 0.5$).

According to the analysis of the above results, it is evident that the adjustment of the coefficients M_{risk} and $M_{benefit}$ changes the preference for risk and benefit in path planning, which leads to significant differences in the parameters of the planning results (average path length, service completion ratio, average risk cost). It also further demonstrates the importance of our proposed path planning approach that considers balancing risk cost and service benefit, which can reflect the process of completing customer service while avoiding risks in the actual operation of logistics drones.

Another critical parameter affecting drone paths in complex urban environments is the acceptable service range for customers. Due to the fact that customer locations often overlap with high risk cost areas such as buildings, pedestrians, and vehicles, part of the customer demand may be discarded if the acceptable service range decreases and the drone path needs to traverse more high-risk cost areas to complete the service. Therefore, we further analyse the impact of acceptable service range R on path planning results.

According to the results shown in Table 6, it can be seen that as the acceptable service range decreases, the overall service completion decreases significantly due to balancing the risk cost and service benefit, and the path risk cost will decrease due to ignoring some customers. The acceptable service range decreases from 200 m to 100 m, and the service completion decreases by 57.00%, while the average path risk cost only increases by 12.26%. This is because the reduction of the acceptable service range causes the drone needs to traverse more high-risk cost areas to complete the service, which is detrimental to the goal of balancing risk cost and service benefit, so the drone discards part of the customer requirement. When $R = 150m$, only the service to Customer 2 was dropped due to balancing risk cost and service benefit, so service completion decreased. However, providing service to Customers 4, 5, and 6 leads to a 5.97% increase in risk cost due to the reduction in the acceptable service range.

Table 6. Sensitivity analysis of the parameter R.

R/m	M_{risk}	Completion Ratio	Average Risk Cost	Unserved Customers
200	20	1	6.20	None
150	20	0.90	6.57	Customer 2
100	20	0.43	5.44	Customers 2, 4, 5, and 6
100	5	0.90	11.35	Customer 2

The variation of completion degree in customer demand shows that the reduction of the acceptable service range does not affect the completion degree for Customers 1 and 3, which do not overlap with the high-risk cost area. Meanwhile, Customers 2, 4, 5, and 6, which overlap with high-risk cost areas, were not served. The comparative experimental results of adjusting the risk cost preference parameter M_{risk} also demonstrate that the purpose of discarding some customer demands is to balance the service benefits and risk costs. For the case that the acceptable service range was 100 m, the drone path accepted a higher risk cost when $M_{risk} = 5$; thus, Customers 4, 5, and 6 that were not served at $M_{risk} = 20$ could be served, and the service completion was improved to 90%. While path risk costs increased by 83.06% due to serving customers whose acceptable service ranges overlap with high risk cost areas.

In summary, it is important to improve the acceptable service range of customers for logistics drone risk management. Logistics drone companies also need to adjust the risk cost and service benefit preferences according to the acceptable service range and customers' location in order to ensure service quality.

4.5. Algorithm Effectiveness Comparison

We propose path point generation rules with an exploration strategy to adapt the scenario for multi-drone logistics operations in urban environments and verify the effectiveness of our algorithm in this section. The A* algorithm [55] is a standard algorithm for path planning by limiting the selectable actions of drone movements. We consider that both the A* algorithm and the algorithm proposed in this paper are node-based optimal algorithms, while the A* algorithm performs local merit by limiting the candidate nodes at the current location when selecting the next location. The algorithm in this paper is a global merit algorithm that selects any point within the map in the form of probability by setting heuristic factors. So, we take the A* algorithm as the benchmark for comparison, which is a more relevant comparison. In addition, we choose the genetic algorithm as another comparison algorithm because it treats paths as individuals and selects individuals with higher fitness through the calculation of individual fitness functions. This approach is consistent with the global merit strategy for node search, and the comparison shows the effectiveness of the method in this paper more clearly.

Based on the A* algorithm, at each step of path planning, a drone can choose among one of a fixed number of equally distributed directions to move one unit step. In our experiments, we set up eight directions for drones and thus have eight candidate nodes p_i for a drone to choose from at each step. To apply the A* algorithm in the constructed environment, it is required to specify the cost-benefit function of performing an optional action at the current location, as shown in Equation (24)

$$A_{iP_i} = d_{i,p_i} + M_{risk}R_{P_i} + \frac{M_{benefit}}{1 + \sum_{P_i \in s(p_j, R)} C_b(C_{demand-j})} \tag{24}$$

where i is the current position, p_i is the candidate point specified for the next step, d_{i,p_i} is the Euclidean distance between i and p_i, R_{P_i} and $C_b(C_{demand-j})$ are risk cost and service benefit consistent with the previous definition.

For genetic algorithms, we need to specify the calculation of individual fitness, as shown in Equation (25)

$$A^i = \frac{1}{d^i} + \frac{1}{M_{risk}R^i} + \frac{C_b^i\left(C_{demand}^0 - C_{demand}^{end}\right)}{M_{benefit}} \tag{25}$$

where i is the i-th path individual, d^i is the path length of individual i, R^i is the total path risk cost of individual i, C_{demand}^{end} is the remaining customer demand after the drone provides service by path, C_{demand}^0 is the initial total customer demand, $C_b^i\left(C_{demand}^0 - C_{demand}^{end}\right)$ is the benefit of the service completed by path i.

According to the comparison with the results of the genetic algorithm, the quality of the results obtained by the algorithm in this paper is basically consistent with the method of directly generating the overall path. There was a difference in path length of about 2% and a difference in risk cost of about 5%. The similarities in the values and trends of the results demonstrate that applying the global merit strategy in the node search process can improve the quality of the results. Service completion is the most important index to measure the result of logistics service path planning in a complex urban environment. As shown in Tables 7 and 8, compared with the A* algorithm, the service completion of our algorithm is significantly higher, with an improvement of 20–40%. The path planning results of the algorithm in this paper have a slightly higher average risk than the A* algorithm. While the difference between the two algorithms' path lengths remains between 1 and 2%, proving that the increase in risk basically comes from the existence of an overlap between the customer location and the risk area.

Table 7. Comparison of the planning results of the two algorithms by varying $M_{benefit}$.

$M_{benefit}$ ($M_{risk}=5$)	Proposed Algorithm			A* Algorithm			Genetic Algorithm		
	CompletionRatio	Path Length	Risk Cost	CompletionRatio	Path Length	Risk Cost	CompletionRatio	Path Length	Risk Cost
0.01	0.66	1.08	1.41	0.45	1.10	0.50	0.65	1.10	1.50
0.05	0.69	1.14	1.57	0.49	1.14	0.51	0.69	1.15	1.64
0.1	0.73	1.14	1.60	0.54	1.13	0.55	0.74	1.15	1.70
0.15	0.74	1.17	1.79	0.54	1.15	0.55	0.74	1.17	1.84
0.2	1.00	1.17	1.83	0.61	1.15	0.54	1	1.18	1.91
0.25	1.00	1.17	1.83	0.61	1.16	0.54	1	1.18	1.91
0.5	1.00	1.21	2.42	0.61	1.19	0.54	1	1.22	2.50
1	1.00	1.21	3.62	0.61	1.19	0.54	1	1.23	3.81
2	1.00	1.21	3.63	0.61	1.19	0.54	1	1.23	3.81
5	1.00	1.21	3.63	0.61	1.19	0.54	1	1.23	3.81

The mentioned indexes show that the proposed search rule promotes the drone's exploration of the environment compared to the A* algorithm. Furthermore, the shortest path could be found based on the guarantee of completing the service. In addition, the algorithm can flexibly respond to the change of risk factor k, which ensures the risk tolerance of drones. It avoids the situation that the original algorithm cannot complete the path planning in the complex environment.

As M_{risk} increased significantly, the drone was more sensitive to risks in the environment. This is equivalent to a more complex risk area in the environment, which requires more detours to avoid, and the A* algorithm fails to find a valid path and deadlocks in this situation. The path search rule proposed in this study still guarantees 100% service completion, while the average path length and average risk have a smooth change. It indicates that the solving ability of our algorithm is still acceptable.

Table 8. Comparing the planning results of the two algorithms by varying M_{risk}.

M_{risk} ($M_{benefit}$=0.5)	Proposed Algorithm			A* Algorithm			Genetic Algorithm		
	CompletionRatio	Path Length	Risk Cost	CompletionRatio	Path Length	Risk Cost	CompletionRatio	Path Length	Risk Cost
1	1.00	1.09	2.86	0.58	1.10	0.73	1	1.12	2.99
2	1.00	1.16	2.72	0.58	1.15	0.62	1	1.18	2.90
3	1.00	1.18	2.50	0.59	1.16	0.61	1	1.13	2.83
4	1.00	1.20	2.45	0.60	1.19	0.61	1	1.21	2.69
5	1.00	1.21	2.42	0.61	1.19	0.54	1	1.25	2.56
10	1.00	1.23	1.65				1	1.27	1.87
20	0.75	1.24	0.60				0.76	1.28	0.73
50	0.60	1.26	0.59				0.58	1.29	0.71
75	0.60	1.26	0.59				0.57	1.29	0.70
100	0.60	1.26	0.58				0.57	1.29	0.70

As for the benefit coefficient $M_{benefit}$, the search rule proposed in this study can better show the change in the preference for customer demand. The service completion was 100% when $M_{benefit} = 0.2$, which increased by 34% compared with $M_{benefit} = 0.01$, and remained at 100% completion. For the A* algorithm, when $M_{benefit} = 0.2$, the service completion was 61%, with an increase of 16%, indicating that the A* algorithm is worse in response to the demand factor $M_{benefit}$. The main reason for this difference is that the path search rule proposed in this paper can guarantee a comprehensive exploration of the environment.

4.6. External Validity Analysis

The effectiveness of the proposed path planning model needs to be validated in balancing risk costs and service benefits when extended to other urban environments. In this work, external validity is performed, and 100 different urban environments are randomly generated.

Randomly generated pedestrian density and vehicle density were in the range $[5, 25] \times 10^3 (people/km^2)$ [56]. The buildings in all environments had randomly generated variance σ. The flight area range was 1000×1000m and was divided into 50×50 grid areas. We set up the customer area to be served and assigned a random initial demand of $d_j \in (0, 10]$. The results of path planning without considering risk and the cost-benefit model proposed in this paper were calculated separately in 100 independent environments. The path risk costs obtained from these two methods were compared to demonstrate the risk mitigation effect of the model in this paper. The total risk cost for each simulation is shown in Figure 14. Among the 100 generated samples (urban model), the average customer service completion rate of the paths planned by the model in this study reached 98.68%, and all showed good risk mitigation effects.

To test the effectiveness of risk mitigation, the results were further statistically analysed to calculate the percentage of risk mitigation at the 95% confidence level. Two sample groups were considered, the risk-mitigated group (Group 1) and the risk-unmitigated group (Group 2). There were 100 samples within each group. Due to the large sample size $(n_1, n_2 \gg 30)$, a normal distribution could be used to calculate confidence intervals. The results of calculating the sample means ($\overline{x_1}$ and $\overline{x_2}$) and sample variances (s_1^2 and s_2^2) for the two groups are shown in Table 9. μ_1 and μ_2 are the population means. $(\mu_2 - \mu_1)/\overline{x_2}$ is the confidence interval for the risk mitigation effect, where $\mu_2 - \mu_1$ was estimated by the following equation: $(\overline{x_2} - \overline{x_1}) \pm Z_{\alpha/2}\sqrt{s_1^2/n_1 + s_2^2/n_2}$.

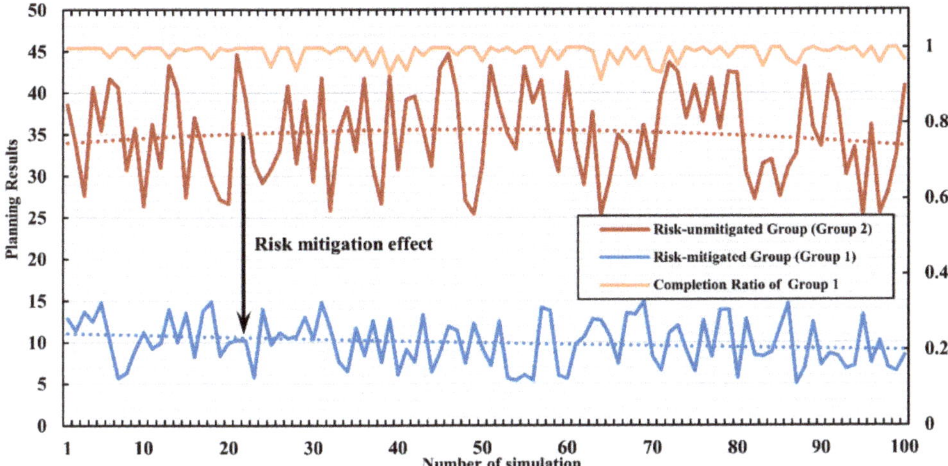

Figure 14. Mitigation effects of path risk in 100 urban environments.

Table 9. Statistical analysis parameters of the risk-mitigated group and the risk-unmitigated group.

	Group 1	Group 2
Sample Size	$n_1 = 100$	$n_2 = 100$
Sample Means	$\overline{x_1} = 10.02$	$\overline{x_2} = 34.99$
Sample Variances	$s_1^2 = 8.14$	$s_2^2 = 31.13$

The results show the 95% confidence interval for the risk mitigation effect $(\mu_2 - \mu_1)/\overline{x_2} \in [0.6962, 0.7312]$. In any urban environment, path planning with the cost-benefit model proposed in this paper mitigates the average total risk by $[69.62\%, 73.12\%]$ at the 95% confidence level and can effectively reduce the risk cost of path planning results for all types of urban environments based on customer service completion.

5. Conclusions

Owing to the complexity of the urban environment, it is still a challenging task to mitigate the security threats from drones while ensuring service completion in logistics drone path planning. To address this issue, we propose a model that couples customers and risk, and guides path planning in logistics drones by means of quantifying and balancing the risk cost and service benefit. The results show that compared to traditional approaches considering only obstacle avoidance, the model proposed in this paper can capture various risks and customers dispersed in all types of urban patterns and mitigate the path risk while ensuring customer service completion. In addition, the different risk and benefit preferences would greatly affect the path planning results, which further demonstrates the importance of our proposed model for balancing risk cost and service benefit. Furthermore, the proposed path search rules with heuristic factors outperform the quality of results in traditional algorithms in complex environments. It is well known that other customer demands and risk areas also exist. For instance, convective weather also has a significant influence on the integrated risk model. In addition, the customer demand model could also consider some more conditions, such as the time window for acceptable service, customer location movement, etc. Therefore, the present work would be further investigated in subsequence research to build a more realistic logistics drone path planning model driven by more customer demands and risk areas.

Author Contributions: Conceptualization, J.L. and Q.S.; methodology, J.L. and Q.S.; formal analysis, J.L. and R.L.; writing—original draft preparation, J.L., R.L. and X.G.; writing—review and editing, J.Z.; supervision, Q.S. All authors have read and agreed to the published version of the manuscript.

Funding: This research was supported by the National Natural Science Foundation of China (Grant no. 71874081), the Natural Science Foundation of Jiangsu Province (Grant no. BK20201296), the Fundamental Research Funds for Nanjing University of Aeronautics and Astronautics (Grant no. NS2022065), and the Qing Lan Project of the Jiangsu Province.

Institutional Review Board Statement: Not applicable.

Informed Consent Statement: Not applicable.

Data Availability Statement: Not applicable.

Conflicts of Interest: The author declares no conflict of interest.

References

1. Placek, M. Global Parcel Shipping Volume between 2013 and 2027. Available online: https://www.statista.com/statistics/1139910/parcel-shipping-volume-worldwide/ (accessed on 2 October 2022).
2. Ranathunga, M.I.D.; Wijayanayake, A.N.; Niwunhella, D.H.H. Simulation-Based Efficiency Assessment of Integrated First-Mile Pickup and Last-Mile Delivery in an E-Commerce Logistics Network. In Proceedings of the 2022 International Research Conference on Smart Computing and Systems Engineering (SCSE), Colombo, Sri Lanka, 1 September 2022; pp. 246–253.
3. Borghetti, F.; Caballini, C.; Carboni, A.; Grossato, G.; Maja, R.; Barabino, B. The Use of Drones for Last-Mile Delivery: A Numerical Case Study in Milan, Italy. *Sustainability* **2022**, *14*, 1766. [CrossRef]
4. Kim, J.J.; Kim, I.; Hwang, J. A change of perceived innovativeness for contactless food delivery services using drones after the outbreak of COVID-19. *Int. J. Hosp. Manag.* **2021**, *93*, 102758. [CrossRef]
5. Eun, J.; Song, B.D.; Lee, S.; Lim, D.-E. Mathematical Investigation on the Sustainability of UAV Logistics. *Sustainability* **2019**, *11*, 5932. [CrossRef]
6. Aurambout, J.-P.; Gkoumas, K.; Ciuffo, B. Last mile delivery by drones: An estimation of viable market potential and access to citizens across European cities. *Eur. Transp. Res. Rev.* **2019**, *11*, 30. [CrossRef]
7. Bányai, T. Impact of the Integration of First-Mile and Last-Mile Drone-Based Operations from Trucks on Energy Efficiency and the Environment. *Drones* **2022**, *6*, 249. [CrossRef]
8. Levasseur, B.; Bertrand, S.; Raballand, N.; Viguier, F.; Goussu, G. Accurate Ground Impact Footprints and Probabilistic Maps for Risk Analysis of UAV Missions. In Proceedings of the 2019 IEEE Aerospace Conference, Big Sky, MT, USA, 2–9 March 2019.
9. Yin, C.; Xiao, Z.Y.; Cao, X.B.; Xi, X.; Yang, P.; Wu, D.P. Offline and Online Search: UAV Multiobjective Path Planning Under Dynamic Urban Environment. *IEEE Internet Things J.* **2018**, *5*, 546–558. [CrossRef]
10. Nikodem, F.; Bierig, A.; Dittrich, J. The New Specific Operations Risk Assessment Approach for UAS Regulation Compared to Common Civil Aviation Risk Assessment. In Proceedings of the DLRK 2018, Friedrichshafen, Deutschland, 4–6 September 2018.
11. Shi, K.X.; Zhang, H.W.; Zhang, Z.Z.; Zhou, X.G. The Algorithm of Terminal Logistics Path Planning Based on TSP Problem. In Proceedings of the International Conference on Artificial Intelligence and Computer Engineering (ICAICE), Beijing, China, 23–25 October 2020; pp. 130–133.
12. Murray, C.C.; Chu, A.G. The flying sidekick traveling salesman problem: Optimization of drone-assisted parcel delivery. *Transp. Res. Part C Emerg. Technol.* **2015**, *54*, 86–109. [CrossRef]
13. Agatz, N.; Bouman, P.; Schmidt, M. Optimization Approaches for the Traveling Salesman Problem with Drone. *Transp. Sci.* **2018**, *52*, 965–981. [CrossRef]
14. de Freitas, J.C.; Vaz Penna, P.H. A variable neighborhood search for flying sidekick traveling salesman problem. *Int. Trans. Oper. Res.* **2020**, *27*, 267–290. [CrossRef]
15. Schermer, D.; Moeini, M.; Wendt, O. *A Variable Neighborhood Search Algorithm for Solving the Vehicle Routing Problem with Drones*; Technical Report BISOR-02/2018; Technical University of Kaiserslautern: Kaiserslautern, Germany, 2018.
16. Mitici, M.; Blom, H.A.P. Mathematical Models for Air Traffic Conflict and Collision Probability Estimation. *IEEE Trans. Intell. Transp. Syst.* **2019**, *20*, 1052–1068. [CrossRef]
17. Bertrand, S.; Raballand, N.; Viguier, F. Evaluating Ground Risk for Road Networks Induced by UAV Operations. In Proceedings of the 2018 International Conference on Unmanned Aircraft Systems (ICUAS), Dallas, TX, USA, 12–15 June 2018.
18. Koh, C.H.; Low, K.H.; Li, L.; Zhao, Y.; Deng, C.; Tan, S.K.; Chen, Y.; Yeap, R.C.; Li, X. Weight threshold estimation of falling UAVs (Unmanned Aerial Vehicles) based on impact energy. *Transp. Res. Part C Emerg. Technol.* **2018**, *93*, 228–255. [CrossRef]
19. Clothier, R.A.; Williams, B.P.; Hayhurst, K.J. Modelling the risks remotely piloted aircraft pose to people on the ground. *Saf. Sci.* **2018**, *101*, 33–47. [CrossRef] [PubMed]
20. Zhang, J.; Yang, S.; Li, Y.; Wu, X. A flight conflict detection model for UAV based on four-dimensional coordinates. *Int. J. Comput. Appl. Technol.* **2019**, *60*, 51–56. [CrossRef]

21. Liu, Z. Collision risk of crossing airlines at the same altitude based on REICH model. *J. Shenzhen University. Sci. Eng.* **2020**, *37*, 136–142. [CrossRef]
22. Brooker, P. Airborne Separation Assurance Systems: Towards a work programme to prove safety. *Saf. Sci.* **2004**, *42*, 723–754. [CrossRef]
23. Zhang, Z.; Shi, R. Research on Collision Risk Model in Free Flight Based on Position Error. In *Geo-Informatics in Resource Management and Sustainable Ecosystem, GRMSE 2015: Communications in Computer and Information Science*; Springer: Berlin/Heidelberg, 2016; Volume 569, pp. 795–803.
24. Bauer, P.; Hiba, A.; Bokor, J.; Zarándy, Á. Three Dimensional Intruder Closest Point of Approach Estimation Based-on Monocular Image Parameters in Aircraft Sense and Avoid. *J. Intell. Robot. Syst.* **2019**, *93*, 261–276. [CrossRef]
25. Chakravarthy, A.; Ghose, D. Generalization of the collision cone approach for motion safety in 3-D environments. *Auton. Robot.* **2012**, *32*, 243–266. [CrossRef]
26. Fan, S.; Wu, G.; Wang, L.; Liu, Y.; Li, L.; Qi, Q. Path Planning for Flight Vehicle by Using Modified Artificial Potential Field Method. *Aerosp. Control* **2018**, *36*, 50–54.
27. Tang, J.; Pan, R.; Zhou, S.; Wang, W.; Zou, R. An Improved Artificial Potential Field Method Integrating Simulated Electric Potential Field. *Electron. Opt. Control* **2020**, *27*, 69–73.
28. Ma, N.; Cao, Y.; Wang, X.; Wang, Z.; Sun, H. A Fast path re-planning method for UAV based on improved A* algorithm. In Proceedings of the 3rd International Conference on Unmanned Systems (ICUS), Harbin, China, 27–28 November 2020; pp. 462–467.
29. Koenig, S.; Likhachev, M. Improved fast replanning for robot navigation in unknown terrain. In Proceedings of the 2002 IEEE International Conference on Robotics and Automation (Cat. No.02CH37292), Washington, DC, USA, 11–15 May 2002; Volume 1, pp. 968–975.
30. De Filippis, L.; Guglieri, G.; Quagliotti, F. Path Planning Strategies for UAVS in 3D Environments. *J. Intell. Robot. Syst.* **2012**, *65*, 247–264. [CrossRef]
31. Nash, A.; Koenig, S.; Tovey, C. Lazy Theta*: Any-Angle Path Planning and Path Length Analysis in 3D. *Proc. AAAI Conf. Artif. Intell.* **2010**, *24*, 147–154. [CrossRef]
32. Hrabar, S. 3D Path Planning and Stereo-based Obstacle Avoidance for Rotorcraft UAVs. In Proceedings of the 2008 IEEE/RSJ International Conference on Intelligent Robots and Systems, Nice, France, 22–26 September 2008; pp. 807–814.
33. Geem, Z.W.; Kim, J.; Loganathan, G.V. A New Heuristic Optimization Algorithm: Harmony Search. *Simulation* **2001**, *76*, 60–68. [CrossRef]
34. Wang, X.; Meng, X. UAV Online Path Planning Based on Improved Genetic Algorithm. In Proceedings of the 38th Chinese Control Conference (CCC), Guangzhou, China, 27–30 July 2019; pp. 4101–4106.
35. Shahidi, N.; Esmaeilzadeh, H.; Abdollahi, M.; Lucas, C. Memetic Algorithm Based Path Planning for a Mobile Robot. *Int. J. Inf. Technol.* **2004**, *1*, 56–59.
36. Foo, J.; Knutzon, J.; Kalivarapu, V.; Oliver, J.; Winer, E. Path Planning of Unmanned Aerial Vehicles Using B-Splines and Particle Swarm Optimization. *J. Aerosp. Comput. Inf. Commun.* **2009**, *6*, 271–290. [CrossRef]
37. Huan, L.; Ning, Z.; Qiang, L. UAV Path Planning Based on an Improved Ant Colony Algorithm. In Proceedings of the 4th International Conference on Intelligent Autonomous Systems (ICoIAS), Wuhan, China, 14–16 May 2021; pp. 357–360.
38. Hassanzadeh, I.; Madani, K.; Badamchizadeh, M.A. Mobile robot path planning based on shuffled frog leaping optimization algorithm. In Proceedings of the 2010 IEEE International Conference on Automation Science and Engineering, Toronto, ON, Canada, 21–24 August 2010; pp. 680–685.
39. Guan-Ting, T.; Jih-Gau, J. Path Planning and Obstacle Avoidance Based on Reinforcement Learning for UAV Application. In Proceedings of the 2021 International Conference on System Science and Engineering (ICSSE), Ho Chi Minh City, Vietnam, 26–28 August 2021; pp. 352–355.
40. Ju-Shan, L.; Hsiao-Ting, C.; Rung-Hung, G. Decentralized Planning-Assisted Deep Reinforcement Learning for Collision and Obstacle Avoidance in UAV Networks. In Proceedings of the 2021 IEEE 93rd Vehicular Technology Conference (VTC2021-Spring), Helsinki, Finland, 25–28 April 2021; pp. 1–7.
41. Park, J.; Jang, S.; Shin, Y. Indoor Path Planning for an Unmanned Aerial Vehicle via Curriculum Learning. In Proceedings of the 2021 21st International Conference on Control, Automation and Systems (ICCAS), Jeju, Republic of Korea, 12–15 October 2021; pp. 529–533.
42. Ghasri, M.; Maghrebi, M. Factors affecting unmanned aerial vehicles' safety: A post-occurrence exploratory data analysis of drones' accidents and incidents in Australia. *Saf. Sci.* **2021**, *139*, 105273. [CrossRef]
43. Neogi, N.A.; Quach, C.C.; Dill, E. A Risk Based Assessment of a small UAS Cargo Delivery Operation in Proximity to Urban Areas. In Proceedings of the IEEE/AIAA 37th Digital Avionics Systems Conference (DASC), London, UK, 23–27 September 2018; pp. 111–119.
44. Enayati, S.; Goeckel, D.L.; Houmansadr, A.; Pishro-nik, H. Privacy-Preserving Path-Planning for UAVs. In Proceedings of the 2022 International Symposium on Networks, Computers and Communications (ISNCC), Shenzhen, China, 19–22 July 2022; pp. 1–6.

45. Bertrand, S.; Raballand, N.; Viguier, F.; Muller, F. Ground Risk Assessment for Long-Range Inspection Missions of Railways by UAVs. In Proceedings of the 2017 International Conference on Unmanned Aircraft Systems (ICUAS), Miami, FL, USA, 13–16 June 2017.

46. Hu, X.; Pang, B.; Dai, F.; Low, K.H. Risk Assessment Model for UAV Cost-Effective Path Planning in Urban Environments. *IEEE Access* **2020**, *8*, 150162–150173. [CrossRef]

47. Primatesta, S.; Rizzo, A.; la Cour-Harbo, A. Ground risk map for Unmanned Aircraft in Urban Environments. *J. Intell. Robot. Syst.* **2020**, *97*, 489–509. [CrossRef]

48. Rappaport, J. Consumption Amenities and City Population Density. *Reg. Sci. Urban Econ.* **2008**, *38*, 533–552. [CrossRef]

49. Yao, Y.; Liu, X.; Li, X.; Zhang, J.; Zhaotang, L.; Mai, K.; Zhang, Y. Mapping fine-scale population distributions at the building level by integrating multi-source geospatial big data. *Int. J. Geogr. Inf. Sci.* **2017**, *31*, 1220–1244. [CrossRef]

50. Zhang, N.; Zhang, M.; Low, K. 3D path planning and real-time collision resolution of multirotor drone operations in complex urban low-altitude airspace. *Transp. Res. Part C Emerg. Technol.* **2021**, *129*, 103123. [CrossRef]

51. Usui, H. Statistical distribution of building lot depth: Theoretical and empirical investigation of downtown districts in Tokyo. *Environ. Plan. B Urban Anal. City Sci.* **2019**, *46*, 1499–1516. [CrossRef]

52. Liu, C.; Yang, S.; Cui, Y.; Yang, Y. An improved risk assessment method based on a comprehensive weighting algorithm in railway signaling safety analysis. *Saf. Sci.* **2020**, *128*, 104768. [CrossRef]

53. Choudhry, A.; Moon, B.; Patrikar, J.; Samaras, C.; Scherer, S. CVaR-based Flight Energy Risk Assessment for Multirotor UAVs using a Deep Energy Model. In Proceedings of the 2021 IEEE International Conference on Robotics and Automation (ICRA), Xi'an, China, 30 May–5 June 2021.

54. Di Franco, C.; Buttazzo, G. Energy-aware Coverage Path Planning of UAVs. In Proceedings of the 2015 IEEE International Conference on Autonomous Robot Systems and Competitions, Vila Real, Portugal, 8–10 April 2015.

55. Sharma, K.; Swarup, C.; Pandey, S.K.; Kumar, A.; Doriya, R.; Singh, K.; Singh, T. Early Detection of Obstacle to Optimize the Robot Path Planning. *Drones* **2022**, *6*, 265. [CrossRef]

56. Wikipedia. List of Cities Proper by Population. Available online: https://en.wikipedia.org/wiki/List_of_cities_proper_by_population_density (accessed on 11 October 2022).

Hypothesis

A Framework to Develop Urban Aerial Networks by Using a Digital Twin Approach

Matteo Brunelli, Chiara Caterina Ditta * and Maria Nadia Postorino *

Department of Civil, Chemical, Environmental and Materials Engineering, Alma Mater Studiorum University of Bologna, 40126 Bologna, Italy
* Correspondence: chiaracaterina.ditt2@unibo.it (C.C.D.); marianadia.postorino@unibo.it (M.N.P.)

Abstract: The new concept of Urban Air Mobility (UAM) and the emergent unmanned aerial vehicles are receiving more and more attention by several stakeholders for implementing new transport solutions. However, there are several issues to solve in order to implement successful UAM systems. Particularly, setting a suitable framework is central for including this new transportation system into the existing ones—both ground and aerial systems. Regulation and definition of aerial networks, but also the characterization of ground facilities (*vertiports*) to allow passengers and freight to access the services are among the most relevant issues to be discussed. To identify UAM transportation networks, suitably connected with ground transportation services, digital twin models could be adopted to support the modelling and simulation of existing—and expected—scenarios with constantly updated data for identifying solutions addressing the design and management of transport systems. In this perspective, a digital twin model applied to an existing urban context—the city of Bologna, in northern Italy—is presented in combination with a novel air transport network that includes the third dimension. The 3D Urban Air Network tries to satisfy the principle of linking origin/destination points by ensuring safe aerial paths and suitable aerial vehicle separations. It involves innovative dynamic links powered by a heuristic cost function. This work provides the initial framework to explore the integration of UAM services into realistic contexts, by avoiding the costs associated with flight simulations in reality. Moreover, it can be used for holistic analyses of UAM systems.

Keywords: vertiports; 3D Urban Aerial Network; dynamic links

Citation: Brunelli, M.; Ditta, C.C.; Postorino, M.N. A Framework to Develop Urban Aerial Networks by Using a Digital Twin Approach. *Drones* **2022**, *6*, 387. https://doi.org/10.3390/drones6120387

Academic Editors: Ivana Semanjski, Antonio Pratelli, Massimiliano Pieraccini, Silvio Semanjski, Massimiliano Petri and Sidharta Gautama

Received: 1 October 2022
Accepted: 28 November 2022
Published: 29 November 2022

Publisher's Note: MDPI stays neutral with regard to jurisdictional claims in published maps and institutional affiliations.

1. Introduction and Background

Nowadays, Unmanned Aerial Vehicles (UAVs) are employed in several fields [1,2], and for this reason Advanced Air Mobility (AAM), also known as Urban Air Mobility (UAM), is expected to turn into reality in the near future. Currently, there are UAV applications in different fields such as smart agriculture, emergency and detection of buildings, bridges and other objects [3–6]. Moreover, UAM services for passengers—performed by electric vertical take-off and landing (eVTOL) aircraft—and freight transport, or some other services such as medical aid, are under development [7–12]. The use cases reported above are only a limited number of UAM application opportunities, but they have attracted the interest for AAM services by many stakeholders. For instance, in the Italian context, UAM services have been considered by the Italian Civil Aviation Authority (ENAC) as a relevant opportunity for civil aviation development [13], and a roadmap to develop UAM passenger services before 2030 has been set. In Europe, public acceptance regarding AAM services has been analysed in a European Union Safety Agency (EASA) report, which points out the importance of operations for medical aid and drone delivery applications as emerged by users' opinions [14].

Most parts of the research efforts on UAM topics are focusing on prototype development [15], but in order to implement an effective UAM service it is necessary to focus on a number of additional aspects, including management of the low airspace, ground facility location and integration with the existing transport systems. Cooperating UAVs or eVTOLs—hereafter named Aerial Cooperative Vehicles (ACVs)—will have to carry out their functions in the low airspace (around 1200 ft), i.e., in the ICAO class G [16], which is currently uncontrolled. Consequently, the implementation of a traffic management system in the Class G airspace is essential to ensure safe and efficient ACV flows—both manned and unmanned. Despite the existence of the air traffic management (ATM) system for traditional aviation, which provides a good basis for the development of a similar system in the lower space, there are technical and management problems to be solved to integrate the UAM system into the current urban multimodal system due to the heterogeneous characteristics of the urban airspace environment.

The National Aeronautics and Space Administration (NASA), in the USA, introduced a Concept of Operations (ConOps) to set safe procedures for Unmanned Aerial Systems (UAS) in the airspace organization, and then UAS Traffic Management (UTM) systems [17,18] have been described. UTM aims to allow safe and successful operations in the low airspace by achieving services such as airspace dynamic configuration, dynamic geo-fencing, weather information, route planning and flight separation [19]. Also in Europe, the U-space concept is being developed with the same goal, i.e., to support low-altitude air traffic management [20] and integrate it with current air traffic. A limited number of studies concerning a framework to develop aspects such as the organization and safety of this emerging transportation system have been proposed [21–23], but the research in this field is still evolving.

Another relevant factor for UAM service development is related to ground infrastructures, also called *vertiports*, which are aerial–ground interchange nodes. Municipalities that want to ensure a successful UAM system should carefully identify the suitable number and location of vertiports that have to be placed in a certain urban area [24]. Moreover, it is important to consider that the introduction of vertiports in urbanized areas would modify the city accessibility and lead to changes in the urban structure [25]. In the literature, some methods have been discussed to locate vertiports as near as possible to people's trip origins and destinations [26,27]. Other studies focus on the minimization of travel times and costs [28], whereas other approaches consider several variables such as the socioeconomic and environmental features of the study area [29]. Solving the vertiport location problem requires a large amount of information (also updated in real time, if possible), concerning both infrastructural and organizational aspects.

The use of a digital twin of the urban area can help finding solutions by taking into account most of the features considered above, both for introducing urban aerial networks and for establishing vertiport locations.

A digital twin can be defined as the digital model of a physical system that is regularly updated by the exchange of information between virtual and physical systems. In recent years, technological innovations have transformed the way to study and simulate processes and systems. Digital twins are a core part of this improvement, and they can be used to simulate, design and process several application scenarios [30]. The digital twin approach has already been explored to test, simulate, design and plan new solutions in the transport field, such as operations of Connected Autonomous Vehicles (CAVs) [31] or airport operations [32,33], as well as to recreate a virtual population in order to realize agent-based simulations [34]. To the best of the authors' knowledge, this is the first paper that explicitly introduces a digital twin model for designing and developing UAM applications, such as vertiport location problems, airspace and air vehicle management. More specifically, the purpose of this paper is to outline a novel framework that aims to improve the use of the digital twin as a tool for developing UAM systems integrated with existing ground transportation systems. A digital twin can be useful to identify the positioning of vertiports—which requires specific conditions—by incorporating a dedicated air network to identify flight paths suitable for ACVs. The paper also proposes a 3D aerial network model, which would benefit from the digital twin model of the urban environment for setting effective and safe air services. In particular, a simulation scenario in a real-world context is reported to show how the digital twin framework would operate to identify vertiports and air corridors on which to set UAM services.

The paper is organized as follows. Section 2 provides the methodological scheme of the paper. Section 3 aims to detail the digital twin concept by referring to the example of the digital twin of a real city (Bologna, in northern Italy). Section 4 provides the possible application of a digital twin for the vertiport location problem, while Section 5 discusses airspace management issues. Finally, Section 6 contains the discussions and the conclusions.

2. Methodology

Figure 1 describes the relationship between Digital Twin, Vertiport Location and 3D Urban Aerial Network (3D-UAN), which are the three interrelated elements here considered.

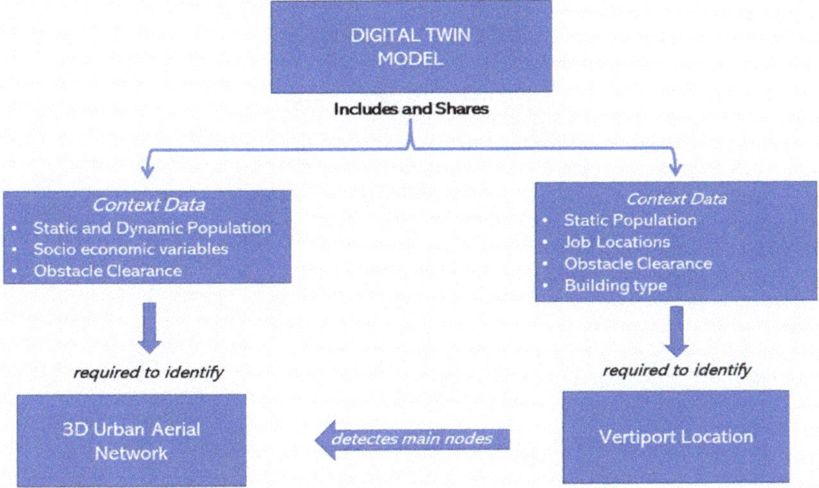

Figure 1. Relationship between Digital Twin, Vertiport Location and 3D Urban Aerial Network.

Digital twin models require data for representing physical systems. As for territorial systems, and urban contexts in particular, data generally refer to socio-economic and anthropic features as well as to the landform. As reported in Figure 1, the "context data" may be different depending on the nature of the problem that the digital twin model would help resolving. Here, the digital twin, and the related data, is thought to support the location of vertiports and the configuration of a suitable 3D-UAN. Some data are similar, although used for different goals. For solving the vertiport location problem, relevant data includes, for example, anthropic features—such as height of buildings—as well as socio-economic characteristics that would identify travel needs and then potential transport demand for UAM services. For the vertiports being located, the further step is setting the aerial network—here, the proposed 3D-UAN. The considered problems (3D-UAM and vertiport location) are not independent, in particular the location of vertiports will affect some of the nodes of the 3D-UAN (defined as "Fixed Nodes", see also Section 5) and then the corresponding links, which are the relationships between nodes and in this case represent urban aerial routes between relevant points. It is worth noting that urban aerial routes must satisfy several constraints within the urban context—such as technical constraints, safety aspects and privacy issues—and then, still based on the use of the digital twin model, additional nodes could be identified in order to set aerial routes that will meet such constraints.

From this perspective, the following sections will discuss the above issues, starting from the potentialities offered by digital twin models (Section 3) and then showing how to apply the framework of Figure 1 for the vertiport location problem (Section 4) and the 3D-UAN features (Section 5). Within the sections, examples are provided for a real context, i.e., the city of Bologna (northern Italy).

3. Digital Twin Perspectives for Transport System Simulations

The employment of digital twin approaches to simulate processes and evolutions of territorial realities has becoming more and more usual, with the aim of describing in a quantitative and detailed way the physical world to simulate and predict the effects of actions. The concept of digital twin was first formalized by Grieves [35] for some industrial contexts, but it is now used in several other fields such as road infrastructures [36], harbours [37], boats [30] and other civil sectors [38] to evaluate noise and air pollution, solar potential, and the effects of future urban development [39], or for simulating, planning and managing urban transportation networks [40,41].

Digital twins of complex systems, such as urban environments, could be fed by off-line data and/or data collected in real time, depending on the aim. With respect to a simple 3D urban model based only on geometrics data, the digital twin of an urban environment considers socio-economic and anthropic information that may be obtained by official sources (e.g., municipalities) or specific surveys—such as population and job densities, median income, working places, recreational areas, building type and points of interest. Such data require only periodic (not real time) updates. On the contrary, some other data—such as traffic flows, pollution levels, traffic noise and meteorological conditions—have to be collected in real time by sensors located suitably in the urban area.

A digital twin model could also be useful for the analysis and simulation of transportation systems that do not currently exist, such as UAM systems in metropolitan areas. In this perspective, the digital twin model may be applied to identify and evaluate the suitable areas for locating UAM ground infrastructures (i.e., vertiports) and also to run realistic simulations and to establish urban air corridors in the lower airspace for ACVs operations. Further details on these two aspects are provided in the following sections.

It worthwhile noting that the amount of gathered information related to the city characteristics affects the UAM scenarios that can be developed. The basic information useful for the vertiport positioning—particularly locations and also integration with ground transportation systems—and to identify aerial corridors in the lower airspace, both size and direction, are:

- Digital Terrain Model (DTM);
- Building height;
- Intended use of the building (or building type, such as schools, hospitals and churches);
- Socioeconomic variables (e.g., population density, workplaces, leisure spaces);
- Ground transportation infrastructures (e.g., roads, rails, terminals);
- Places where people can gather (e.g., stadium, main squares, pedestrian areas).

The use of these variables to locate vertiports and aerial corridors will be discussed in detail in the next sections.

As an example for the application of the proposed framework depicted in Figure 1, the digital twin model of the city of Bologna has been realized. Bologna is located in northern Italy, with a population of almost 390,000 inhabitants. Some characteristics that may represent relevant challenges for the introduction of UAM services are: (i) the relevant number of medieval towers in the 3D city centre (see the digital twin model in Figure 2); (ii) the land morphology, i.e., the northern part of the city is on a flat area, while the southern one is placed on a hilly surface (see Figure 3). These characteristics, which introduce a high number of obstacles for ACVs operations and vertiport location, make the city of Bologna an interesting case study.

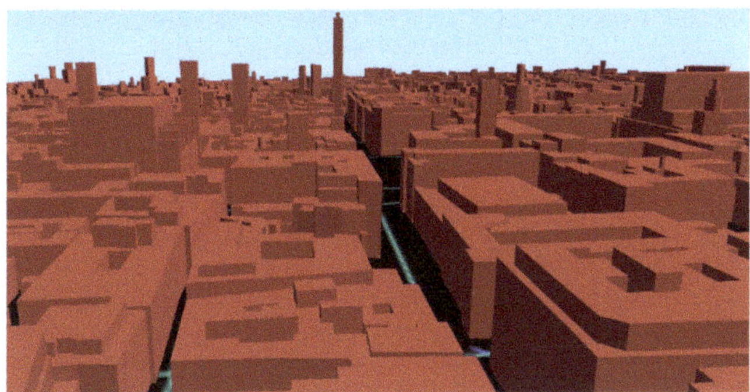

Figure 2. Part of the digital twin of Bologna city centre.

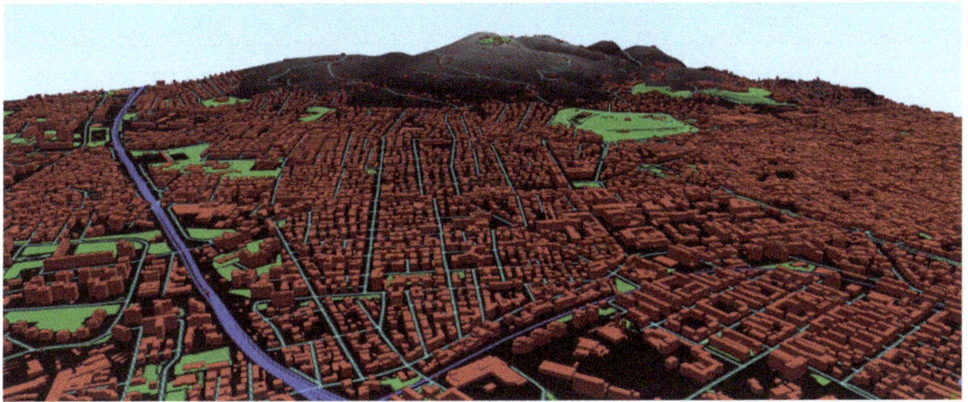

Figure 3. Example of different information in the Bologna digital twin.

To realize the city digital twin, a GIS tool has been used to merge and process a high amount of information that has been gathered by different sources (i.e., municipalities and cadastral data). In this first attempt, the model is based on freely available data. First, a DTM for the entire city was used as a base. Data on buildings such as height, type (e.g., schools, hospitals, churches, sport and leisure facilities), shape and size have been collected and added together with information on population density referring to the different city districts. Green areas such as parks have been modelled explicitly because they are places where people group. The main ground transportation infrastructures (roads, railways), the public transport stops and the airport area have then been added to the city model. Finally, uncovered rivers or streams have been included because ACVs could overfly them to minimize the risk of collision.

Figure 3 shows a graphic representation of the city digital twin, included the DTM on the background, where buildings are in red, roads and railways are in light blue and blue, respectively, and parks are in green. More data, included real time data, may be included depending on availability.

4. Vertiport Location

Vertiports are important infrastructures of UAM systems, and their location plays a crucial role for the development of the entire 3D-UAN in a given area. The location of vertiports depends on a wide range of factors, which include: (i) existing/planned obstacles, which have to be avoided by ACVs; (ii) available space, depending on vertiport layout; (iii) procedures, which are required to set suitable connections between the ground and the aerial transportation networks; (iv) ground travel times (accessibility); (v) externalities produced in the neighbourhood.

In the European context, the first regulation for vertiport design and placement has been published by EASA [42]. However, it mainly focuses on vertiport design and only takes into account the potential obstacles for ACV flight procedures in urban area. In the Cologne Airport case study [43], considerations such as the ones in the EASA report on obstacle clearance have been used to identify potential vertiport locations. The *Geofences* data, i.e., virtual volumes around obstacles (such as buildings) set to avoid collisions, especially for high urbanized cities, are considered among the most important ones to support obstacle avoidance by ACVs and vertiport design suitability [44]. Furthermore, as in current airport plan procedures, analyses of wind intensity and direction as well as externalities produced by noise emissions are considered main criteria to identify vertiport localization [45]. As for this latter, some studies suggest placing vertiports next to roads, railways or airports [29] or in cloverleaf interchanges [46] to reduce noise impacts. In the European context, people seem worried about UAM safety and noise pollution, which might represent a relevant limit to implement UAM services in the EU [14].

The proximity between vertiports and the main ground transportation nodes could be an important element to guarantee a good connection between the ground and aerial transportation networks, which is a successful element for the introduction of UAM services [47]. Other features that have been explored to choose vertiport locations involve rooftop flatness and its shape and size. As an example, these analyses, coupled with the socio-economic data of the study area, have been carried out for the city of New York [48], whereas other applications—such as for Munich and Los Angeles—mainly focus only on the socioeconomic characteristics of the area [29].

In some other analyses, obstacles or other physical hindrances have not been considered, while the aim has been to optimize some variables under specific criteria. For example, some studies have developed optimization algorithms to place vertiports in order to minimize the total travel time [49,50], maximize the number of passengers that can use UAM services [51,52], maximize the service provider revenues [51] and maximize the travel time saving [53].

All the studies cited above consider, in turn, only a limited number of variables useful to identify vertiport locations. However, an effective vertiport location based on several points of view requires a holistic approach that takes into account the different aspects, such as urban environment, rules and impacts.

From this perspective, a digital twin of a city could give significant support to design, place and manage vertiports as it contains detailed information that allows one to identify space occupancy, obstacle clearance, acoustic impacts and accessibility to ground services, but also potential demand based on socio-economic features as well as limitations for safety and security issues. In more detail, information on building heights allows the identification of potential spaces for locating vertiports, or, on the contrary, the areas that cannot be used for this aim. Similarly, obstacles (e.g., antennas) would prevent the use of such areas for locating vertiport. At the same time, population density, position of interchange ground nodes and suitable socio-economic indexes (e.g., high-income areas) would facilitate the identification of spaces potentially allocable to host vertiports. Integrating socio-economic and trip data, in particular concerning passengers' origin and destination points inside the study area, is useful for optimal vertiport positioning.

High-density areas associated with high-income, which are attractive for starting UAM services and then locating vertiports for accessing such services, generally correspond to densely urbanized areas, where space is rarely available. Furthermore, space availability does not refer simply to the area occupied by the vertiport infrastructure, but also to a suitable volume around it for allowing safe landing and take-off manoeuvres. The advantage of the digital twin city in this context is to use multi-criteria approaches for identifying vertiport location also facilitated by the visual representation.

Finally, externalities generated by the vertiport can be assessed based on some aerial traffic hypotheses. Moreover, as the vertiport will be accessed mainly by ground transportation modes, additional externalities due to the multi-modal ground transportation systems can be considered and estimated—e.g., noise impacts, which is the most perceived one by the population, can be assessed, and it might represent another criterion used to identify the optimal vertiport location with respect to residents.

As an example, Figure 4 reports a preliminary solution—still referring to the city of Bologna—where the vertiport location has been identified based on the information of its digital twin. Particularly, vertiport suitability has been identified by Fadhil [29], based on the weighted overlay of different features. In more detail, the suitability, S, may be defined as [54]:

$$S = \sum w_i x_i \cdot \prod C_j \tag{1}$$

where w_i is the weight assigned to factor i, X_i is the criterion score of factor i and C_j is a constraint. Factors and constraints are identified in the digital twin database of the case study (see also Section 3); in particular, the main factors are population density, job density, median income, the characteristics of ground transportation and the main point of interests, while the constraints are elements that prevent the location of a vertiport in a given area (e.g., space unavailability, obstacles, proximity to schools, and similar). As for weights, in this first attempt they have been identified based on some preliminary results, also in the work by Fadhil [29]. The selected factors for the weighted overlay are summarized in Table 1. A different radius for points of interest with respect to public transport stops—respectively 1000 m and 500 m—has been identified, as points of interest are expected to be less frequent than public transport stops, which generally are 500 m from each other in city central areas.

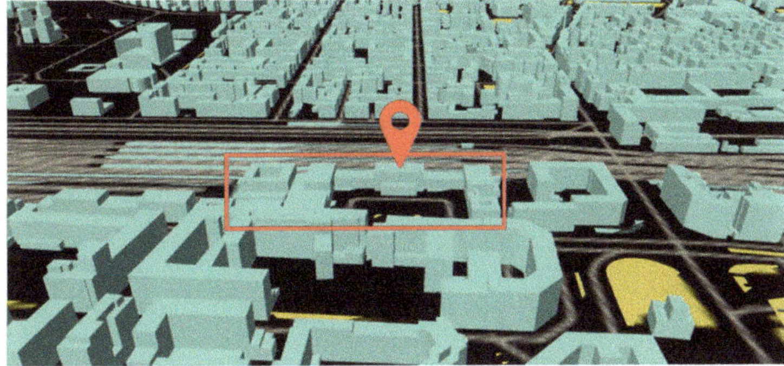

Figure 4. Example of vertiport location (Bologna digital twin model).

Table 1. Factors used in the vertiport location procedure.

Factors	Type of Information
Population Density	Residential population per km^2 for each census unit
Job Density	Number of job activities for each census unit
Median Income	Annual median income in a census unit
Ground Transportation	For each census zone, number of public transport (train, bus, sharing mobility) stops within a radius of 500 m
Points of Interest	For each census zone, number of points of interest (especially tourist attraction) within a radius of 1000 m

From Equation (1), a preliminary vertiport location has been found close to the main train station, which is characterized by a large square that would facilitate safe manoeuvres.

The identified area is also close to both the city centre, with several points of interest in the surroundings, and an important interchange node, with train and bus services that would improve intermodality between ground and aerial services. It is worth noting that the assigned weights and constrains used in Equation (1) reflect, from one side, the importance given to several factors and from the other side the unfeasibility of some locations based on criteria such as building *geofence*, obstacles and so on, as it emerges from the digital twin database. The area is also characterized by a high value of population and job density. Although high population and job densities might be considered a constraint for vertiport location, the considered *geofence* data would guarantee suitable distances from relevant buildings and at the same time a suitable, potential demand for UAM services. Based on the chosen factors, weights and constraints, the airport area, which is also close to the city, has not been identified as a potential solution for locating the vertiport. For example, in this case the population density is not so high as in the city centre, and intermodality is not as relevant as in the location close to the train station.

This brief example has shown how the digital twin framework depicted in Figure 1 could support the location of vertiports. While this solution has been found by considering simple constraints and weights coming from the literature, more precise solutions may be found by improving the set of data in the digital twin and by adding further constraints related to environmental impacts or minimization of access time to the vertiports, which in this example have not been considered as not all the data were available at this stage. As for environmental impacts in particular, several hypotheses on the expected traffic at the vertiport could provide different solutions, thus also suggesting capacity limits to guarantee good life quality levels to the community living around the vertiport.

Finally, it is worth noting that airports are expected to host vertiports inside or close to their neighbourhood [55], especially for aerial airport shuttle services. Mainly for airports close to urban areas (city airports)—but generally for airports serving a given territorial system—the digital twin should also include information on aircraft take-off and landing trajectories (and the consequent obstacle limitation surfaces) in order to consider the potential risks due to ACVs movements in airport area and the related possible interferences with traditional aviation. *Geofencing* data would also be a suitable solution in this case.

5. Urban Aerial Network Model

Ensuring effective and safe connectivity between trip origin and destination pairs by UAM services requires setting a suitable aerial network where ACVs will move. Particularly, vertiports represent access/egress points of UAM services, landing/take off spaces for ACVs and initial/end nodes of aerial links, which may also be combined for providing more complex aerial routes between some relevant vertiports. As depicted in Figure 1, the location of vertiports has a direct effect on the urban aerial network that would guarantee the aerial services in the area.

Among the considered factors and constraints, the appropriate location of vertiports also has to take into account the local features of the low airspace—e.g., fixed and dynamic obstacles, climatic conditions, protected zones. Again, the digital twin helps in identifying such aerial routes, which should be suitably high above buildings, natural or artificial obstacles and should guarantee safe flight conditions. In addition to the third aerial dimension, it is necessary to define a structure for the management of routes and in-flight operations. For this purpose, an aerial network model has been proposed in order to enable ACV [23]—i.e., drones, UAVs, eVTOLs—operations in uncontrolled airspace (class G). This model requires a corresponding set of data that should be included in the digital twin model.

5.1. Main Features of the Aerial Network Model

The aerial network model integrates the concept of dynamic air corridors (e.g., DDCs) [56] in a multilayer structure [57] in order to define a three-dimensional graph that involves the vertical dimension for ensuring trips between origin/destination points. Here a 3D-UAN model is proposed, given by the union of bi-dimensional graphs, G_L, in multiple layers, L, which includes the set of Fixed Nodes ($N_{F,L}$), a set of Transition Nodes ($N_{T,L}$) and a set of Dynamic Links (D_L) connecting the nodes and the layers.

The 3D Graph (Θ) model [23] is summarized by the following formulation:

$$\Theta = \cup_{L=\{1,\dots,n\}} G_L \cup D_{v,L} \tag{2}$$

where $G_L = (N_{F,L}, N_{T,L}, D_{h,L})$.

In more detail, Fixed Nodes correspond to vertiports, which are the access to egress from UAM services, and are located based on the digital twin of the urban system, as described in Section 3. Transition Nodes are set to allow horizontal crossings and shifting to an upper or lower layer; some of them may be located at the same coordinates as Fixed Nodes, except for the vertical coordinate. Again, the location of Transition Nodes—or just their vertical position if they correspond to Fixed Nodes—is obtained by the digital twin model, particularly by using the information related to height of buildings, potential obstacles, and no-fly zones, among others. Pairs of nodes (both fixed and transition) are connected by Dynamic Links, $d_{m,L}$, which belong to the set $D_L = \{d_{m,L}\} \mid m = \{1, 2 \dots P_L\}$, where P_L is the total number of links for layer L. More specifically, the dynamic link set consists of horizontal and vertical link subsets, respectively, $D_{h,L} = \{h_{mL}\} \subset D_L \mid m = \{1, 2 \dots\}$ and $D_{v,L} = \{v_{mL}\} \subset D_L \mid m = \{1, 2 \dots\}$. Such links are identified by ensuring safe flight conditions with respect to the external features, i.e., urban or extra-urban environment.

The proposed 3D-UAN model also includes a cost function, defined on each link, with the aim of providing minimum cost origin/destination connections, suitable ACV separation and in-flight safety. The following link cost function $c(T_t, T_g)$ has been defined for each link belonging to D_L:

$$c(T_t, T_g) = \begin{cases} T_{t_j} \text{ for } j = 1 \\ T_{t_j} + T_{g(j,\,j-1)} \; \forall \, j > 1 \end{cases} \tag{3}$$

where j is the j-th ACV using the dynamic link $d_{m,L}$ at a given time period; T_{t_j} is the travel time of j on $d_{m,L}$; $T_{g(j,\,j-1)}$ is the time gap between j and $j - 1$.

Depending on $d_{m,L}$—horizontal or vertical link—the travel time, $T_{t_{ij}}$, will change. By considering vertical links, if $j = 1$ the $T_{t_{ij}}$ may be climbing (T_{a_j}) or descent (T_{f_j}) time depending on the link direction—i.e., to upper layers or to lower layers. If there are several ACVs on the same link, i.e., $j > 1$, the time gap $T_{g(j,j-1)}$ guarantees suitable separation between two following ACVs along vertical links. For horizontal links, T_{t_j} is the running time, T_{r_j}, if $j = 1$, while if $j > 1$, the time gap $T_{g(j,j-1)}$ guarantees appropriate separation between two following ACVs. To assess the waiting time component, time gap $T_{g(j,j-1)}$ is assigned at a fixed node before ACV departure, to keep safe travel conditions among them.

Equation (3) may be re-written for different types of links as:

$$c_{h,L}(T_r, T_g) = \begin{cases} T_{r_j} \text{ for } j = 1 \\ T_{r_j} + T_{g(j,\,j-1)} \; \forall \, j > 1 \end{cases} \tag{4}$$

$$c_{v,L}(T_{a,f}, T_g) = \begin{cases} T_{a_j} \text{ for upper layer transitions, } j = 1 \\ T_{a_j} + T_{g(j,\,j-1)} \text{ for upper layer transitions, } j > 1 \\ \\ T_{f_j} \text{ for lower layer transitions, } j = 1 \\ T_{f_j} + T_{g(j,\,j-1)} \text{ for lower layer transitions, } j > 1 \end{cases} \tag{5}$$

Data regarding travel, climbing and descendent times; physical obstacles; day time (related to social economics habits); overflight zones; and environmental conditions—such as wind phenomena or meteorological conditions, which affect flight conditions—should be included in the digital twin model, together with real time information on aerial traffic flows in more advanced versions.

The cost function (4) is used to compute the minimum cost paths that each ACV will use to realize the trip between the origin and destination vertiports, and to ensure appropriate ACV safety separation (defined by $T_{g_{(j,j-1)}}$). Shortest paths based on such link cost function may be found by iterative search algorithms—such as Dijkstra [58] or A* [59].

Flight operations within the same layer, L, occur along horizontal dynamic links (connecting fixed and transition nodes) belonging to $D_{h,L}$. The vertical links belonging to $D_{v,L}$ guarantee some specific procedures—i.e., landing and take-off operations, as well as layer transitions to/from the upper or lower layers. As for the dynamic nature of both horizontal and vertical links, it consists of enabling/disabling them in compliance with data on environmental conditions and traffic capacity (e.g., operational delays, unfavourable weather conditions). In addition, the features of the dynamic links may change according to ACV size, which requires different features of $d_{m,L}$ cross sections. For example, large ACVs require a greater distance between the layers and vertical link length increase and changes in transition node positions in order to ensure suitable protection volumes around them [60].

5.2. Aerial Network Model Simulation Scenario

By considering the digital twin model of the metropolitan area of Bologna—which includes a wider area with respect to the example in Section 3 and also factors and constraints for implementing the 3D-UAN in the city—three vertiports have been identified (Figure 5), which are, respectively, at the airport (vertiport 1), at the main station (vertiport 2) and in a city area with high population, job density and median income (vertiport 3). It is worth noting that this result has been obtained by using still freely available data. More accurate results and possibly different vertiport locations and/or number might be obtained by using a more detailed database.

Figure 5. Vertiport location supporting 3D-UAN—case study.

Starting from the vertiport locations, the routes linking them have been set using Equation (1) and the information provided by the digital twin model about local orography and barriers. Where available, data about local environmental features have also been considered. Particularly, some constraints to avoid conflicts with traditional aviation have been introduced in the digital twin model. As for the urban environment, the presence of high buildings—such as historical towers and modern skyscrapers—has prevented the connection of vertiport 1 with the other remaining two with a straight line (see Figure 5).

Figure 6 depicts the scheme of the obtained 3D-UAN network, while link and vertiport features are reported in Tables 2 and 3, respectively. Due to the information contained in the digital twin model (i.e., DTM and building heights) and the introduced constraints to avoids obstacles along the dynamic air corridors, the altitude of the first layer (L_1) resulted at 200 m ASL (the highest building roof is at 184 m), while the following one (L_2) is at 250 m ASL (distance between L_1 and L_2 = 50 m).

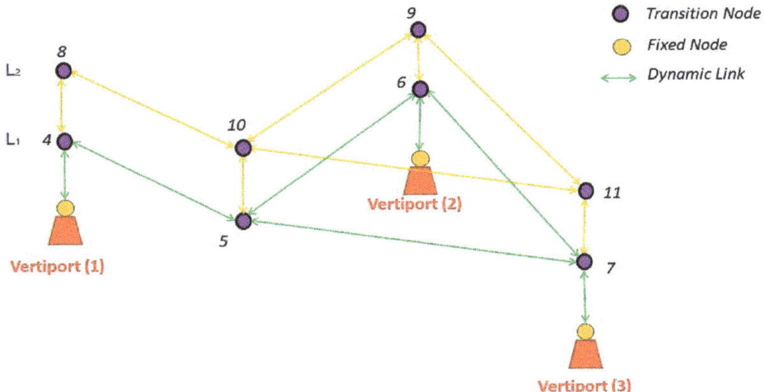

Figure 6. The 3D-UAN model for the case study.

Table 2. Dynamic link features in the 3D-UAN case study.

Bidirectional Link			
Horizontal	**Length (m)**	**Vertical**	**Length (m)**
4–5	3310	1–4	160
8–10	3310	2–6	140
5–7	6670	3–7	110
10–11	6670	4–8	50
5–6	1700	5–10	50
10–9	1700	6–9	50
6–7	5190	7–11	50
9–11	5190		

Table 3. Vertiport elevation.

Vertiport	Elevation ASL (m)
Node 1 (Airport)	40
Node 2 (Main Station)	60
Node 3 (Populated–High income area)	90

To provide a preliminary application of the 3D-UAN for the case study, the following hypotheses have been considered, particularly the existence of j on the $d_{m,L}$; $T_{g(j,\,j-1)}$ is the time gap between ACVs j and $j-1$., i.e., to the upper layers or to the lower layers. If there are multiple ACVs on the same link, i.e., $i > 1$, the gap time, $T_{g(i,i-1)}$, guarantees suitable separation between two following ACVs along vertical links.

For horizontal links, T_{t_i} is the running time, T_{r_i}, if $i = 1$, while if $i > 1$, the gap time is $T_{g(i,i-1)}$. To assess the waiting time component, the gap time, $T_{g(i,i-1)}$, the following are utilized: (i) advanced communication technologies (V2V, V2I/V2X) [61,62]; (ii) high precision ACVs on-board sensors, such as radars, Detection and Avoidance (DAA) [63]; (iii) data transmission networks (i.e., FANET system [64]).

The average cruising speed allowed on the horizontal links is considered equal to 90 km/h, while on the vertical links (included take-off and landing phases) it is 50 km/h. This average speed value was chosen based on the data distributed by Volocopter on their first prototypes [65], which should also support aspects related to flight safety.

Two scenarios have been investigated, which have been set as follows:

(a) scenario S_1: time gap between successive departures T_g = 180 s
(b) scenario S_2: time gap between successive departures T_g = 120 s.

For both scenarios, the dynamic link features are reported in Table 4. Furthermore, for S_1 and S_2, six ACVs have been considered connecting specific origin/destination pairs, and their shortest paths have been computed (Table 5). Additionally, a T_g value has been assigned to fixed nodes, particularly for each scenario: ACV $_{(1)}$ and ACV $_{(2)}$ takes-off at the same time t_0 and T_{g1} = 0; ACV $_{(3)}$ and ACV $_{(4)}$ takes-off at time t_1 and T_{g2} = $t_0 + T_{g(1,2)}$; ACV $_{(5)}$ and ACV $_{(6)}$ takes-off at time t_2 and T_{g3} = $t_1 + T_{g(3,4)}$. Table 6 reports a detailed analysis of the two scenarios over different periods of time

Table 4. Dynamic link features.

Dynamic Link Features	
Maximum link capacity (number of ACVs)	2
Average cruise speed on horizontal links	90 km/h
Average cruise speed on vertical links	50 km/h

Table 5. O/D nodes, travelled distances and path costs in the two scenarios.

	Origin Node	*Destination Node*	*S_1: Travelled Distance (m)*	*S_1: Path Cost (s)*	*S_2: Travelled Distance (m)*	*S_2: Path Cost (s)*
ACV $_{(1)}$	3	1	10,250	418.64	10,250	418.64
ACV $_{(2)}$	2	1	5310	222	5310	222
ACV $_{(3)}$	3	1	10,250	418.64	10,250	418.64
ACV $_{(4)}$	2	1	5310	222	5310	222
ACV $_{(5)}$	3	1	10,250	418.64	10,350	605.84
ACV $_{(6)}$	2	1	5310	222	5310	222

Table 6. Dynamic link status and ACV position in the two tested scenarios.

	Scenario 1: Tg = 180 s						*Scenario 2: Tg = 120 s*					
	0	120 s	180 s	240 s	360 s	420 s	0	120 s	180 s	240 s	270s	360 s
ACV $_{(1)}$	N * 3	(7-5) **	(7-5)	(7-5)	(5-4)	\	N 3	(7-5)	(7-5)	(7-5)	(7-5)	(5-4)
ACV $_{(2)}$	N 2	(5-4)	(5-4)	\	\	\	N 2	(5-4)	(5-4)	\	\	\
ACV $_{(3)}$	\	\	N 3	(7-5)	(7-5)	(7-5)	\	N 3	(7-5)	(7-5)	(7-5)	(7-5)
ACV $_{(4)}$	\	\	N 2	(6-5)	(5-4)	\	\	N 2	(6-5)	(6-5)	(5-4)	\
ACV $_{(5)}$	\	\	\	\	N 3	(7-5)	\	\	\	N 3	(11-10)	(11-10)
ACV $_{(6)}$	\	\	\	\	N 2	(6-5)	\	\	\	N 2	(6-5)	(5-4)

* N = node. ** = link.

From Tables 5 and 6, it can be seemed that, in a small urban context (such as the city of Bologna), a time gap T_g = 180 s between successive departures would not generate air traffic congestion, even if there are crossing routes. A different result is obtained in the case of T_g = 120 s. In this scenario, the ACV $_{(5)}$ has to exploit the dynamism of the links and switches to the next layer to ensure flight safety and avoid link congestion. In fact, thanks to the information stored regarding the air traffic conditions, by both the ACV and the control centre, the link (7-5) is disabled for transit, while the vertical links (7-11) and (8-4), and the horizontal links (11-10) and (10-8) are enabled, ensuring a suitable air transport service.

6. Discussion and Conclusions

The previous sections described two main aspects of UAM systems—i.e., vertiport locations and 3D UAN—that may benefit from the digital twin approach. The presented framework, together with the proposed 3D-UAN model, has been applied to the real context of a medium size city and its metropolitan area. Although this is preliminary research, the results obtained are very encouraging.

The first step, i.e., the location of the vertiport close to the main train station in the city area, confirmed some of the suggestions and preliminary results in the literature. The identified area has some interesting features, such as high levels of population and job densities—which would generate demand levels suitable for supporting UAM services—and good ground connections—which would assure great accessibility to the vertiport from the remaining part of the city.

As for the 3D-UAN structure, in the case study the metropolitan area has been considered, which is more suitable to this aim. Three vertiports have been identified and, by using the digital twin information on the most important factors and constraints to set safe aerial routes, a preliminary 3D-UAN structure has been identified. Furthermore, a preliminary simulation of the aerial traffic flows has been provided, based on Equations (2)–(4). In a real operational context, detected traffic data could also be added to the digital twin of the system as time series data useful for figures and off-line scheduling purposes. In fact, in the case of scheduled services, the computation of the shortest path and assignment of ACVs at specific enabled links is performed before the departure of each aerial vehicle, based on pre-trip information regarding the origin and destination points of ACVs. Scheduled minimum paths and vehicle separations may be computed based on Equation (2).

It is worth noting that the data used for feeding the digital twin of the case study were limited, because only freely available data have been used at this stage, and some other data should be added for improving the nature of the information provided by the digital twin, both static and dynamic data. For example, information regarding static and dynamic populations could be relevant for dynamically adapting the 3D-UAN in order to avoid overflying crowded locations. Similarly, information on dynamic and static population density could be used to adjust the dynamic corridors in order to reduce the externalities produced by ACVs, e.g., noise emissions during the day or night, respectively. In the case study of Bologna, the information about the dynamic population is particularly relevant because the city hosts one of the most important universities in Italy, with a student population of about 90,000, which is a high number compared to the whole population of the city. Particularly, many of them often live in Bologna for a limited number of months. In this context, dynamic population density data are probably the best option for safety evaluations of aerial corridors. Moreover, information on how much a location is busy—which might be obtained by several sources (e.g., Google popular times)—can also be useful to design and adapt the 3D network, while information on areas subjected to urban canyon effects may help in refining the optimal routes.

Another important aspect that affects the location of the vertiports and the design and management of the 3D-UAN is the energy consumption of each ACV. While for small urban areas—as, for example, the city of Bologna—this kind of analyses is not necessary because current aerial vehicle prototypes could realize several trips with a single charge; however, for larger cities such as Rome, Paris or London the cost function of the 3D-UAN model should include the energy consumption factor so that the computation of paths between vertiports will also consider the vehicles autonomy. The maximum autonomy range can then be included as relevant information for setting suitable locations of vertiports and routes between them, as ACV energy autonomy affects the length of the links, also depending on the vertiport (fixed node) where the recharging facilities has been located [66].

To summarize, the opportunity to use a digital twin approach at a high detail level will help system designers and urban planners to evaluate and implement procedures to realize successful aerial networks which can support the existent ground transportation system.

38. Jiang, F.; Ma, L.; Broyd, T.; Chen, K. Digital twin and its implementations in the civil engineering sector. *Autom. Constr.* **2021**, *130*, 103838. [CrossRef]
39. Schrotter, G.; Hürzeler, C. The Digital Twin of the City of Zurich for Urban Planning. *PFG—J. Photogramm. Remote Sens. Geoinf. Sci.* **2020**, *88*, 99–112. [CrossRef]
40. Schweizer, J.; Poliziani, C.; Rupi, F.; Morgano, D.; Magi, M. Building a large-scale micro-simulation transport scenario using big data. *ISPRS Int. J. Geo-Inf.* **2021**, *10*, 165. [CrossRef]
41. Sánchez-Vaquerizo, J.A. Getting Real: The Challenge of Building and Validating a Large-Scale Digital Twin of Barcelona's Traffic with Empirical Data. *ISPRS Int. J. Geo-Inf.* **2022**, *11*, 24. [CrossRef]
42. EASA. Prototype Technical Specifications for the Design of VFR Vertiports for Operation with Manned VTOL-Capable Aircraft Certified in the Enhanced Category. 2022. Available online: https://www.easa.europa.eu/downloads/136259/en (accessed on 1 September 2022).
43. Feldhoff, E.; Soares Roque, G. Determining infrastructure requirements for an air taxi service at Cologne Bonn Airport. *CEAS Aeronaut. J.* **2021**, *12*, 821–833. [CrossRef]
44. Kim, N.; Yoon, Y. Regionalization for urban air mobility application with analyses of 3D urban space and geodemography in San Francisco and New York. *Procedia Comput. Sci.* **2021**, *184*, 388–395. [CrossRef]
45. Otte, T.; Metzner, N.; Lipp, J.; Schwienhorst, M.S.; Solvay, A.F.; Meisen, T. User-centered Integration of Automated Air Mobility into Urban Transportation Networks. In Proceedings of the 2018 IEEE/AIAA 37th Digital Avionics Systems Conference (DASC), London, UK, 23–27 September 2018. [CrossRef]
46. Antcliff, K.R.; Moore, M.D.; Goodrich, K.H. Silicon Valley as an Early Adopter for on-Demand. In Proceedings of the 16th AIAA Aviation Technology, Integration, and Operations Conference, Washington, DC, USA, 13–17 June 2016; p. 3466. [CrossRef]
47. Preis, L. Quick Sizing, Throughput Estimating and Layout Planning for VTOL Aerodromes—A Methodology for Vertiport Design. In Proceedings of the AIAA Aviation 2021 Forum, American Institute of Aeronautics and Astronautics, Reston, VA, USA, 2–6 August 2021; p. 2372.
48. Delgado Gonzalez, C.J. Rooftop-Place Suitability Analysis for Urban Air Mobility Hubs: A GIS and Neural Network Approach. Master's Thesis, Universida de Nova de Lisboa, Lisbon, Portugal, 2020.
49. Rothfeld, R.; Fu, M.; Balać, M.; Antoniou, C. Potential Urban Air Mobility Travel Time Savings: An Exploratory Analysis of Munich, Paris, and San Francisco. *Sustainability* **2021**, *13*, 2217. [CrossRef]
50. Shin, H.; Lee, T.; Lee, H. Computers and Operations Research Skyport location problem for urban air mobility system. *Comput. Oper. Res.* **2022**, *138*, 105611. [CrossRef]
51. Rath, S.; Chow, J.Y.J. Air taxi skyport location problem with single-allocation choice-constrained elastic demand for airport access. *arXiv* **2021**, arXiv:1904.01497v4. [CrossRef]
52. Rimjha, M.; Hotle, S.; Trani, A.; Hinze, N. Commuter demand estimation and feasibility assessment for Urban Air Mobility in Northern California. *Transp. Res. Part A Policy Pract.* **2021**, *148*, 506–524. [CrossRef]
53. Daskilewicz, M.; German, B.; Warren, M.; Garrow, L.A.; Boddupalli, S.-S.; Douthat, T.H. Progress in vertiport placement and estimating aircraft range requirements for evtol daily commuting. In Proceedings of the 2018 Aviation Technology, Integration, and Operations Conference, Atlanta, GA, USA, 25–29 June 2018. [CrossRef]
54. Eastman, J.R. Multi-criteria evaluation and GIS. *Geogr. Inf. Syst.* **1999**, *1*, 493–502.
55. Desai, K.; Al Haddad, C.; Antoniou, C. Roadmap to early implementation of passenger air mobility: Findings from a delphi study. *Sustainability* **2021**, *13*, 612. [CrossRef]
56. Lascara, B.; Lacher, A.; DeGarmo, M.; Maroney, D.; Niles, R.; Vempati, L. *Urban Air Mobility Airspace Integration Concepts: Operational Concepts and Exploration Approachs*; Technical Report; MITRE: Bedford, MA, USA, 2019.
57. Samir Labib, N.; Danoy, G.; Musial, J.; Brust, M.R.; Bouvry, P. Internet of unmanned aerial vehicles—A multilayer low-altitude airspace model for distributed UAV traffic management. *Sensors* **2019**, *19*, 4779. [CrossRef] [PubMed]
58. Dijkstra, E.W. A note on two problems in connexion with graphs. *Numer. Math.* **1959**, *1*, 269–271. [CrossRef]
59. Fu, L.; Sun, D.; Rilett, L.R. Heuristic shortest path algorithms for transportation applications: State of the art. *Comput. Oper. Res.* **2006**, *33*, 3324–3343. [CrossRef]
60. Healy, R.; Misiorwoski, M.; Gandhi, F. A systematic cfd-based examination of rotor-rotor separation effects on interactional aerodynamics for large evtol aircraft. *Lead. Edge* **2019**, *2*, 4.
61. Demba, A.; Möller, D.P. Vehicle-to-vehicle communication technology. In Proceedings of the 2018 IEEE International Conference on Electro/Information Technology (EIT), Rochester, MI, USA, 3–5 May 2018; pp. 459–464.
62. Djahel, S.; Jabeur, N.; Barrett, R.; Murphy, J. Toward V2I communication technology-based solution for reducing road traffic congestion in smart cities. In Proceedings of the 2015 International Symposium on Networks, Computers and Communications (ISNCC), Hammamet, Tunisia, 13–15 May 2015; pp. 1–6.
63. Lyu, H. Detect and avoid system based on multi sensor fusion for UAV. In Proceedings of the 2018 International Conference on Information and Communication Technology Convergence (ICTC), Jeju Island, Republic of Korea, 17–19 October 2018; pp. 1107–1109.
64. Chriki, A.; Touati, H.; Snoussi, H.; Kamoun, F. FANET: Communication, mobility models and security issues. *Comput. Netw.* **2019**, *163*, 106877. [CrossRef]

65. Boelens, J.-H. *Pioneering the Urban Air Taxi Revolution 1.0*; Volocopter: Bruchsal, Germany, 2019; Available online: https://press.volocopter.com/images/pdf/Volocopter-WhitePaper-1-0.pdf (accessed on 1 September 2022).
66. Wu, Z.; Zhang, Y. Optimal eVTOL charging and passenger serving scheduling for on-demand urban air mobility. In Proceedings of the AIAA Aviation 2020 Forum, American Institute of Aeronautics and Astronautics, Reston, VA, USA, 15–19 June 2020; p. 3253.

Article

Fast Obstacle Detection System for UAS Based on Complementary Use of Radar and Stereoscopic Camera

Luca Bigazzi, Lapo Miccinesi, Enrico Boni, Michele Basso, Tommaso Consumi and Massimiliano Pieraccini *

Department of Information Engineering, University of Florence, via Santa Marta 3, 50139 Firenze, Italy
* Correspondence: massimiliano.pieraccini@unifi.it

Abstract: Autonomous unmanned aerial systems (UAS) are having an increasing impact in the scientific community. One of the most challenging problems in this research area is the design of robust real-time obstacle detection and avoidance systems. In the automotive field, applications of obstacle detection systems combining radar and vision sensors are common and widely documented. However, these technologies are not currently employed in the UAS field due to the major complexity of the flight scenario, especially in urban environments. In this paper, a real-time obstacle-detection system based on the use of a 77 GHz radar and a stereoscopic camera is proposed for use in small UASs. The resulting system is capable of detecting obstacles in a broad spectrum of environmental conditions. In particular, the vision system guarantees a high resolution for short distances, while the radar has a lower resolution but can cover greater distances, being insensitive to poor lighting conditions. The developed hardware and software architecture and the related obstacle-detection algorithm are illustrated within the European project AURORA. Experimental results carried out employing a small UAS show the effectiveness of the obstacle detection system and of a simple avoidance strategy during several autonomous missions on a test site.

Keywords: UAS; radar; computer vision; obstacle detection and avoidance; autonomous navigation

Citation: Bigazzi, L.; Miccinesi, L.; Boni, E.; Basso, M.; Consumi, T.; Pieraccini, M. Fast Obstacle Detection System for UAS Based on Complementary Use of Radar and Stereoscopic Camera. *Drones* **2022**, *6*, 361. https://doi.org/10.3390/drones6110361

Academic Editor: Anastasios Dimou

Received: 29 October 2022
Accepted: 16 November 2022
Published: 18 November 2022

Publisher's Note: MDPI stays neutral with regard to jurisdictional claims in published maps and institutional affiliations.

1. Introduction

In recent years, there has been growing interest in autonomous UAS, and this is also reflected in the increase in scientific publications on the topic [1–5]. The importance of this research area is fundamental since the development of autonomous technologies in drones can guarantee significant benefits for society in the near future. For instance, implementations for the search of missing people and the exploration of post-disaster or inaccessible environments to humans where the use of drones with autonomous decision-making capabilities could play a decisive factor in terms of the number of lives saved [6,7].

Nowadays, the topic of human transport in urban contexts through the use of autonomous drones is also becoming more and more interesting, precisely because it would allow to limit ground traffic by moving part of it to the sky. In this regard, the work presented in this article was developed within the European project AURORA (Safe Urban Air Mobility for European Citizens), which has as its ultimate goal the development of autonomous technologies aimed at this type of use. In this regard, a fundamental element that an autonomous driving system must implement on board is the obstacle-detection and avoidance part [8,9]. In fact, without it, the aircraft cannot modify the global trajectory to avoid unexpected obstacles and therefore would not be able to be used for applications that instead require high safety and flexibility in terms of automatic recalculation of the desired trajectory. In the literature, there are papers that deal with the development of such systems; many of these are based on the use of stereoscopic cameras capable of depth perception, allowing to carry out the detection of obstacles [10,11].

Recently, the scientific community has begun to experiment with the use of radar systems, which have multiple advantages, such as longer detection ranges and being insensitive to light and visibility conditions [12,13]. Compared to optical systems, it is important to note that radar technology also has disadvantages, such as a lower spatial resolution and the almost total absence of resolution in height [12,14]. However, the two technologies have complementary characteristics and therefore lend themselves well to being used together; in fact, they are able to compensate for each other's disadvantages. Automotive applications of simultaneous radar and vision systems for obstacle detection [15–17] are widely documented in the literature; however, this is not common in the UAS field. In fact, except for some implementations where the 3D perception is obtained by merging together the radar data with the monocular vision [14], until now and to the authors' knowledge, there are no documented techniques that use radar and stereoscopic vision in a complementary way.

In this article, an approach for UAS applications that exploits the complementary features of an automotive-derived radar and a stereoscopic optical sensor to increase the reliability of the detection algorithm is presented.

The two systems have been kept completely independent in such a way that a malfunction of one cannot affect the functioning of the other. As it will be detailed in the following sections, the avoidance strategy comes into operation as soon as one of the two systems detects a dangerous obstacle according to the detection policy implemented.

The paper is organized as follows: Section 2 describes the hardware and software architecture, explains the method of creating environmental maps and defines the algorithm that has the task of detecting dangerous obstacles inside them. Furthermore, Section 2 also discusses the implemented avoidance strategy which comes into operation only when obstacles considered dangerous are highlighted. Section 3 shows the results obtained from field tests carried out in an open environment that simulates an urban context. Finally, Section 4 reports the discussion on results and future developments.

2. Materials and Methods

2.1. UAS Architecture

The platform used for the development of the obstacle detection and avoidance system consists of a DJI Matrice 300 RTK drone to which a ZED 2 stereoscopic camera, an AWR1843BOOST radar and an Nvidia Jetson Xavier AGX board have been added as a payload. The obstacle detection algorithm runs on the Jetson board, which is directly interconnected with the camera and the radar. Moreover, through a UART connection, the board is also able to pilot the UAS autonomously.

Figure 1 shows the DJI Matrice 300 RTK equipped with the additional components. In particular, at the top, it is possible to observe the Nvidia Jetson Xavier AGX board, while at the bottom, there are the radar and the optical sensors used for the implementation of the obstacle detection algorithm.

2.2. Stereoscopic Vision

To obtain three-dimensional information on the surrounding environment through the use of computer vision, there are basically two technologies that can be used. The first is based on RGB-D technology, where an RGB optical sensor is used alongside a *TOF* (time of flight) depth sensor. The second one uses the optical flow coming from two RGB cameras that needs to be processed on board the companion computer to provide 3D data. The TOF sensor mounted on RGB-D cameras uses a laser beam matrix, very sensitive to ambient light conditions, and does not guarantee high operating ranges.

Figure 1. DJI Matrice 300 RTK drone equipped with the additional hardware components.

To comply with the specifications imposed by the AURORA project, classical stereoscopy based on two RGB optical streams was chosen, which provides more stable performance, especially in outdoor environments. In particular, the ZED 2 stereoscopic camera by StereoLabs was used, which in addition to providing the point cloud of the surrounding environment defined in the fixed frame, is able to generate an estimate of the trajectory traveled by the camera in 3D space. This estimate was obtained from the SLAM algorithm implemented in the SDK (Software Development Kit) supplied with the camera. Unlike other optical systems (see Intel t-265), the ZED 2 camera does not perform calculations on board, leaving all the computational load to the Nvidia Jetson module. This is a significant limitation since a considerable amount of the hardware resources is already reserved for the camera SDK, leaving less space for user applications. For these limitations, it is important to pay particular attention to the optimization of the detection algorithm in order to minimize the computational cost necessary for the detection of dangerous obstacles. Despite the presence of these problems, currently the ZED 2 camera represents the state of the art for stereoscopic vision systems. In fact, it guarantees an operating range of up to 40 m and performance superior to those offered by other products. Figure 2 shows a functional example of this camera, where the RGB optical flow (a) and the depth flow, processed on board the Jetson module (b), are visible.

Figure 2. Functional example of a ZED 2 stereoscopic camera. Image (**a**) shows an RGB frame coming from the camera, while image (**b**) shows a depth frame processed by the ZED SDK on board the Jetson module.

2.3. Radar

The radar used for this work is an AWR1843BOOST by Texas Instruments [12]. A radar detects the distance of the target by sending and receiving an electromagnetic signal through at least a couple of antennas. Using a multiple input multiple output array, it is also able to retrieve the direction of arrival. The sensor used is equipped with 3 TX and 4 RX antennas, which correspond to 12 virtual antennas, disposed as shown in Figure 3. In Figure 3, the z axis represents the altitude, while the x axis is left to right, and λ is the wavelength of the electromagnetic signal. This arrangement of virtual antennas achieves a good azimuth resolution and a poor elevation resolution, the elevation resolution being related to the inverse of the z-distance between the antennas.

For the current application, the elevation resolution is used only as an angular cut-off. In other words, the value of elevation measured by the radar is not used for mapping, and it is set equal to zero for each target, but it is used as a spatial filter for rejecting the target outside of a selected angular area. We set the angular field as ±45 deg in azimuth and [0,20] deg in elevation. Therefore, all targets outside this interval are not used for mapping.

Since the radar is not able to provide the elevation of the target, we decided to always consider zero as the elevation of the object. This is equivalent to assuming that each target is on a horizontal plane at the same height of the drone. This hypothesis is not as restrictive as it seems since usually a target at $z = 0$ m (at the same height of the UAS) is the most reflective. Under this hypothesis, the map generated using the radar is bi-dimensional. It becomes three-dimensional by changing the drone altitude [12].

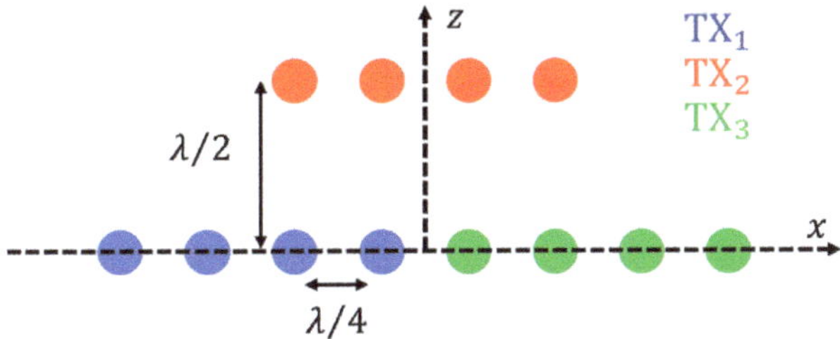

Figure 3. Virtual antennas position of AWR1843BOOST [12].

In order to reduce the number of false alarms, the radar signal was filtered using two constant false alarm rate (CFAR) algorithms [12]. The first CFAR algorithm discriminates the physical targets from the thermal noise and from possible clutter using a moving threshold on range direction. The second CFAR algorithm was applied in the Doppler direction (which means on the speed) to further discriminate the possible targets from false alarm. The radar provides the coordinates of each target that exceed the CFAR thresholds.

In the current application, the radar range resolution was 0.5 m with a maximum range of 120 m and 10 Hz frame periodicity.

2.4. Framework

The whole framework runs on board the Jetson card, and it was developed through the *ROS* (robot operating system) environment. The main advantage offered by ROS is to guarantee extreme flexibility and modularity: it is possible to interconnect multiple software packages, called nodes, through a publisher–subscriber scheme. The framework developed is shown in Figure 4, where the nodes are visible in blue, the topics in green, and the functions implemented within the nodes in black. The DJI OSDK node provides a communication interface to the Matrice 300 RTK. In particular, through the telemetry function, it is possible to obtain all the data of interest, while through the control function, it is possible to autonomously fly the drone sending a set-point velocity vector. In the current implementation, the telemetry data that are actually used in the framework are the *GNSS* (global navigation satellite system) coordinates and the attitude vector of the drone. This data are then used to build the radar maps, which will be explained in Section 2.5.

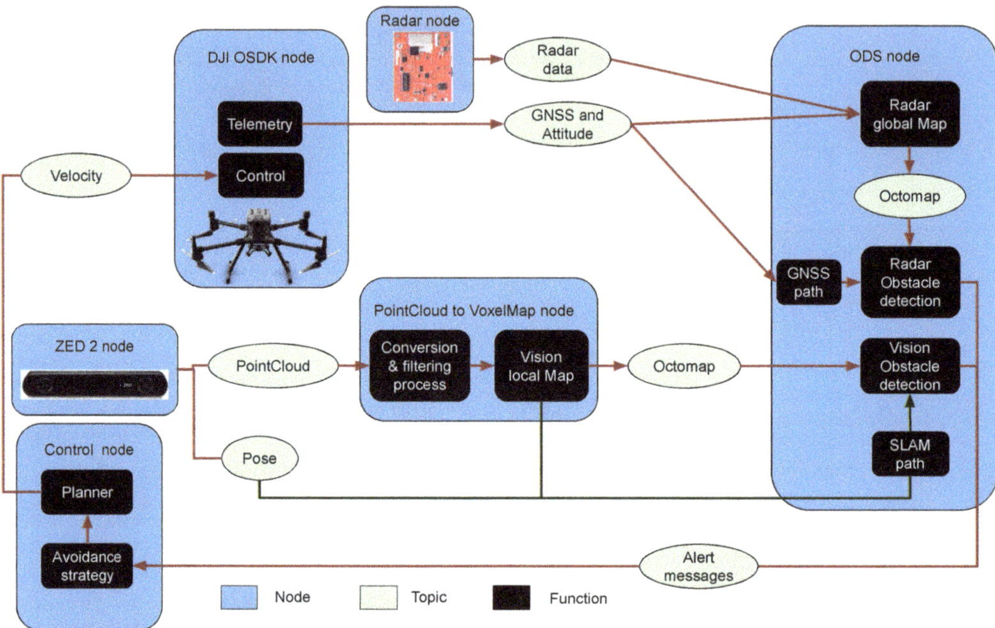

Figure 4. Block diagram of the framework that implements the obstacle detection and avoidance algorithm. In particular, the framework is able to communicate with the drone and to receive data from the sensors to search for obstacles in the surrounding environment.

Regarding the radar node, it provides the topic containing the 2D coordinates of the obstacles detected in the radar frame. In fact, as already explained in Section 2.3, the AWR1843BOOST radar does not reliably provide the relative elevation coordinate. The ZED 2 node creates the interface toward the stereoscopic camera: this node provides the fixed-frame position estimation by processing the vision data, the camera attitude and the topic point cloud that contains the 3D image of the environment. As will be explained in Section 2.6, by suitably processing this topic, it is possible to create an obstacle-detection system. The *ODS* node (obstacle-detection system) implements various functions: in addition to the obstacle-detection task, it is responsible for the creation of the maps [12] explained in Section 2.5, exploiting the absolute paths generated from the GNSS data [12] via the DJI OSDK node and from the SLAM (simultaneous localization and mapping) estimated by the vision process [18–21].

Since through the point cloud topic, it is not possible to create vision maps directly, a specific node called "point cloud to voxel map" was developed. This node converts the point cloud topic into a usable data structure [22] and the processed data are used for the creation of the vision maps.

The last node implemented in the framework is the control node, through which it is possible to define the generating policy (planner) of the desired set points for the autonomous driving algorithm. In addition, the control node implements the avoidance strategy, which is enabled in the event that a potentially dangerous obstacle is detected. In this case, the function that realizes the avoidance strategy bypasses the planner function, which is normally enabled during the autonomous mission.

2.5. Mapping Algorithm

Both vision and radar maps exploit the Octomap library [23,24] to create and manage volume maps (defined by sets of voxels) in an efficient and fast way, which is a requirement when performed on embedded devices, such as Jetson boards. Two different type of maps can be employed: local and global maps, both using the same Earth reference system (ENU in this work). In general, we define a map as local when it only contains the obstacles detected at the current time, whereas a global map includes all the objects detected during the entire mission. In either cases, the task for updating the maps is computationally expensive since it requires a transformation of each detected point from the sensor frame to Earth frame, given the UAS attitude and position (pose). As said, for updating a local map, it is necessary to destroy the previous information at each new frame. Conversely, for the creation of the global map, it is sufficient to define a global container, where at each frame, all detected targets are inserted, without, however, deleting previously detected objects, even if they may no longer be present. Figure 5 shows the differences between the global map built using the radar sensor and the global map built starting from the optical one.

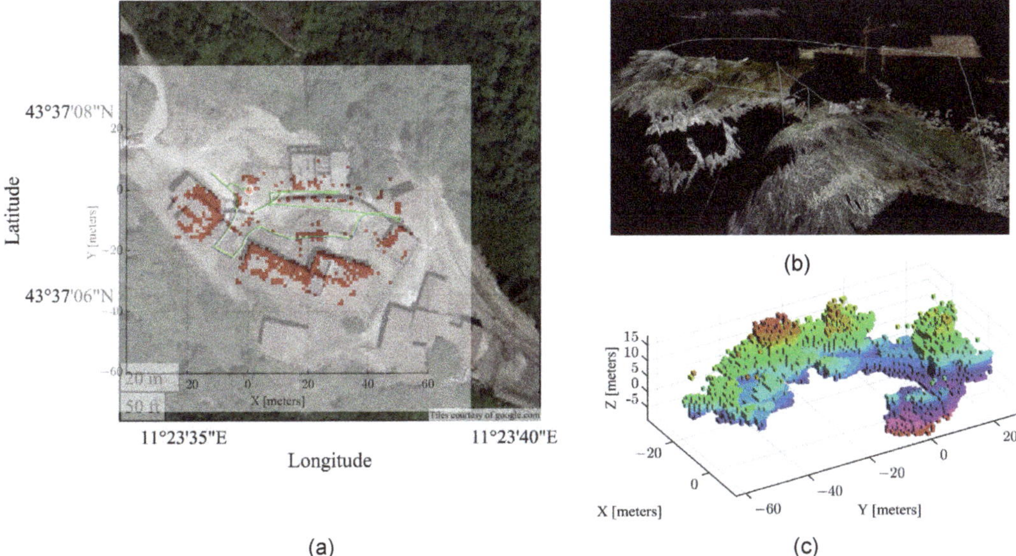

Figure 5. Differences between vision and radar global maps: left (**a**) is the radar map where detected obstacles are reported on the same plane as the drone; top right (**b**) is the optical point cloud, while on the bottom right (**c**), there is the corresponding conversion into a set of voxels used for the detection of dangerous obstacles. In this latter case, the map obtained also allows one to know the altitude of the detected obstacles.

In order to meet project AURORA specifications, which require independent and complementary sources for the obstacle detection subsystem, the proposed solution exploits two specific maps: a Vision local map and a Radar global map, as shown in the diagram of Figure 4. In particular, the vision maps are computed using the position estimate obtained from the vision process as a reference. Due to the camera operating range, it is quite common to have no visible objects during a mission for certain time intervals: in these conditions the SLAM process can diverge, creating inconsistencies between the new coordinates entered at the current time and the old ones, giving origin to distortions within the global map (see Figure 6). For this reason, the global map is to be considered unreliable, while the local map is the best choice for the task since in the latter, there are only the current obstacles, and these are always consistent with the position estimate provided by the SLAM process.

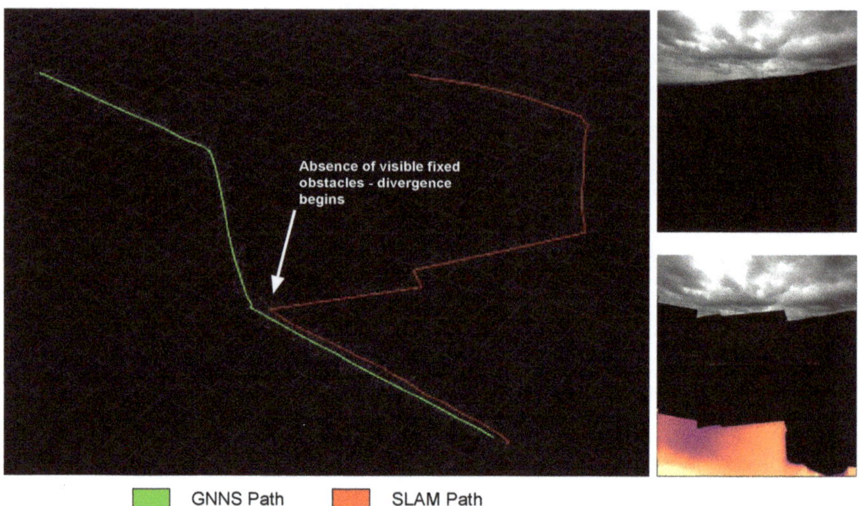

Figure 6. Path divergence issues: estimated trajectories starting from data provided by the RTK system (**green**) and by the SLAM algorithm that uses the vision data (**red**). The latter diverges if there are no fixed objects in the operating range of the camera. This situation is shown in the image on the top right, where, after the avoidance of the building, all the fixed obstacles are out of the range of the optical sensor.

As for the radar obstacles detection, the reasoning for choosing a global map is different: unlike the stereoscopic vision system, the radar has a much lower resolution, so some obstacles may not always be reflective and consequently they could be detected only in certain time intervals. Therefore, a global map reporting all the targets detected during the mission allows for greater safety. It should be also noted that the radar maps, unlike the vision ones, are created using the path received by the RTK system as a reference [12], which has no particular divergence problems, being based on GNSS localization.

2.6. Obstacle Detection Methodology

Obstacle detection refers to the task of searching for dangerous obstacles in specific regions of the maps introduced in Section 2.5 in order to guarantee safe UAS autonomous navigation. As already mentioned, the vision system guarantees high resolution for distances within 40 m, while the radar system provides a lower resolution but an operating range up to 120 m [12]. Therefore, the visual detection algorithm is typically more effective in urban/indoor environments or in lower UAS speed ranges, whereas in larger open environments, and with few obstacles, the radar detection system guarantees safer operations, allowing for faster flight. In addition to this, it must be considered that in urban or indoor environments, there may be poor or incomplete GNSS coverage. In these conditions, the visual detection system is more reliable, not only in terms of resolution, but also because the vision maps are computed using the position provided by the SLAM process, which does not require geolocation data.

In order to be able to detect the presence of dangerous obstacles starting from the two maps used (vision and radar), it is necessary to define a region of interest (ROI) [25] as a subset of the maps where the obstacles are searched. The approach followed in this article considers any object (static or dynamic) found inside the ROI to be potentially dangerous. The shape of the ROI is defined to ensure that the global trajectory produced a priori by the planner can be considered safe under the condition that no obstacles are inside the ROI. In order to simplify the computational cost of the obstacle search, we make some assumptions about the UAS navigation which are quite reasonable:

(i)   The flight occurs with the sensors (radar and vision) facing the direction of motion;
(ii)   The altitude is kept constant during the flight.

Under these operating conditions, we define the ROI as a cylindrical sector with radius d_m (maximum search depth along the flight direction), a FOV angle ψ_R, and vertical height h_m. This region can be first computed with respect to a reference frame B centered in the UAS position $\vec{P}$ and with Euler angles $(0,0,\psi)$, i.e., with the y^b axis aligned with the drone heading angle ψ, as shown in Figure 7. The figure shows a schematic of the detection process applied to the local map: the sensor FOV where obstacles are detected is shown in gray. Instead, the yellow area defines the ROI such that the presence of obstacles in this sub-region is considered dangerous.

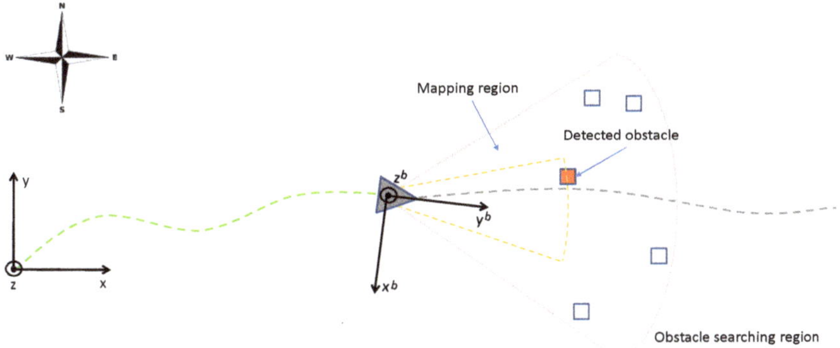

Figure 7. Mapping region of the local map (**gray**); searching region where the presence of an obstacle is considered dangerous (**yellow**); occupied voxels by obstacles (**blue**).

As the drone moves along the planned trajectory, the coordinates of the ROI in the fixed reference frame must be updated at each iteration of the algorithm, taking into account the UAS current position and yaw angle through the relation

$$\vec{F}_i = R_z(\psi)^{-1}\vec{F}_i^b + \vec{P}, \quad i = 1, \cdots, N \tag{1}$$

where $\vec{F}_i$ and $\vec{F}_i^b$ are the coordinates of the i-th point of the ROI in ENU and B frames, respectively, and

$$R_z(\psi)^{-1} = \begin{bmatrix} \cos(\psi) & \sin(\psi) & 0 \\ -\sin(\psi) & \cos(\psi) & 0 \\ 0 & 0 & 1 \end{bmatrix} \in \mathbb{R}^{3\times3} \tag{2}$$

is the corresponding rotation matrix.

This operation can be expensive, and its burden depends on the number N of points representing the ROI. For this reason, in this article, an efficient algorithm to generate the minimum number of points belonging to the ROI volume, given the map resolution m_r, is reported in Appendix A.

Once all the iterations necessary to update the entire ROI are performed, the result shown in Figure 8 is obtained, where the relative voxel can be seen for each calculated point. Finally, the last step of the obstacle detection task is to query the radar and vision maps (via the Octomap libraries) using the coordinates of the ROI vectors $\vec{F}_i$ in Equation (1): for each point of the ROI, an occupation probability is returned such that for probabilities greater than a given threshold (0.5 in our experiments), the relative voxel is considered occupied, and an obstacle-avoiding strategy needs to be activated.

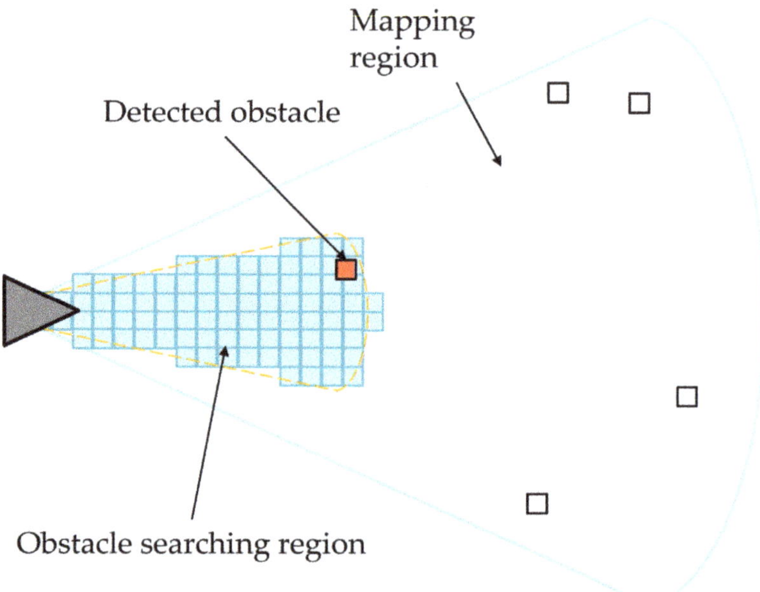

Figure 8. The figure shows the voxels calculated by the detection algorithm within the search region. In particular, the number of points computed by the algorithm depends both on the map resolution and ROI volume.

It is important to underline that the ROI update algorithm should start calculating the coordinates of the points from those closest to the drone. This is preferable to make sure that the search for dangerous obstacles starts from the most critical points, i.e., those that are closest to the drone.

2.7. Simplified Avoidance Strategy

To carry out the experiments required by the AURORA project, it was useful to design and implement an obstacle avoidance strategy on board the drone. Although in the literature, there are many refined techniques [26–29] that could have been used, the main objective of the work package assigned to the University of Florence unit was the development and testing of a fast complementary obstacle detection system exploiting radar and optical sensors. In this context, it was chosen to implement a very simple avoidance strategy able to correctly operate in an open environment.

In Figure 9, a flowchart shows the workflow of the algorithm: during the autonomous mission, the planner and the obstacle detection (OD) blocks are enabled. As long as no obstacles are detected inside the ROI, the OD block keeps the planner generating a trajectory through a velocity reference $\vec{V}_1$. When the OD task detects the presence of one or more obstacles, it disables the planner by interrupting the loop connection and activates the obstacle avoidance (OA) block, which remains active as long as there are obstacles inside the ROI. As already mentioned, a simplified avoidance strategy was implemented by switching to a velocity reference vector $\vec{V}_2$, which corresponds to an increase in altitude. This state persists until no more obstacles are found in the ROI such that the standard mission planner can be re-enabled.

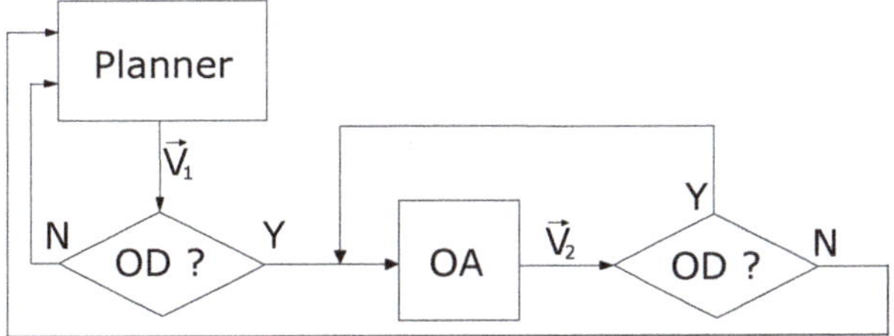

Figure 9. Flowchart of the obstacle detection (OD) and avoidance algorithm (OA).

Figure 10 illustrates the different steps of the simplified avoidance strategy.

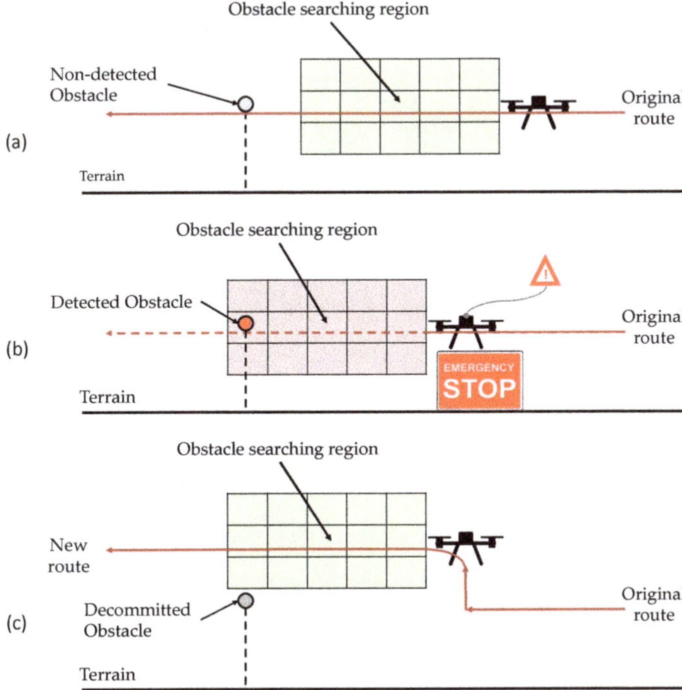

Figure 10. Avoidance strategy of dangerous obstacles. The voxels in front of the drone define the ROI where the algorithm searches for dangerous obstacles (**a**). When an object is detected inside it (**b**), the avoidance strategy is activated (**c**).

Notice that this strategy, once enabled, does not allow the subsequent lowering of the altitude for realignment with the global path. This means that the mission will be completed at a higher altitude than that of the path defined a priori. However, this is not a problem since, as already mentioned, the avoidance strategy has been implemented only for testing the detection system.

3. Results

The tests were carried out at the Celle castle at Incisa Valdarno (Italy) visible in Figure 11. To test the correct functioning of the obstacle detection and avoidance system in the planner, an autonomous mission was preloaded inside the control node, where the global trajectory consists of a straight path that collides with one of the buildings present on the site. Considering the type of environment in which the tests were carried out, to prevent the avoidance strategy from immediately starting to work, the desired depth of the ROI was set at 7 m for the vision and 10 m for the radar.

The trajectory that the drone needs to follow is shown in Figure 12. This mission foresees an automatic take-off from point 1, an arrival in point 2 and a return to the take-off point with consequent landing. All phases of the mission were managed on board the Jetson, which triggered the avoidance strategy explained in the previous section as soon as the ODS node detected the presence of a building.

Figure 11. Celle castle site used to carry out the tests on the obstacle detection and avoidance system.

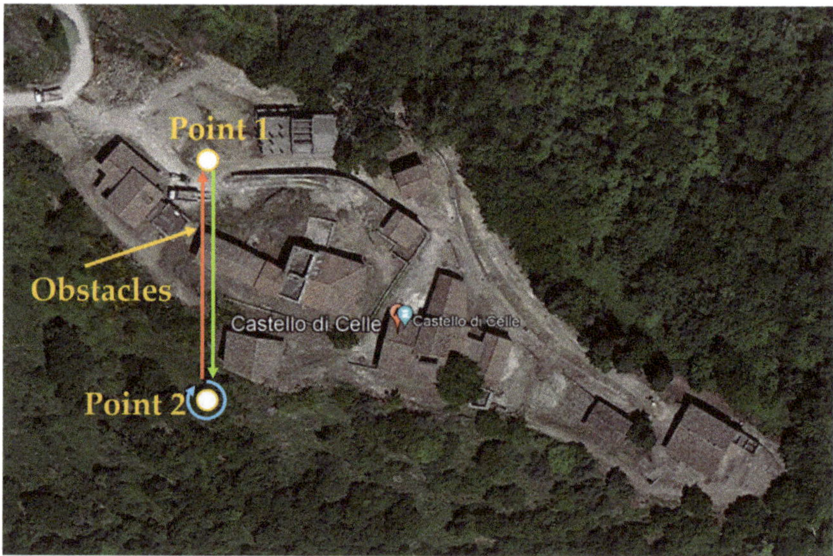

Figure 12. Example of mission carried out to test the correct functioning of the obstacle detection and avoidance system. This mission involves a trajectory that takes the drone on a collision course with a building on the site.

Since the two detection systems (radar and optical) are able to work independently, several missions were carried out in order to separately test each sensor. Figure 13 presents the results obtained by enabling first the radar detection system and then the optical one. On the top row of Figure 13a, three snapshots of a mission for the radar detection system are shown together with the voxels relating to the global map and the voxels that make up the ROI. In the central image, the ROI turns red since the algorithm detects an obstacle (the edge of a building). Instead, the bottom row shows the text area, where the ODS node starts publishing the warning messages with the coordinates of the detected obstacles (central image) and the function that implements the avoidance strategy (right image). This leads the drone to climb in altitude and subsequently to pass the building, avoiding the collision. A similar behavior can also be observed for the optical detection system, visible in Figure 13b.

As already explained, the resolutions that the two sensors can provide are very different: in fact, for the global radar map, only the voxels relating to the edges of the building are displayed, being more reflective, while in the optical local map, the entire visible wall of the building is filled with voxels. In this regard, the resolution of the radar maps is set at 1 meter, while for the vision maps, it is equal to 0.5 m. Since the radar has less resolution, it is safer to increase the volume associated with the single obstacles detected.

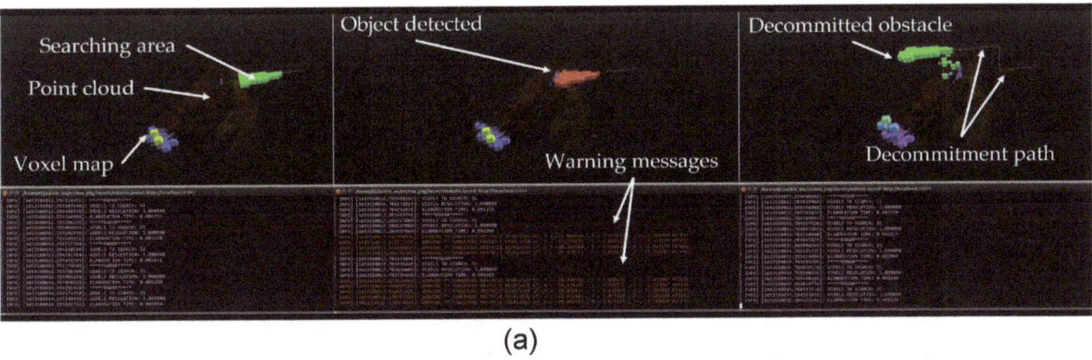

(a)

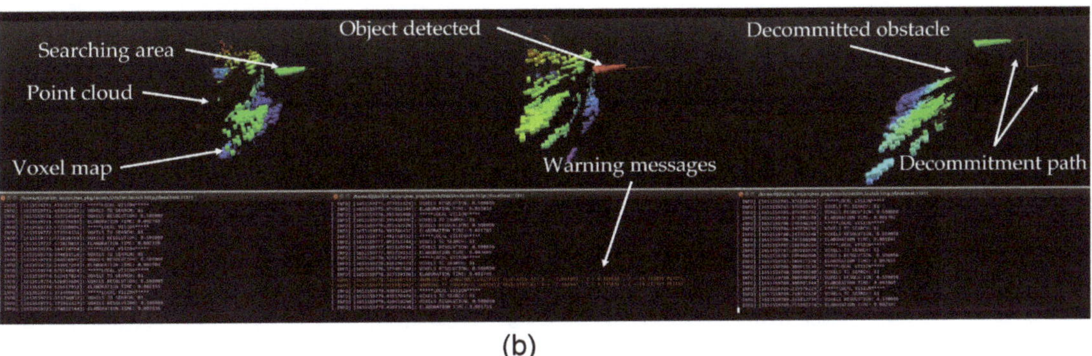

(b)

Figure 13. Screenshots of the building detection and avoidance phases through the separate use of the two sensors: radar detection system (**a**), and optical detection system (**b**).

Table 1 shows the average update times for the coordinates of the ROI as a function of different desired depths. Obviously, at greater depths, the algorithm must calculate a larger number of voxels, and consequently, the complexity increases. These timings were achieved by setting a map resolution of 0.5 m and a FOV of 24 degrees.

Table 1. Number of voxels automatically calculated by the update algorithm of the ROI, the resolution and the computation time as a function of various desired depths.

Depth (m)	Resolution (m)	Calculated Voxels	Computation Time (s)
7	0.5	81	0.0032
10	0.5	139	0.0053
20	0.5	443	0.0081
30	0.5	915	0.0095
50	0.5	2361	0.0184
70	0.5	4477	0.0240
90	0.5	7261	0.0346
120	0.5	12699	0.0462

As can be seen, even at the maximum search depth (120 m), the average update time is only 46.2 ms. This allows for the effective real-time detection of dangerous obstacles, thus maximizing the time available for the implementation of the avoidance strategy.

4. Discussion

The experimental tests presented in this paper have shown how in an urban scenario, the complementary obstacle detection and avoidance system is always able to detect and avoid dangerous static obstacles along the UAS path. Having adopted a redundant architecture consisting of a radar sensor and an optical sensor, which possess complementary operational features, it is possible to guarantee the detection of dangerous obstacles in different operating conditions, significantly increasing the safety of the entire system. In fact, as already discussed in the previous sections, in addition to having different operating ranges and resolutions between the radar and the optical sensor, in low light conditions, the radar can effectively replace the vision, even for short ranges. In conditions of acceptable brightness, vision can guarantee higher resolutions and therefore greater safety in detecting short-range obstacles. Moreover, redundancy can also be found in the different map reference systems employed: radar maps are based on GNSS geolocation data, while the vision maps use an independent position reference obtained from the optical SLAM process. In this way, if the satellite coverage fails, the vision maps will continue to function correctly and vice versa.

It is important to note that research associated with radar maps can involve larger regions of interest than optical ones, as the radar sensor is able to reach longer ranges. This implies that by appropriately scaling the resolutions of the vision and radar maps as a function of the specific sensor range, it is possible to obtain equivalent calculation times for the related regions of interest. This approach allows to add greater complementarity for the two detection systems.

Future research work will focus on the implementation of more complex avoidance strategies, which, eventually with the help of GPUs, can simultaneously compute multiple alternative trajectories in order to find the optimal one, also avoiding moving obstacles. Another interesting aspect that deserves additional work concerns the study of possible different strategies for the creation and management of radar and vision maps, where, for example, the RTK and SLAM references can be merged or replaced if necessary.

Author Contributions: Conceptualization, M.P. and M.B.; methodology, M.B. and E.B.; software, L.B.; validation, L.B., L.M. and T.C.; investigation, L.B, L.M. and T.C.; formal analysis, L.B.; data curation, L.B.; writing–original draft preparation, L.B.; writing—review and editing, M.P.; funding acquisition, M.P. All authors have read and agreed to the published version of the manuscript.

Funding: This research was co-funded by Horizon 2020, European Community (EU), AURORA (Safe Urban Air Mobility for European Citizens) project, ID number 101007134.

Institutional Review Board Statement: Not applicable.

Informed Consent Statement: Not applicable.

Data Availability Statement: The data presented in this study are available on request from the corresponding author.

Conflicts of Interest: The authors declare no conflict of interest.

Abbreviations

The following abbreviations are used in this manuscript:

FOV	Field Of View
ROI	Region Of Interest
SLAM	Simultaneous Localization And Mapping
UAS	Unmanned Aerial System
GNSS	Global Navigation Satellite Systems
RTK	Real-Time Kinematic
UART	Universal Asynchronous Receiver-Transmitter
ROS	Robot Operating System
ODS	Obstacle Detection System
OSDK	Onboard Software Development Kit
OD	Obstacle Detection
OA	Obstacle Avoidance

Appendix A. ROI Generation Algorithm

This appendix presents an algorithm to define a ROI geometry with the shape of a cylindrical sector, depending on the parameters:

- m_r: map resolution;
- d_m: maximum depth to investigate (ROI radius);
- ψ_R: ROI field of view [rad];
- h_m: ROI height.

The ROI is computed as the following set of N coordinates $\vec{F}^b$ in the reference frame B

$$ROI = \left\{ \vec{F}^b \in \mathbb{R}^3 \text{ s.t. } \vec{F}^b = \begin{bmatrix} n_i m_r \sin\left(n_j/n_i\right) \\ n_i m_r \cos\left(n_j/n_i\right) \\ n_k m_r \end{bmatrix} \right\} \tag{A1}$$

where the indexes n_i, n_j, n_k are integers depending both on the ROI geometry and map resolution:

$$n_i = 1, \cdots, \left\lceil \frac{d_m}{m_r} \right\rceil, \quad -\left\lceil n_i \frac{\psi_R}{2} \right\rceil \leq n_j \leq \left\lceil n_i \frac{\psi_R}{2} \right\rceil, \quad -\left\lfloor \frac{h_m}{2m_r} \right\rfloor \leq n_k \leq \left\lfloor \frac{h_m}{2m_r} \right\rfloor.$$

In brief, the computation of the N points of the ROI proceeds along the radius (using the n_i index) from the origin up to the maximum depth, with a step resolution equal to m_r. For any generic depth, the n_j index spans the FOV angle ψ_R to compute all the points along the corresponding arc of circle. Finally, the third index n_k computes a copy of the above planar geometry along the vertical dimension. The number N of the points computed by the algorithm can be approximately estimated by the ratio between the ROI geometrical volume and the voxel volume, such that

$$N \approx \frac{\psi_R d_m^2 h_m}{2m_r^3}.$$

Figure A1 shows two examples of ROI with a common FOV angle of 24 degrees but different depths.

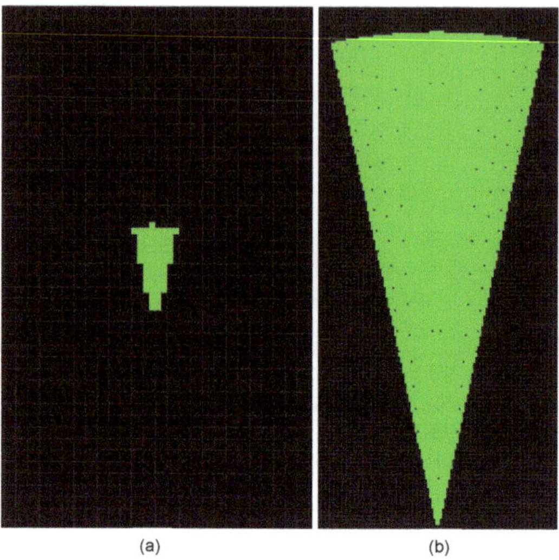

Figure A1. On the left (**a**) is the ROI relative to a desired depth of 7 m; on the right (**b**) is the ROI relative to a desired depth of 120 m. FOV angle is equal to 24 degrees.

References

1. Samaras, S.; Diamantidou, E.; Ataloglou, D.; Sakellariou, N.; Vafeiadis, A.; Magoulianitis, V.; Lalas, A.; Dimou, A.; Zarpalas, D.; Votis, K.; et al. Deep Learning on Multi Sensor Data for Counter UAV Applications—A Systematic Review. *Sensors* **2019**, *11*, 4837. [CrossRef] [PubMed]
2. Palossi, D.; Loquercio, A.; Conti, F.; Flamand, E.; Scaramuzza, D.; Benini, L. A 64-mW DNN-Based Visual Navigation Engine for Autonomous Nano-Drones. *IEEE Internet Things J.* **2019**, *5*, 8357–8371. [CrossRef]
3. Lo, L.Y.; Yiu, C.H.; Tang, Y.; Yang, A.S.; Li, B.; Wen, C.Y. Dynamic Object Tracking on Autonomous UAV System for Surveillance Applications. *Sensors* **2021**, *21*, 7888. [CrossRef] [PubMed]
4. Bigazzi, L.; Basso, M.; Boni, E.; Innocenti, G.; Pieraccini, M. A Multilevel Architecture for Autonomous UAVs. *Drones* **2021**, *6*, 55. [CrossRef]
5. Bigazzi, L.; Gherardini, S.; Innocenti, G.; Basso, M. Development of Non Expensive Technologies for Precise Maneuvering of Completely Autonomous Unmanned Aerial Vehicles. *Sensors* **2021**, *21*, 391. [CrossRef] [PubMed]
6. Zhaocheng Niu; Hui Liu; Xiaomin Lin; Junzhao Du Niu, Z.; Liu, H.; Lin, X.; Du, J. Task Scheduling With UAV-Assisted Dispersed Computing for Disaster Scenario. *IEEE Syst. J.* **2022**, *2*, 1–12. [CrossRef]
7. Alex, C.; Vijaychandra, A. Autonomous cloud based drone system for disaster response and mitigation. In Proceedings of the 2016 International Conference on Robotics and Automation for Humanitarian Applications (RAHA), Amritapuri, India, 18–20 December 2016.
8. Yu, X.; Marinov, M. A Study on Recent Developments and Issues with Obstacle Detection Systems for Automated Vehicles. *Sustainability* **2020**, *12*, 3281. [CrossRef]
9. Gageik, N.; Benz, P.; Montenegro, S. Obstacle Detection and Collision Avoidance for a UAV With Complementary Low-Cost Sensors. *IEEE Access* **2015**, *3*, 599–609. [CrossRef]
10. Yu, Y.; Tingting, W.; Long, C.; Weiwei, Z. A Stereo vision based obstacle avoidance strategy for quadcopter UAV. In Proceedings of the 2018 Chinese Control And Decision Conference (CCDC), Shenyang, China, 9–11 June 2018.
11. Xiao, Y.; Lei, X.; Liao, S. Research on UAV Multi-Obstacle Detection Algorithm based on Stereo Vision. In Proceedings of the 2019 IEEE 3rd Information Technology, Networking, Electronic and Automation Control Conference (ITNEC), Chengdu, China, 15–17 March 2019.
12. Miccinesi, L.; Bigazzi, L.; Consumi, T.; Pieraccini, M.; Beni, A.; Boni, E.; Basso, M. Geo-Referenced Mapping through an Anti-Collision Radar Aboard an Unmanned Aerial System. *Drones* **2022**, *3*, 72. [CrossRef]
13. Safa, A.; Verbelen, T.; Keuninckx, L.; Ocket, I.; Hartmann, M.; Bourdoux, A.; Catthoor, F.; Gielen, G. A Low-Complexity Radar Detector Outperforming OS-CFAR for Indoor Drone Obstacle Avoidance. *IEEE J. Sel. Top. Appl. Earth Obs. Remote Sens.* **2021**, *8*, 9162–9175. [CrossRef]
14. Yu, H.; Zhang, F.; Huang, P.; Wang, C.; Yuanhao, L. Autonomous Obstacle Avoidance for UAV based on Fusion of Radar and Monocular Camera. In Proceedings of the 2020 IEEE/RSJ International Conference on Intelligent Robots and Systems (IROS), Las Vegas, NV, USA, 10 February 2021.

15. Zhang, X.; Zhou, M.; Qiu, P.; Huang, Y.; Li, J. Radar and vision fusion for the real-time obstacle detection and identification. *Ind. Robot* **2019**, *46*, 391–395. [CrossRef]
16. Tsai, Y.; Chen, K.; Chen, Y.; Cheng, J. Accurate and Fast Obstacle Detection Method for Automotive Applications Based on Stereo Vision. In Proceedings of the 2018 International Symposium on VLSI Design, Automation and Test (VLSI-DAT), Hsinchu, Taiwan, 16–19 April 2018.
17. Kim, J.; Han, D.S.; Senouci, B. Radar and Vision Sensor Fusion for Object Detection in Autonomous Vehicle Surroundings. In Proceedings of the 2018 Tenth International Conference on Ubiquitous and Future Networks (ICUFN), Prague, Czech Republic, 3–6 July 2018.
18. Basso, M.; Bigazzi, L.; Innocenti, G. DART Project: A High Precision UAV Prototype Exploiting On-board Visual Sensing. In Proceedings of the ICAS 2019: The Fifteenth International Conference on Autonomic and Autonomous Systems, Athens, Greece, 2–6 June 2019.
19. Bigazzi, L.; Basso, M.; Gherardini, S.; Innocenti, G. Mitigating latency problems in vision-based autonomous UAVs. In Proceedings of the 2021 29th Mediterranean Conference on Control and Automation (MED), Puglia, Italy, 22–25 June 2021.
20. Memon, S.A.; Ullah, I. Detection and tracking of the trajectories of dynamic UAVs in restricted and cluttered environment. *Expert Syst. Appl.* **2021**, *11*, 115309. [CrossRef]
21. Kim, M.; Memon, S.A.; Shin, M.; Son, H. Dynamic based trajectory estimation and tracking in an uncertain environment. *Expert Syst. Appl.* **2021**, *177*, 114919. [CrossRef]
22. Huang, M.; Wei, P.; Liu, X. An Efficient Encoding Voxel-Based Segmentation (EVBS) Algorithm Based on Fast Adjacent Voxel Search for Point Cloud Plane Segmentation. *Remote Sens.* **2019**, *11*, 2727. [CrossRef]
23. Rho, E.; Jo, S. OctoMap-based semi-autonomous quadcopter navigation with biosignal classification. In Proceedings of the 2018 6th International Conference on Brain-Computer Interface (BCI), Gangwon, Republic of Korea, 15–17 January 2018.
24. Hornung, A.; Wurm, K.M.; Bennewitz, M.; Stachniss, C.; Burgard, W. OctoMap: An efficient probabilistic 3D mapping framework based on octrees. *Auton. Robot.* **2013**, *34*, 189–206. [CrossRef]
25. Al-Kaff, A.; García, F.; Martín, D.; De La Escalera, A.; Armingol, J. Obstacle Detection and Avoidance System Based on Monocular Camera and Size Expansion Algorithm for UAVs. *Sensors* **2017**, *5*, 1061. [CrossRef] [PubMed]
26. Yasin, J.N.; Mohamed, S.A.; Haghbayan, M.H.; Heikkonen, J.; Tenhunen, H.; Plosila, J. Unmanned Aerial Vehicles (UAVs): Collision Avoidance Systems and Approaches. *IEEE Access* **2020**, *6*, 105139–105155. [CrossRef]
27. Ferrick, A.; Fish, J.; Venator, E.; Lee, G.S. UAV obstacle avoidance using image processing techniques. In Proceedings of the 2012 IEEE International Conference on Technologies for Practical Robot Applications (TePRA), Woburn, MA, USA, 23–24 April 2012.
28. Budiyanto, A.; Cahyadi, A.; Adji, T.B.; Wahyunggoro, O. UAV obstacle avoidance using potential field under dynamic environment. In Proceedings of the 2015 International Conference on Control, Electronics, Renewable Energy and Communications (ICCEREC), Bandung, Indonesia, 27–29 August 2015; Volume 11.
29. [CrossRef] [PubMed] Wang, L.; Lan, Y.; Zhang, Y.; Zhang, H.; Tahir, M.N.; Ou, S.; Liu, X.; Chen, P. Applications and Prospects of Agricultural Unmanned Aerial Vehicle Obstacle Avoidance Technology in China. *Sensors* **2019**, *2*, 642. [CrossRef] [PubMed]

Article

Urban Air Mobility: Systematic Review of Scientific Publications and Regulations for Vertiport Design and Operations

Karolin Schweiger [1,*] and **Lukas Preis** [2]

1   German Aerospace Center (DLR), Institute of Flight Guidance, Lilienthalplatz 7, 38108 Braunschweig, Germany
2   Bauhaus Luftfahrt e.V., Willy-Messerschmidt-Straße 1, 82024 Taufkirchen, Germany; lukas.preis@bauhaus-luftfahrt.net
*   Correspondence: karolin.schweiger@dlr.de; Tel.: +49-531-295-3835

Abstract: Novel electric aircraft designs coupled with intense efforts from academia, government and industry led to a paradigm shift in urban transportation by introducing UAM. While UAM promises to introduce a new mode of transport, it depends on ground infrastructure to operate safely and efficiently in a highly constrained urban environment. Due to its novelty, the research of UAM ground infrastructure is widely scattered. Therefore, this paper selects, categorizes and summarizes existing literature in a systematic fashion and strives to support the harmonization process of contributions made by industry, research and regulatory authorities. Through a document term matrix approach, we identified 49 *Scopus*-listed scientific publications (2016–2021) addressing the topic of UAM ground infrastructure with respect to *airspace operation* followed by *design, location and network, throughput and capacity, ground operations, cost, safety, regulation, weather* and lastly *noise* and *security*. Last listed topics from *cost* onwards appear to be substantially under-represented, but will be influencing current developments and challenges. This manuscript further presents regulatory considerations (Europe, U.S., international) and introduces additional noteworthy scientific publications and industry contributions. Initial uncertainties in naming UAM ground infrastructure seem to be overcome; *vertiport* is now being predominantly used when speaking about vertical take-off and landing UAM operations.

Keywords: urban air mobility; UAM; eVTOL; vertiport; literature review

Citation: Schweiger, K.; Preis, L. Urban Air Mobility: Systematic Review of Scientific Publications and Regulations for Vertiport Design and Operations. *Drones* **2022**, *6*, 179. https://doi.org/10.3390/drones 6070179

Academic Editors: Ivana Semanjski, Antonio Pratelli, Massimiliano Pieraccini, Silvio Semanjski, Massimiliano Petri and Sidharta Gautama

Received: 19 June 2022
Accepted: 12 July 2022
Published: 19 July 2022

Publisher's Note: MDPI stays neutral with regard to jurisdictional claims in published maps and institutional affiliations.

1. Introduction

"To take off, flying vehicles first need places to land" [1]

The interest in suitable VTOL ground infrastructure is rising due to the growing amount of small UAS applications and the thriving topic of UAM introducing a new mode of passenger transport and on-demand deliveries inside urban areas. UAM is striving for revolutionizing the status quo of ground transportation, aircraft design, ATM processes and the principles of multi-modality. Furthermore, UAM seeks to connect residential areas and airports to city centres, to attract as many residents as possible by promising immense time savings under affordable conditions. UAM is setting the scene for new approaches, new technologies and new potential markets. However, UAM is describing a new mode of aerial transportation which will be implemented in very challenging urban environment in which VTOL capabilities and early considerations of infrastructure design specifications are expected to be crucial. This is supported by EASA "Study on societal acceptance of Urban Air Mobility in Europe" which concluded with infrastructure being the biggest challenge for UAM [2].

These days, the topic of UAM is thriving, the number of published contributions is large, but those who focus specifically on UAM ground infrastructure are widely scattered and are addressing different business cases, time horizons and technological readiness. This manuscript provides a detailed and systematic review of 49 *Scopus*-listed, scientific publications about ground infrastructure in the context of UAM and published between the years 2016 and 2021 (including). The publications were selected through a text mining approach: if the abstract of a publication contained both "urban air mobility" and at least one keyword related to ground infrastructure (see the list of keywords in Section 1.1) it was included in the selection. The various text mining techniques used in the analysis are explained in Sections 1.2 and 1.3. These encompass database overlap analysis, document term matrix and document classification. All scripts were written by the authors using the following *Python 3.8* packages: *pandas, nltk, stop_words* and *statistics*. A comprehensive introduction into the text mining approaches used in this review can be found in [3].

The review predominantly focuses on VTOL operations and subsequently calls UAM ground infrastructure: *vertiports*. Furthermore, additional noteworthy contributions made by research, regulatory authorities and industry are presented. This review complements already existing UAM review publications of Garrow et al. [4] and Straubinger et al. [5] and contributes thereto by focusing explicitly on ongoing research, regulatory and industrial contributions as well as intermediate achievements in the field of UAM VTOL ground infrastructure. We are aware that the term "urban air mobility" indicates a limited view compared to "advanced air mobility" (AAM) as proposed by NASA [6]. Yet NASA continues to use the term UAM as a subset of AAM, as do comprehensive reviews of the field [4,5]. For this reason we will use the term UAM, but we do not intend to exclude other applications of AAM, such as regional or rural air mobility.

Throughout the review, eleven research topics were identified: *airspace operation, design, location and network, throughput and capacity, ground operations, cost, safety, regulation, weather, noise* and *security* (sequence: descending prominence), which shaped the following structure of the manuscript. Section 1 provides an overview and a systematic trend analysis (text mining) of already used UAM ground infrastructure terminology and classifications. Section 2 elaborates a summary of current heliport design guidelines and introduces first drafts and prototypes of vertiport design specifications focusing mainly on European and American contributions. The subsequent Sections 3–5 summarize and discuss the contributions of 49 publications based on the trend analysis introduced in Section 1. Additional noteworthy scientific, regulatory and industry contributions are discussed. Section 3 examines the development of vertiport networks considering different operating environments and groups of customer. Section 4 summarizes vertiport design proposals, analyzes different approaches of developing vertiport airside air and ground operations and collects initial investment estimations for specific vertiport designs. Section 5 concludes the review by providing initial evaluations of weather impacting UAM and vertiport operations. Finally, Section 6 conducts a critical evaluation of all sighted contributions and highlights pending and under-represented research questions.

1.1. Taxonomy of UAM Ground Infrastructure

One might ask the question, why is there a need to define a new class of ground infrastructure specifically for UAM when we already have a distinct set of thoroughly practiced design guidelines covering aerodromes, airports and heliports?

Assuming affordable access to UAM flights is targeted, high numbers of throughput need to be achieved which will require larger and probably more complex ground infrastructure topology and access management as it is currently available for helicopter/heliport operations [7]. This may include ground taxiing of VTOL aircraft, reduced separation, simultaneous/automatic/autonomous operations as well as steep/vertical approach and departure profiles in order to operate in densely populated and built-up urban environment. For comparison, basic flight maneuvers for rotorcraft address a typical descent profile of 8 to 12 degrees whereas a steep approach is defined by approx. 15 degrees descent angle [8]. Moreover, UAM being considered on-demand, following high dispatch frequencies and mainly operating in urban scenery with shortly changing flight phases are characteristics of significant difference compared to current aviation operations.

As to understand with what UAM ground infrastructure is associated with and what considerations are stated in terms of classification and definition, the following Sections 1.1.1 and 1.1.2 will provide an overview of historic and current developments.

1.1.1. Regulatory and Standardization Context

Both well-established and novel aircraft manufacturers, research facilities, local and public authorities, regulatory agencies, CNS providers, air navigation service providers, consulting companies and many more all around the world are currently contributing to the development of UAM. A considerable inconsistency was found in the classification of such UAM VTOL ground infrastructures throughout different (scientific) publications addressing UAM.

Starting with already familiar aviation ground infrastructure and according to ICAO, the *aerodrome*, is "a defined area on land or water (including any buildings, installations and equipment) intended to be used either wholly or in part for the arrival, departure and movement of aircraft" [9]. In the European certification specification for aerodrome design *CS-ADR-DSN*, EASA follows ICAO's guidelines but added the specification of being located "on land or water or on a fixed offshore or floating structure" [10]. This also includes small general aviation airfields, heliports, commercial airports and military airbases [11]. A distinct version for rotorcraft, the *heliport*, is defined by ICAO's *Annex 14* and EASA's *CS-HPT-DSN* as "an aerodrome or a defined area on a structure intended to be used wholly or in part for the arrival, departure and surface movement of helicopters" [9,12]. For completion, an airport has terminal(s) and car parks additional to the infrastructure used by the aircraft itself, thus the aerodrome is part of an airport [11]. Consequently, the heliport extends the characteristic of an aerodrome by the definition of an area on structure which includes the possibility of elevated areas. Also, the heliport is exclusively used by helicopters, whereas the aerodrome can be used by both vehicles. It needs to be highlighted that EASA's *CS-HPT-DSN* only provides design certification specification for heliports located at aerodromes that fall under scope of *Regulation (EU) 2018/1139*.

Transitioning from "traditional" aviation towards initial serious considerations of inter-city aerial transportation, in 1983, the National Rotorcraft Program analyzed how the national inter-urban transportation market in the U.S. can be improved [13]. Among others, the report determined that conventional helicopters did not satisfy the stated requirements due to lack of capacity, high operational costs and high noise levels. The recommendation of considering tiltrotor aircrafts offered higher speed and range and vertical take-off and landing capabilities.

Followed by this recommendation, in 1985, the FAA, NASA and the Department of Defense conducted a joint civil tiltrotor study in order to identify the potential of the commercial tiltrotor transport market [13] . Several studies followed covering the topics civil tiltrotor missions and applications, potential risk areas, market evaluations, ground infrastructure planning and development, air traffic control and public acceptance (see [13–16]).

Driven by those civil tiltrotor developments generated by industry, military and government, in 1991, the FAA developed an *AC 150/5390-3* guiding vertiport design [17]. The terminologies *vertiport* and *vertistop* were first introduced describing respectively "an identifiable ground or elevated area, including any buildings or facilities thereon, used for takeoff and landing of tiltrotor aircraft and rotorcraft" and "a vertiport intended solely for takeoff and landing of tiltrotor aircraft and rotorcraft to drop off or pick up passengers or cargo". This AC paved the way for the term vertiport and the general idea of creating classes of ground infrastructure to describe different characteristics and operational capabilities. Those considerations were never put into practice since military tiltrotor technologies were never used commercially therefore causing the cancellation of *AC 150/5390-3* in July

2010 [18]. However, years later, those former developments serve as important precedent being now adjusted and refined for modern UAM operations.

First, the generic term *UAM aerodrome* was introduced by FAA's first version of a UAM ConOps [19] addressing foundational principles, roles and responsibilities, scenarios and operational threats. It describes "a location from which UAM flight operations depart or arrive. [...] UAM aerodrome is used explicitly when the context indicates functionality to support UAM operations that is not present in NAS [National Airspace System] operations" [19].

NASA is following FAA's approach by using the term *UAM aerodrome* in the first version of the published UAM Vision ConOps in 2020 [6], addressing a UAM operation of medium density and complexity. The term *UAM aerodrome* is further specified by addressing operational UAM characteristics such as VTOL capabilities and ground movement leading into the definition of a "specifically defined area that is intended for the arrival, departure, and ground movement of UAM aircraft. Because of the VTOL nature of many UAM aircraft, most UAM aerodromes look more like today's heliports with landing pads as opposed to long runways" [6]. In a follow-up ConOps addressing high-density automated vertiports [20], NASA again further specified the classification and defined the term *vertiport* in correspondence to the aircraft design (VTOL and rotorcraft) and its propulsion unit (eVTOL). Also, the physical location of a vertiport (ground-based or elevated) is now part of the definition which resulted into "an identifiable ground or elevated area, including any buildings or facilities thereon, used for the takeoff and landing of eVTOL and rotorcraft".

Responding to the rising requests claiming for a consolidated UAM ground infrastructure design guideline, in March 2022 the FAA published an engineering brief on the subject of vertiport design limited to piloted and VFR VTOL operations in order to capture early UAM VTOL operations [18]. In [18], UAM ground infrastructure is now following the initial classification of [17], but clearly stating propulsion characteristics, VTOL capabilities and the specific use of co-located buildings for passenger handling and other UAM services. Consequently, the *vertiport* is defined as "an area of land or a structure, used or intended to be used, for electric, hydrogen, and hybrid VTOL landings and takeoffs and includes associated buildings and facilities" and the *vertistop* as "an area similar to a vertiport, except that no charging, fueling, defueling, maintenance, repairs, or storage of aircraft are permitted" [18].

Transitioning to European UAM applications, EASA introduced the term *vertiport* in the first draft of the SC SC-VTOL-01 [21] in 2019. It provides an initial description naming the vertiport "an area of land, water, or structure used or intended to be used for the landing and take-off of VTOL aircraft". There is no specific requirement attached to that definition addressing the VTOL aircraft's propulsion unit, passenger handling and service facilities providing e.g., charging/refuelling and maintenance. This rather generic definition was picked-up by EASA's Prototype Technical Specification *(PTS-VPT-DSN)* for VFR Vertiports [22] published in 2022.

Since regulatory authorities are working closely together with standardization bodies, it is noteworthy mentioning them in this context. The EUROCAE, operating as a non-profit organization, is dedicated to the elaboration of aviation standards since 1963. The development of UAM operations is incorporated in working group 112 "Vertical Takeoff and Landing" which is developing several standards such as vertiport operations (ED-299 currently under development [23]), and VTOL aircraft ConOps (ED-293 [24]). Important groundwork for [22] was provided by EUROCAE. In [24], EUROCAE makes use of the term *vertiport* following the definition stated in EASA's SC-VTOL-01.

On an international standardization level, the International Organization for Standardization *ISO*, is currently developing a vertiport standard *ISO/AWI 5491* under the technical committee *ISO/TC 20/SC 17 Airport Infrastructure* [25]. A publication is still pending. Further, *ASTM International* initiated already in 2017 the work item of "New Specifications for Vertiport Design" which also indicates the usage of *vertiport* and *vertistop* and providing the following, sofar most precise definition: "Vertiport means a generic reference to the area of land, water, or structure used, or intended to be used, for the landing and takeoff of VTOL aircraft together with associated buildings and facilities. Vertistop means a minimally developed VTOL aircraft facility for boarding and discharging passengers or cargo. The vertiport/vertistop relationship is comparable to a bus terminal-bus stop relationship with respect to the extent of services provided or expected" [26]. It is also highlighted that vertiports are expected to serve both civil VTOL aircraft and civil helicopters and the extension for electric driven VTOL aircraft should be considered carefully [26].

1.1.2. Commercial and Research Context

In 2016, when *UBER Elevate* published the whitepaper "Fast-Forwarding to a Future of On-Demand Urban Air Transportation" [27], the topic short range metropolitan air transportation including the vertiport "came back to life". Several whitepapers followed addressing among others "The Roadmap towards scalable urban air mobility" [28], "The New Digital Era of Aviation" [29] and a "Concept of Operations: Autonomous UAM Aircraft Operations and Vertiport Integration" [30].

Ref. [27] picks up the terminologies introduced by [17] but focusses on layout and charging characteristics. The infrastructure which supports urban VTOL operations is defined as *vertiports*, described as "VTOL hubs with multiple takeoff and landing pads, as well as charging infrastructure" and as *vertistops* "a single TLOF pad with minimal infrastructure". This whitepaper together with the following *UBER Elevate* Summits in the years 2017, 2018 and 2019 received considerable attention and significantly pushed forward the topic of UAM. This trend is also depicted by the number of publications related to the topic UAM ground infrastructure in Figure 1. When investigating all publications listed in the online database *Scopus* from the year 2000 onwards, which are displaying a connection to the keyword UAM ground infrastructure, it appears that the number of publications is increasing explicitly with the year 2016.

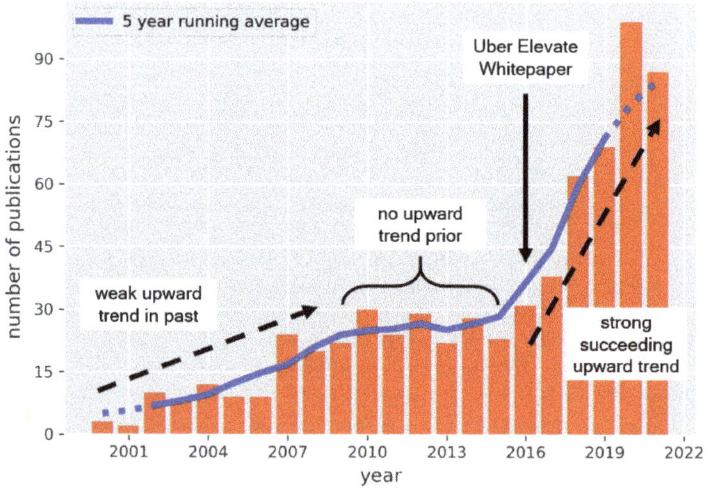

Figure 1. Publications related to UAM ground infrastructure as listed in *Scopus* after the year 2000 and in relation to the publication of *UBER*'s whitepaper in 2016 [27].

The consulting companies *Deloitte* [31] and *McKinsey and Company* [1] both established a UAM ground infrastructure classification with multiple sub-categories addressing varying features, capabilities and local implementation. The generic term of the physical infrastructure is termed as *vertiplaces* [31] and *VTOL ports* [1], respectively. The largest archetype is defined by both as *vertihub*. Ref. [31] describes it as small airports for eVTOL aircraft, mainly located on the periphery of urban or suburban areas because of their large footprint including the availability of MRO infrastructure, whereas, ref. [1] envisions it as a stand-alone building implemented in central and high-traffic areas providing charging/refueling capabilities for VTOL aircraft and distinct services for passenger. The second archetype is termed as *vertiport* and *vertibase*, respectively. Based on [31], the *vertiport* is located at points of interests ideally integrated with other modes of ground transportation. Multiple eVTOL aircraft can be accommodated, fast-charging, refueling and minor MRO services are provided. Security check-points, passenger waiting lounges, systems for fire safety and real-time surveillance are highlighted as well. According to [1], *vertibases* are medium size, located at medium-traffic areas and are either newly built or retro-fitted. As third archetype depicting the smallest footprint, Refs. [1,31] use the term *vertistation* and *vertipad* respectively. On the one hand a *vertistation* provides only one or two pads for which the use of existent helipads can be considered. On the other hand, *vertipads* are assigned to a "spoke" in a hub-and-spoke network. Both share the characteristic of smaller footprints and lower costs which could enable an easy implementation as peripheral infrastructure in suburban or rural locations.

Following the approach of multiple archetypes but based on aircraft performance and UAM ground infrastructure capabilities, ref. [7] uses the term *UAM aerodrome* by [19] as hypernym for UAM ground infrastructure. With regard to a UAM aircraft's performance, VTOL or STOL capabilities are distinguished resulting into different UAM aerodrome classes. The term *vertidrome* was used for VTOL operations and *stoldrome* for STOL operations only. Two additional flavors of vertidromes are used, *vertiport* and *vertistop*, in order to distinguish between operational and technical capabilities like charging, refueling, MRO and passenger handling.

Numerous terms for novel take-off and landing ground infrastructure were found by [32], such as *vertiport, vertipad, pocket airport, skypark, sky node* and *sky port*. To avoid the definition of a specific term and therefore limiting ground infrastructure to a specific characteristic, ref. [32] uses the generic term *TOLA*, take-off and landing area, for on-demand mobility operations, which describes any location an aircraft, VTOL or STOL aircraft, can depart from or arrive at. Additional terms were found such as *Verti-X* [33], *skyports* [34] and *airpark* [35] if super STOL (SSTOL) and STOL aircraft are being considered to serve metropolitan areas and intra-city operations.

But towards what terminology is the UAM community trending? The next section will run a systematic analysis of what terminologies are used in the scientific context, based on the set of *identified terms* introduced in this section.

1.2. Trends in Research and Scientific Publication

In this section, the use and prominence of the above-mentioned terms or keywords (both words used synonymously) will be analyzed. The goal is to illuminate the usage of different keywords in the past and present and help the community become more aware of current developments in the field of UAM ground infrastructure. As hinted in the title of this paper, we believe "vertiport" to be the most prominent keyword and it is therefore used throughout this manuscript.

In Section 1.1, a total of 19 keywords were discussed. A search in the publication database *Scopus* (find the *Scopus* publication portal under https://www.scopus.com/; accessed on 11 July 2022) yielded that 11 of 19 keywords were used at least once in the listed scientific literature (equals database "ground infrastructure"); this means 8 keywords were not used at all. A limitation of this approach was that the keywords needed to occur in the title or abstract of the publication, as *Scopus* only searches the meta data of publications. This database was chosen as a compromise between a wide range of publications (e.g., *Web of Science* does not list conference proceedings) and quality of publications (e.g., *Google Scholar* has no transparent mechanism of selecting papers).

To gain a feeling for the trend of each keyword (see Table 1), three time spans and sub-databases were looked at in particular: the last two decades, the last 10 years, and the years after 2016 which marked a turning point due to the publication of the UAM white paper by *UBER Elevate* [27] (see Figure 1). The number of publications in each sub-database is shown as well in Table 1. The size of the database does not have to match the sum of the occurrences of all keywords for various technical reasons: for example, one paper could contain multiple keywords from the list.

Table 1. Prominence of "ground infrastructure" related keywords.

Keyword	*Scopus* All Years	Past Two Decades (2000–2021)	Past 10 Years (2012–2021)	Since *UBER Elevate* (2016–2021)
aerodrome	662	536	383	296
airpark	30	27	12	9
pocket airport	2	2	2	1
skynode	23	23	13	8
skypark	6	5	5	4
vertidrome	1	1	1	1
vertihub	2	2	2	2
vertipad	4	6	6	6
vertiport	82	63	62	60
vertistop	2	2	2	2
verti-x ine	1	1	1	1
Size of database without duplicates	810	689	500	396

As the focus of this review is ground infrastructure in the context of UAM, the same search was then applied to the keyword of "urban air mobility" (equals database "urban air mobility"). The goal of this analysis is to find the best-fitting keyword for ground infrastructure in the context of UAM, which is done by comparing the two databases derived from *Scopus*. In Figure 2, the overlap of these two databases is visualized. Set A and B represent the UAM and ground infrastructure database, respectively. The comparison of two sets is conducted by looking at DOIs as unique identifier. As not all listed entries carry a DOI, these entries are removed, yielding the sets C and D representing the cleaned databases for UAM and ground infrastructure, respectively.

Entries in *Scopus* that do not carry a DOI number can be proceedings, workshop summaries or other material, but also conference papers and articles. There are other ways of comparing entries such as using the title, but this might lead to problems with consistency. Excluding all entries that do not carry a DOI number is therefore a way of dataset quality control, while we acknowledge that this might create a bias within the dataset.

The combination of both databases is labeled as set E, the papers exclusively occurring in the urban air mobility database as set F and the papers exclusively occurring in the ground infrastructure database as set G. Our set of interest are those papers shared by both databases, which are labelled as set H. In Table 2, a brief description of all sets is given, including the size of each set and their relation to one another. Comparing the databases of both searches showed that only 8 keywords (see Table 3) of the 11 keywords shown in Table 1 are used in the context of UAM throughout 49 scientific publications listed by *Scopus*.

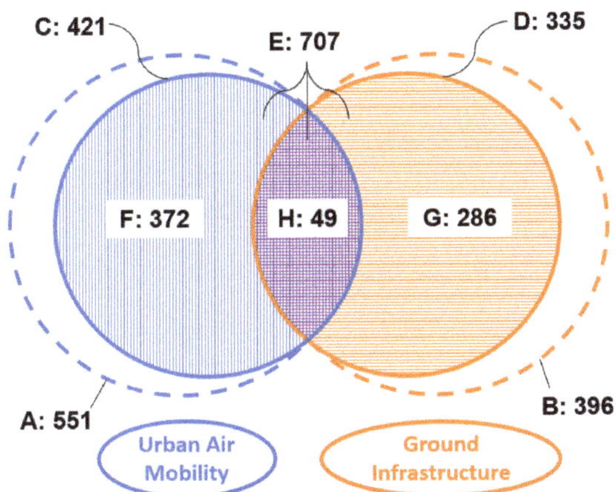

Figure 2. Overlap between databases derived from the keyword "urban air mobility" and 11 keywords related to "ground infrastructure".

Table 2. Size of sets from database overlap analysis: 49 shared papers including keyword "urban air mobility" (UAM) and keywords related to "ground infrastructure" (GI).

Set	Descripton	Size	Mathematical Relation of Sets
A	UAM all publications	551	$A \supseteq C \supseteq F$
B	GI all publications	396	$B \supseteq D \supseteq G$
C	UAM only publications with DOI	421	-
D	GI only publications with DOI	335	-
E	UAM and GI combined	707	$E = C \cup D$
F	UAM exclusive	372	$F = C \setminus H$
G	GI exclusive	286	$G = D \setminus H$
H	UAM and GI shared	49	$H = C \cap D$

Table 3. Keyword occurrences describing UAM ground infrastructure (set H).

Keyword	Hits
aerodrome	1
airpark	4
vertidrome	1
vertihub	1
vertipad	4
vertiport	40
vertistop	1
verti-x	1
total hits	**49**

Applying a document term matrix approach, the number of occurrences of each keyword in the final database can be highlighted (see Table 3). A document term matrix shows how often each keyword occurs in each publication. The number of hits shown are the sum of occurrences across all 49 publications. It can be seen that "vertiport" occurs in 40 of the total 49 publications and therefore covers over 80%. *Vertiport is the most prominently used term or keyword to describe UAM ground infrastructure.* This is in direct contrast to the wider field of aerospace research where "aerodrome" (296 occurences in the last 6 years) appears to be the more prominent keyword while "vertiport" is used less often (60 occurences in the last 6 years) as can be seen in Table 1. Yet, the keyword "aerodrome" has negligible relevance in the UAM community (1 occurence in the context of "urban air mobility", see Table 3). The process of selecting keywords describing ground infrastructure and finding the overlap with the body of research concerned with UAM is summarized in Figure 3.

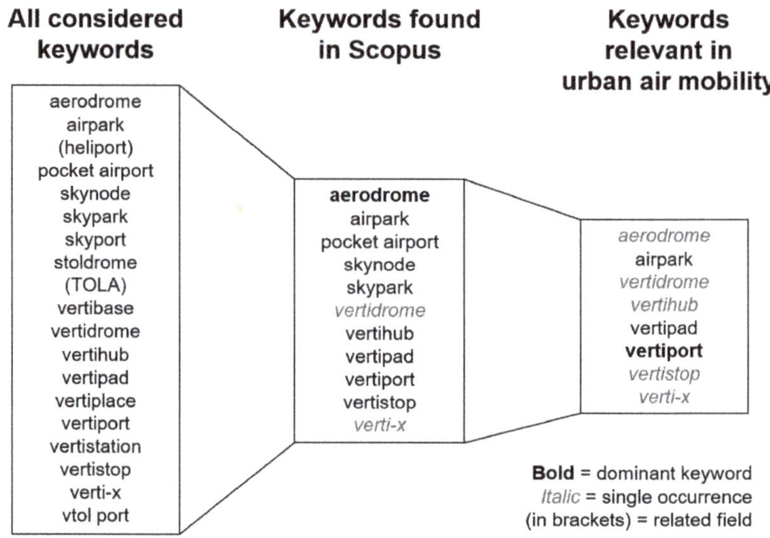

Figure 3. Selection process of keywords used to describe "ground infrastructure" in the context of "urban air mobility".

An analysis of the full UAM database (set A displayed in Figure 2) shows, that only two papers have been published before 2016, wherefore the assumption to start our analysis with the publication of *UBER*'s whitepaper [27] in 2016 is justified. We are aware that searching for the keyword "urban air mobility" may neglect former UAM-like contributions covering intra-city air travel. The focus of this manuscript, however, is to specifically cover the recent trend of UAM addressing novel eVTOL aircraft and airspace designs as well as the concept of on-demand and multi-modal mobility.

An exponential growth in UAM related publications can be seen after the year 2018. Analyzing the vertiport database (set H displayed in Figure 2) also shows a rising trend in publications. Both trends are visualized in Figure 4. Using a data analytics approach the most frequent authors are listed in the Appendix A. Similarly, the conference proceedings and journals which published most often about the topic of vertiports are identified (see Figure A1a and Figure A1b, respectively). Finally, a list of the top ten papers with the highest impact according to number of citations is shown in the Appendix A in Table A1. This overview is supposed to give the reader an idea of which publications and authors impacted the research community; and where to search for articles and submit personal contributions to.

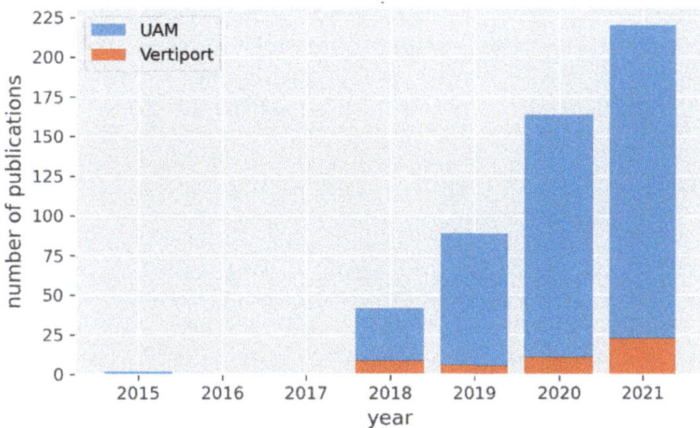

Figure 4. Trends of publication in the fields of UAM and vertiports.

1.3. Classification of Vertiport-Related Topics

Reading through the 49 scientific publications extracted from *Scopus* as explained in Section 1.2, we identified eleven topics which will be proposed as a classification of the current vertiport research. The topics and their prominence across those 49 publications are displayed in Figure 5. The sizes of the rectangles correspond to the weight of each topic. The larger the area of a topic the more attention it received so far. The weight of a topic was determined via weighted sum analysis. First, for each publication it was analyzed if the topic played no role (0 points), a minor role (1 point), or a major role (2 points). This was applied for all topics giving each publication a sum of points. Second, the amount of points for the topic was divided by the sum of all points of that particular publication to create a normalized point-score (so that the sum of point-scores for each publication is 1). Third, the normalized point-scores of the topic were added up across all publications.

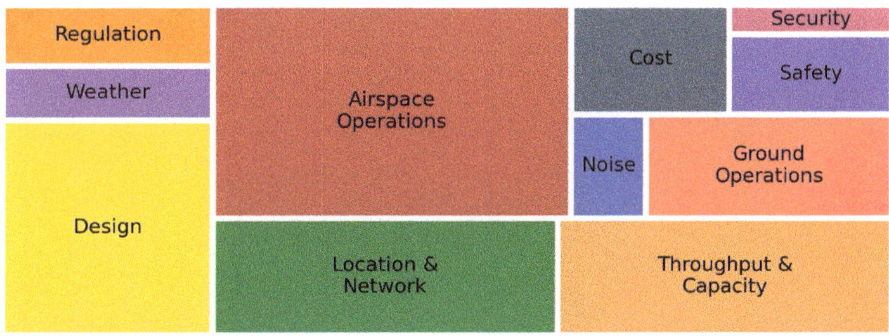

Figure 5. Classification of vertiport-related topics and their weight in the reviewed scientific literature (49 publications); size of rectangle corresponds to prominence of the topic.

1.4. Summary

Reviewing different publications addressing the description of UAM ground infrastructure resulted into a collection of various approaches, classifications and terminologies used for UAM ground infrastructure (cf. Section 1.1). UAM ground infrastructure is often classified based on the operating vehicle's performance (VTOL, STOL, civil helicopter), propulsion characteristics (electric, hybrid, hydrogen, LNG), operational features (charging, refueling, MRO), entertainment services (passenger, residents) and training capabilities. Additionally, the overall footprint (large, middle, small), the way of implementation (newly built, retro-fitted) and the location where UAM ground infrastructure is going to be placed (city-center, urban, sub-urban, periphery, connected to other modes of transport) play an important role when establishing UAM ground infrastructure and its specific services. Based on the individual perspective, 19 different terms have been identified. Searching in the database *Scopus* for "ground infrastructure" in connection to "urban air mobility" and trough database overlap analysis, we found 49 publications building the basis for this manuscript (cf. Section 1.2). Using a document term matrix, we were able to show that "vertiport" is the most commonly used term occurring in over 80% of the sighted publications. Additionally, we found a rising trend of vertiport publications starting in the year 2018; this affirms our assumption to only include recent years in our analysis. Lastly in Section 1.3, we identified eleven research areas in the vertiport domain presently addressed with varying significance. This includes airspace operation, design, location and network, throughput and capacity, ground operations, cost, safety, regulation, weather as well as noise and security.

2. Heliport and Vertiport Design Guidelines

"Heliports provide the most analogous present-day model for VTOL vertiports. However, despite the similarities between the two types of aircraft, there are design differences between traditional helicopters and VTOL aircraft. VTOL aircraft come in varied configurations and propulsion systems, with and without wings, and with varied landing configurations." [18]

Merging aerial transportation with our daily lives would often require vertiports to be located in densely populated areas and inside city boundaries which is currently more a vision than a reality. If future vertiports are going to play an eligible part of a multi-modal transportation network already following certain standards, they have to be additionally aligned with aviation safety standards in order to operate in the first place. *Skyports*, a globally acting developer of UAM ground infrastructure, demands that "national and international aviation rules and industry standards must be changed rapidly to enable the introduction of new VTOL aircraft and associated ground infrastructure" [36]. Driven by these demands, national aviation agencies who are responsible for providing and regulating safe flight conditions are now working on adjusting current design guidelines and regulations, and where necessary, to develop and implement new ones. Since the UAM community is still lacking a comprehensive understanding of how VTOL operations are changing ground infrastructure design specifications and requirements, it is frequently referred to already existent heliport and rotorcraft terminologies, approaches and procedures. Figure 6 depicts the terminology typically used in the context of UAM and vertiports.

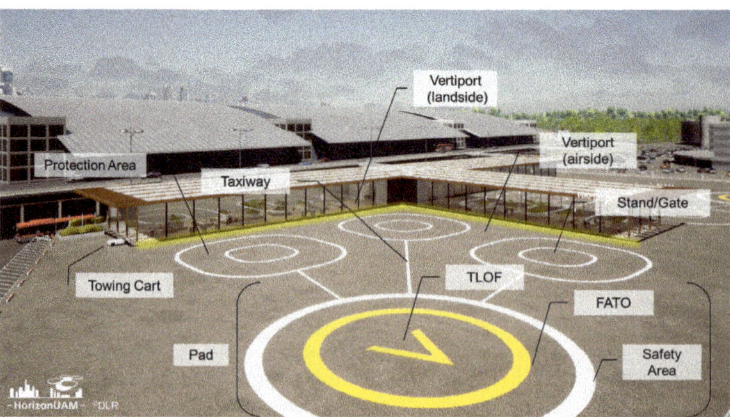

Figure 6. Vertiport topology terms used in the context of UAM.

Depending on different time horizons, maturity levels and traffic densities, vertiports can differ in elements, capability, size and throughput. One key element is the TLOF of specific size, pavement, marking, load-bearing and drainage, etc. in order to withstand dynamic forces during touchdown. At the TLOF, the VTOL aircraft initiates take-off and conducts final touchdown. The FATO is a defined area of specific size over which the VTOL aircraft is completing its final phase of approach or initial phase of departure. A dedicated safety area surrounds the FATO to specific extent and provides an extended obstacle free area. Additional stands of specific size and protection area can be used for parking and passenger handling. They are connected by a taxi route in order to provide a safe transition from one element to another. Taxi routes must follow pre-defined requirements and have to provide protection areas to ensure a safe operation. Various operational modes of taxiways can be considered, such as moving the vehicle through air or on the ground resulting into different size and safety margins (see Section 4.2.2).

In the following two sections, a summary of historic and current regulatory design guidelines will be provided with the focus on European and American contributions.

2.1. Europe

Ongoing vertiport research and regulatory work is driven by EASA's drone and VTOL operation initiative.

In 2020, a first issue of a proposed means of compliance *MOC SC-VTOL* was published focusing primarily on basic VTOL aircraft design topics such as minimum handling qualities and CFP [37]. A thorough definition of a vertiport's role and minimum requirements was missing. EASA's second publication of proposed *MOC-2 SC-VTOL* [38] started to address the airside operation of a vertiport such as approach and departure paths, operating volumes, FATO dimension and climb gradients, for which a final publication is expected in 2022.

Based on those developments, a Prototype Technical Specification for the design of VFR vertiports accommodating manned eVTOL aircraft, *PTS-VTP-DSN*, was published in March 2022 and is leading the way for a first European regulatory framework [22].

2.1.1. Operation Classes

In Europe, UAS operations are grouped in different operation classes based on the performance involved and the operational risk addressed. Its categories are *open*, *specific* and *certified*. Operations in the open and specific category address (leisure) operations with low and medium level of risks for which we already have a European regulatory framework for (Open: [39], Specific: [40]). Lastly, the certified category caters for the highest level of risk, therefore asking for the highest safety standards compared to other operation classes. According to [41], certified operations need to meet aircraft standards for manned aviation requiring a type certificate and a certificate of airworthiness. The dependency between type certificate, risk-levels and operational requirements including the use of designated UAM ground infrastructure was developed in the first issue of *SC-VTOL-01* in 2019 [21]. "VTOL aircraft that are certified in the Category Enhanced would have to meet requirements for continued safe flight and landing, and be able to continue to the original intended destination or a suitable alternate vertiport after a failure. Whereas for Category Basic only controlled emergency landing requirements would have to be met, in a similar manner to a controlled glide or auto-rotation" [21]. In order to better understand the European approach of classiyfing UAS operations, a structured overview of its setup is depicted in Figure 7. European regulation for certified UAS operations is currently under development under the rule making task *RMT.0230(C)* which initially defines three types of operation [42]. Operation type #1, IFR cargo UAS operations in class A-C airspace. Operation type #2, UAS operation in congested environment in *U-space* airspace including unmanned passenger and cargo transport. Completed by operation type #3 following characteristics of type #2 but with pilot on-board and considering also operations outside of *U-space* airspace. For further description of the topic *U-space*, please visit Section 2.1.5.

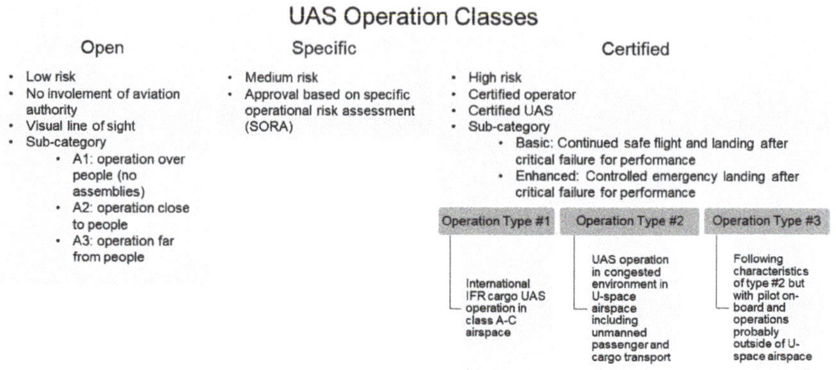

Figure 7. European UAS operation classes, subcategories and types based on [41,43,44].

Later on, when operating volumes and contingency procedures at vertiports are being defined, the corresponding operation class and operation type will determine performance and therefore vertiport footprint requirements.

2.1.2. D-Value

Following former heliport design guidelines such as [12], the D-value has been used to dimension a heliport's airside topology, safety margins and operating constraints. The D-value defines "the largest overall dimension of the helicopter when rotor(s) are turning measured from the most forward position of the main rotor tip path plane to the most rearward position of the tail rotor tip path plane or helicopter structure" [12]. Comparing novel VTOL aircraft designs (cf. [45]), ref. [46] found that the smallest enclosing circle being equally to the D-value for rotorcraft can be off by 15%. A thorough mathematical derivation is provided in Appendix 1 of [22]. In order to secure sufficient obstacle clearance, EASA re-defined the D-value for VTOL aircraft by changing it into "the diameter of the smallest circle enclosing the VTOL aircraft projection on a horizontal plane, while the aircraft is in the take-off or landing configuration, with rotor(s) turning if applicable. [...] If the VTOL aircraft changes dimension during taxi or parking (e.g., folding wings), a corresponding D_{taxi} and $D_{parking}$ should also be provided" [38].

2.1.3. Vertiport Design Guidelines

Taking into account the new D-value definition specifically fitting VTOL aircraft designs, key elements of a vertiport (airside ground) can be dimensioned in order to establish an operating environment. Please re-visit Figure 6 to refresh specific heliport/vertiport design elements and terminologies used.

According to [22], a vertiport has to offer at least one FATO, in order to provide a designated area free of obstacles and with sufficient surface and load-bearing qualities. The dimension of a FATO is driven by the vehicle with the largest D-value intending to operate on the designated ground infrastructure. Furthermore, at least one TLOF needs to be provided at a vertiport. It can be located within a FATO or co-located with a stand. An additional safety area (solid/non-solid) exceeding the FATO and a protection side slope should protect the operation from penetrating obstacles. The vertiport might also offer taxiways and stands for additional operation. Both can be designed to meet either ground or hover movement capabilities of the VTOL aircraft resulting in higher footprints for the latter. Stands can be used simultaneously, sequentially, by turning in a hover or by taxiing-through without a need to turn. Depending on the intended operation, different requirements need to be met. Furthermore, EASA's *PTS-VTP-DSN* proposes a lightning vertiport identification marking of a letter "V" inside a blue circle, a D-value marking to clearly state those aircraft designs being able to be accommodated at the vertiport, a FATO identification number, as well as a marker for the maximum allowable mass. Additional proposals for approach lighting systems and flight path alignment guidance markings and lights were elaborated, defining the location, characteristics, and configurations of each system. It is expected, as a second step, that a full regulatory framework will be developed in the context of the rule making task *RMT.0230* "Introduction of a regulatory framework for the operation of unmanned aircraft systems and for urban air mobility in the European Union aviation system" [42] in the near-term.

For further details, the reader is pointed to EASA's certification specification for VFR heliports *CS-HPT-DSN* [12] and VFR vertiports *PTS-VTP-DSN* [22].

2.1.4. Proposed Reference Volume for VTOL Procedures

After examining the design requirements for a vertiport's airside ground topology, the airspace directly attached to the vertiport accommodating among others approach and departure paths (airside air) needs to be structured. Reviewing different regulatory proposals and guidelines, in the second publication of the proposed *MOC-2 SC-VTOL* [38], VTOL take-off and landing procedures are building on existing regulations for helicopters of category A. "Category A with respect to helicopters' means a multi-engined helicopter designed with engine and system isolation features specified in the applicable airworthiness codes and capable of operations using take-off and landing data scheduled under a critical engine failure concept that assures adequate designated surface area and adequate performance capability for continued safe flight or safe rejected take-off in the event of engine failure" [47]. Novel VTOL aircraft designs are expected to offer advanced vertical take-off and landing capabilities in order to meet the needs of emerging VTOL operations in urban environment. Therefore, a novel take-off path was elaborated addressing explicitly vertical take-off. It consists of a significant vertical climb segment until the take-off decision point is reached. Additionally, at least two take-off/climb and approach surfaces with a separation of at least 135° (ideally 180°) should be provided. Furthermore, obstacle clearance in terms of protection surfaces apply with respect to the virtual elevated vertiport which describes the top of the vertical climbing segment until positive rate of climb is achieved and the VTOL aircraft is starting the acceleration into forward flight. VTOL aircraft can either follow conventional landing or a newly developed vertical landing procedure while complying with the requirements of obstacle separation. For this purpose, vehicle performance as well as navigation and communication performance requirements need to be elaborated in order to define the maximum allowed deviation from the nominal landing path. The required landing distance provides a safe environment if a CFP event is recognized at the landing decision point (LDP). For additional details please refer to Figures 1 and 2 of [38].

Due to the variety of VTOL designs, a first "Reference Volume Type 1" was proposed by *MOC-2 SC-VTOL* providing standardized parameter values for vertical take-off and landing procedures [38]. This proposed reference volume for VTOL procedures led into EASA's so called obstacle free volume (OFV) proposed in [22]. It describes a protection volume above take-off/landing pads in order to create a safe environment for UAM operations especially in congested and obstacle-rich environment (see left visualization in Table 4). In order to qualify as a OFV, certain criteria and dimensions must be met. Considering different accumulations of approach and departure surfaces to fit different obstacle characteristics can lead into bi-directional or omni-directional OFVs. A standardized reference volume Type 1 was developed and is displayed in Table 4. Manufacturer of VTOL aircraft may voluntarily comply with the reference volume type 1, and if required, additional reference volumes can be defined. It needs to be highlighted that the reference volume type 1 displayed in *PTS-VPT-DSN* [22] was enlarged compared to what was proposed initially in *MOC-2 SC-VTOL* [38].

Next to the design dimensions of a VTOL-specific operating volume, VTOL aircraft manufacturer and certification authorities need to agree jointly on an operating procedure and minimum performance requirements. This also includes strategies and measures if non-nominal situations occur during different flight phases.

During the flight, ref. [22] introduced the concept of alternate vertiports assigned to the flight prior take-off in cases of a critical failure. Whereas, if an individual take-off procedure needs to be aborted, the vertiport needs to provide a suitable FATO extension (rejected take-off distance) for the VTOL aircraft to complete a rejected take-off under a CFP at the take-off decision point. This results into bigger vertiport footprints in order to accommodate those contingency procedures. Similar to the aborted take-off procedure, a vertiport needs to offer a safe operating volume when balked landing is conducted due to CFP and a go-around procedure needs to be in place guiding the VTOL aircraft from LDP back to LDP in order to start a second approach.

Table 4. VTOL reference volume type 1 according to *PTS-VPT-DSN* [22]; visualization (left) extracted from [22], ©EASA.

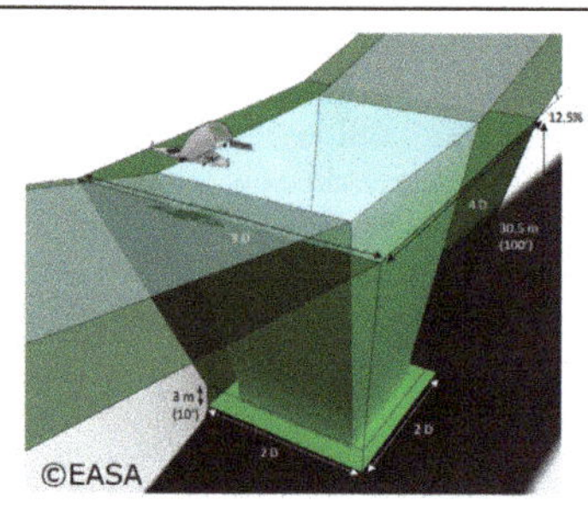

Parameter	Short Description	Reference Volume Type 1
D	D-Value	VTOL aircraft specific
h_1	Low hover height	3 m
h_2	High hover height	30.5 m
TO_{width}	Width at h_2	3 D
TO_{front}	Front distance at h_2	2 D
TO_{back}	Back distance at h_2	2 D
$FATO_{width}$	Width of the FATO	2 D
$FATO_{front}$	Front distance of the FATO	1 D
$FATO_{back}$	Back distance of the FATO	1 D
α_{app}	Slope of approach surface	12.5%
α_{dep}	Slope of departure surface	12.5%

2.1.5. Airspace Structure and Traffic Management

Latest European UAM development show, that urban passenger-carrying operations are considered to operate first under current ATM procedures and most probably under visual flight rules, but are targeting an operation inside the European UTM system *U-space* in the mid- and long-term. *U-space* was elaborated initially in form of a ConOps (see [48,49]) providing a first set of operational practices and rules, predominantly addressing drones and small UAS. Those insights contributed to the recent regulation describing the *U-space* framework, its foundational structure and mandatory services [50]. Furthermore, a corresponding draft of acceptable means of compliance and guidance material was developed in accordance with the *U-space* framework [51]. However, the peculiarities of passenger-carrying operations were not considered during the initial *U-space* ConOps, consequently a vertiport's role, responsibility and participation in *U-space* is not defined yet on a ConOps or regulatory basis. In addition, *U-space* is currently limited to very low-level airspace up to 500 ft (150 m) AGL which might be re-evaluated considering passenger-carrying UAM traffic. As UAM is considered to grow over time, the *U-space* system is assumed to mature in levels of connectivity and automation as well (*U-space* services U1 to U4). Starting from foundational services like e-identification and traffic information, it targets a full set of strategic and tactical operating *U-space* services in order to accommodate the complexity and dynamic behaviour of UAM including passenger-carrying VTOL operation. The basis of the *U-space* framework and its corresponding ConOps asks for a detailed analysis of stakeholders, roles, required services and a thorough ground and air risk evaluation. In 2021, the European standardization organisation *EUROCAE* published the second volume of an eVTOL ConOps *ED-293* [24], in which the vertiport was highlighted as an essential stakeholder and operational procedures such as ground handling processes were proposed. Further details including the distinct definition of roles and responsibilities within a vertiport's organisation are currently finalized in ED-299 [23] and are expected to be published this year.

For vertiport operations, a thorough traffic management analysis is still pending. What information is required by the *U-space* community during the course of different flight phases? How is a vertiport integrated into urban airspace? Who is responsible for the air traffic management at a vertiport and how do multiple *U-space* service provider interact in the vicinity of a vertiport In the next years, *U-space* will be re-evaluated and expanded in order to fit UAM demands in the mid-and long-term. The completion of several European *U-space* research projects including but not limited to CORUS-XUAM developing an extended *U-space* ConOps [52], TINDAiR investigating the safe integration of UAM as an additional airspace user [53], DACUS developing demand and capacity balancing strategies [54] and PJ34-W3 AURA developing a ATM *U-space* interface [55]) will support essentially this development.

2.2. USA

In the U.S., heliport design guidelines have an extensive history and impacted regulatory efforts worldwide. According to the World Factbook of the Central Intelligence Agency, over 80% of all heliports worldwide are located in the U.S. [56]. The current FAA heliport design guideline published in *AC 150/5390-2C* in 2012 [57] is building the basis for most ongoing vertiport research.

2.2.1. Heliport Design Guidelines

The FAA heliport design guidelines describe the dimensions of the airfield elements, approach and departure paths, safety related questions and the heliport facility as a whole. In the current version, general aviation heliports, transport heliports and hospital heliports are treated individually. As general aviation heliports are most closely related to anticipated early UAM operations, the following descriptions will focus on this application. The dimensions for TLOF, FATO and safey area of pads are defined, as well as widths of taxiways and safety zones around parking positions. The slope of approach and departure operations should be 8:1 and two FATOs need to be at least 200 ft (61 m) apart to be operated simultaneously. The safety area of the pad needs to be obstruction free, but can expand over the rim of a building for elevated heliports. In Figure 8, two key figures from FAA's vertiport engineering brief can be seen.

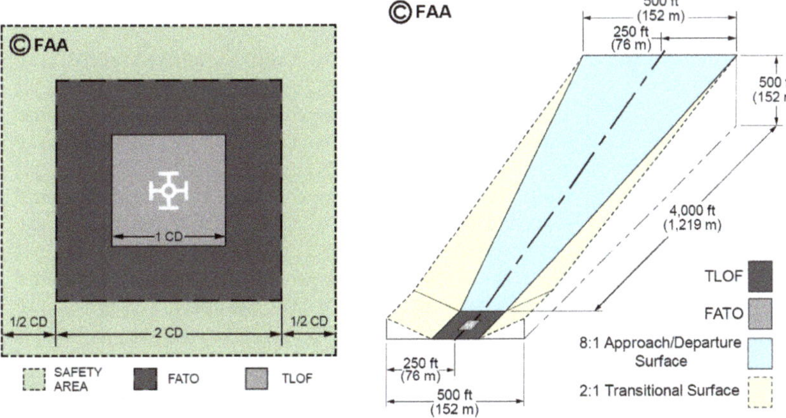

Figure 8. Dimensions of pad and approach/departure slope according to FAA engineering brief on vertiport design [18], ©FAA.

Various reports have been published containing considerations for updating heliport guidelines to fit future vertiport requirements. As there are no vertiport guidelines in effect today, heliport guidelines are the closest scenario. An update of the FAA heliport design guidelines, *AC 150/5390-2D*, is currently drafted [58]. The National Air Transportation Association published a review of UAM related literature in 2019 and finds that "there is no comprehensive canon of policy guidance or regulatory mandates governing vertiport operations" [59]. The report goes on to address regulatory gaps in passenger facilitation, ground handling, security, (ground) marking, design and planning and first response. A similar view on regulatory aspects, but with a stronger focus on building codes is taken in an article written by Zoldi [60]. Here, building codes around fire, health, safety, electricity, plumbing, air circulation and sustainability standards are listed, which are not heliport specific, but must be considered in the process of designing the facilities. A more operations-related perspective is taken by [61], who describes a safe helicopter approach path to be at a slope of 500 ft (150 m) per nautical mile for helicopter-carrying sea vessels.

In 2020 the FAA published a ConOps for UAM with an emphasis on novel airspace structures in the national airspace [19]. Vertiports are viewed as "location[s] from which UAM flights arrive and depart". New "corridors" or tubes in the air are established through which eVTOL aircraft travel. This airspace is designed to be shared by manned and unmanned transport.

Lastly, NUAIR has recently published a ConOps for high-density automated vertiport operations with the perspective of having hundreds of vehicles airborne simultaneously in a metropolitan area [20]. Similar to the FAA ConOps, vertiports are defined as nodes at the end of airspace corridors: "identifiable ground or elevated area used for the takeoff and landing of VTOL aircraft". In the NUAIR ConOps the NASA UAM maturity level 4 as defined by [62] is treated. Vertiport operations are conceptualized as (1) a wider vertiport operations area, (2) a smaller vertiport volume and (3) surface operations. A comprehensive list of vertiport stakeholders is provided. The ConOps claims that "no vertiport exists and operates today", that "heliports are the most analogous current-state model for vertiports of the future" and that early vertiports might be retro-fitted heliports [20]. Together, the FAA and NUAIR ConOps show maturing thoughts towards creating future vertiport design guidelines.

2.2.2. Historic and Future Regulatory Considerations for Vertiports

In the past, there have been attempts to formulate distinct vertiport design guidelines. While they were discontinued they still form the historic root for current vertiport design guidelines. Some things have changed dramatically, in particular aircraft technology, automation and the electrification of aviation. Selected vertiport considerations will be presented in this section.

In 1970 a vertiport study was published by [63] looking at intra-city air travel with tilt-wing configurations using conventional fuels. The study already considered similar aspects as today's efforts, among others passenger processing, air traffic management and design of vertiport airfields. One remarkable point is that noise and community acceptance had already been identified as a key constraint. In 1991, the FAA launched efforts to investigate vertiport design using larger tilt-propeller configurations for inter-city air travel [17]. The design of approach and departure slopes and other regulations resemble today's heliport regulations, except for the sizes of take-off and landing pads, which are larger due to the different vehicle sizes and configurations. Various studies followed, such as [64] designing a single-FATO, eight-gate vertiport layout to be built at the Hudson river. In order to operate the vertiport sufficient demand would be necessary and small access and egress times were identified as essential to meet this goal. In a follow-up study, 13 vertiport locations nationwide were investigated for passenger transport from the suburb to the city center [65]. It was concluded that only about half of the 14 cities have the demand structure to build a profitable vertiport. Only one vertiport was built, namely in Dallas. The FAA *AC 150/5390-3*, responsible for those efforts, was cancelled in 2010 [17].

Most recently the FAA released a pre-print of a new edition of vertiport design guidelines to be published in June 2022, which were already mentioned in Section 1.1.1. Many aspects are identical to the current FAA heliport design guidelines and the authors acknowledge that the guidelines will be subject to continuous change in the near future. Yet, one of the novelties is the explicit treatment of charging for electric vehicles and the question of vertiport placement in the proximity of airport runways. The report uses the term "controlling dimension" *CD* to describe the maximum dimension of the vehicle. The dimensions of a pad are defined as TLOF (1 *CD*), FATO (2 *CD*) and safety area (3 *CD*) depending on the maximum dimension of the vehicle, as can be seen in Figure 8 (left).

2.2.3. Air Traffic Management

Regulations for ATM are not exclusive to vertiports, but they overlap and, in particular, NASA has espoused ATM for UAS as part of their focus. First thoughts on how to integrate high numbers of UAS into the national airspace were presented by [66]. Here, it was already clear that "UAS operations today challenge the ATM system in several ways", seeing that human air traffic controller would quickly experience overwhelming workload. In 2014, NASA then coined the term UTM, which will "support safe and efficient UAS operations for the delivery of goods and services" [67]. A range of new concepts are introduced, such as dynamic geo-fencing, new flight rules and tactical de-confliction with improved CNS capabilities. In 2017, NASA published their ConOps for the UTM system [68,69], while the FAA released in parallel the ConOps for a Low Altitude Authorization and Notification Capability [70]. Another noticeable effort is the *ATM-X* project done by NASA, who started asking the question of how to integrate in particular UAM passenger services into the national airspace [71].

Finally, in the year 2018, the ConOps for UTM was published by the FAA in cooperation with the Department of Transportation under the umbrella of "NextGen" [72]; also under this umbrella the above-mentioned ConOps for UAM has been published in 2020 [19]. In the UTM ConOps the airspace class G below 400 ft (122 m) AGL is proposed for operations. Various principles are introduced, e.g., a hybrid of private/public partnership and guarantee of equal access to the airspace by all participants. Further, the UAS service suppliers or providers of services for UAM (PSU) are introduced and take on a central role in the envisioned architecture. In contrast to the initial European U-space ConOps (see Section 2.1.5), where vertiports are not specifically addressed yet, the U.S. UTM system explicitly includes vertiports in its concept.

2.3. International

Next to the U.S. and Europe, there are considerations around vertiport design worldwide, which also play a role in the current effort to draft first vertiport design guidelines. ICAO released its *Heliport Manual Doc 9261-AN/903* in the fifth edition in 2021 [73]. Yet, this document is not open to public and follows generally speaking the guidelines set by the FAA [57]. Airbus released a blueprint [74] sketching out principles for UTM and stakeholders involved in UTM. In this report, next to UTM efforts in Europe and the U.S., China [75,76] and Japan [77,78] are mentioned to have started investigating UTM. It is not clear if these investigations yielded mentionable results or were further pursued beyond 2018. Further, there where efforts in Australia in 2020 to define a ConOps for UTM involving the Airservices Australia and Embraer [79]. In this report the relevance of vertiport capacities was highlighted and an example for a vertiport network in Melbourne was presented.

Lastly, a most recent report by the *Organisation for Economic Co-operation and Development* (OECD) should be mentioned on the question of integrating drones into the transport system [80]. The report considers both cargo and passenger drones. Noise and the environmental impact are identified as key challenges, which will be important aspects to be considered while drafting future vertiport design guidelines.

2.4. Summary: Selective Comparison

Different approaches to formulating vertiport design guidelines in Europe, the U.S. and internationally have been described in the previous sections. Across these approaches there are many similarities which reflect the desire to integrate UAM into existing airspace regulations and structures. At the same time there are variations. A comparative summary of various design guidelines is contrasted in Table 5. This is a selective list and only reflects a momentary snapshot since the elaboration of vertiport design guidelines is still an ongoing worldwide development.

Table 5. Selection of diverging characteristics between various design guidelines.

Description	FAA	EASA	International
UTM airspace	below 400 ft (AGL) [72]	up to 500 ft (AGL) [51]	-
Main focus of reviewed reports	UAS/UAM [19,72]	(s)UAS/(UAM) [48]	UAS/(UAM) (see Section 2.3)
First mention of vertiports in the context of UTM	2020 [19]	2019 [48]	(2018/20) [74,79]
VTOL aircraft dimensions	Control dimension CD [18] (historically tip-to-tip span TTS [57])	Enclosing circle D [22]	maximum dimension MD [73]
Pad dimensions (references same as aircraft)	TLOF $= 1CD$ FATO $= 2CD$ Safety Area $= 3CD$	FATO $= 2D$	TLOF $= 2$ under-carriage FATO $= 1.5 - 2MD$ Safety Area $= +6\,m$
Pad symbol	cross [18]	letter "V" [22]	-
Approach/departure slope	7.1° (8:1) [18]	12.5° [22]	-
Vertical segment as part of approach/departure path	no [18]	yes [22]	-

3. Vertiport Location and Networks

"Ground infrastructure and planning decisions at this stage of the project development carry significant project risk, and hence, decision makers and stakeholders need to be able to make well-considered business and operations decisions." [81]

According to [82], the following factors make a location favorable for placing UAM ground infrastructure: less densely built-up cities with substantial amount of free and undeveloped land; access to water like lakes and rivers; no existing strong and efficient public transportation network; large commercial airport located nearby. Furthermore, a city's climate degrades initial UAM operations if reduced visibility, wind and icy conditions are faced frequently. Therefore, an initial setup is recommended in consistent weather patterns and mild climate until more operational experience is gained. In addition, the wealth of the city and its population has to be considered since early implementation of UAM and on-demand mobility services require high investments and will create high initial operating costs.

How scientific publications are addressing the question of vertiport location and how they propose to solve the optimization problem of finding the best place for a vertiport and the right size of a corresponding network, is discussed in the following chapter.

For additional orientation, the reader is pointed to Figure 9 which shows the operating areas of those vertiport networks discussed throughout the sections addressing the use-cases: commuting, airport shuttle, holistic UAM system, other covering delivery and STOL operations, and mixed.

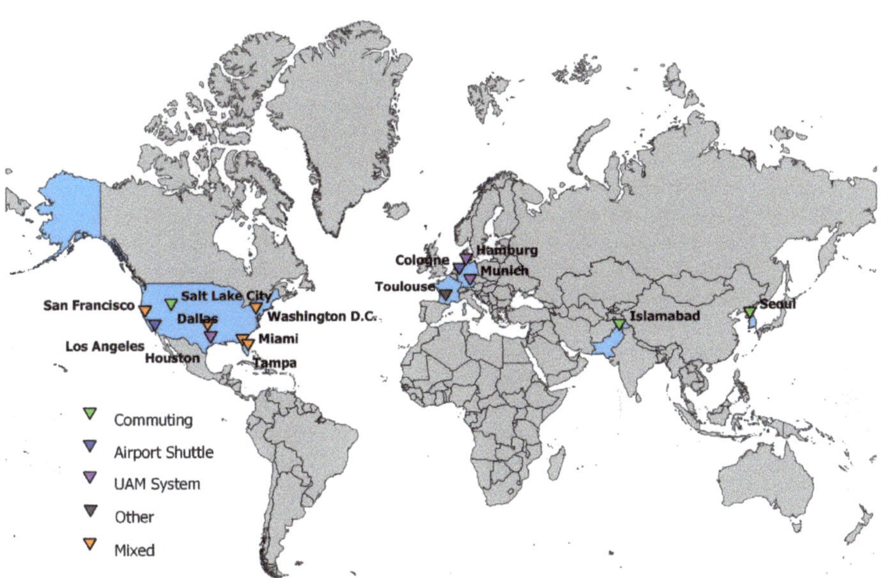

Figure 9. Vertiport network locations covered by selected scientific publications. Use-case is expressed through color-code.

3.1. Vertiport Networks Based on Commuting Trends

Air mobility operations may be differentiated between urban air mobility inside city limits, sub-urban air mobility connecting city and surrounding metropolitan areas (trip exceeds 20 miles (32 km)), and regional air mobility providing city to city transport [81]. Depending on the operation type, different repercussions on vertiport location, size, resource provision and operating concept may be expected. Historic commuting behavior can be used as a starting reference to evaluate where and to what extend air mobility may serve mobility needs. Once the need and potential demand is evaluated, a suitable location has to be defined for each vertiport of the network; on the one hand a vertiport needs to be conveniently reachable, on the other hand the amount of vertiports should be reduced to the most needed.

Developing theoretically a vertiport network may consider "uncapacitated" and "capacitated" facilities. The use of "uncapacitated" facilities makes sure that individual vertiports are not causing any operational bottlenecks during analysis and, therefore, are able to serve unlimited demand (see e.g., [81]). Instead, "capacitated" vertiports only serve limited demand (see e.g., [81,83]).

For the U.S areas *San Francisco Bay area*, and *Salt Lake City-Provo-Orem, Dallas-Fort Worth* and *Washington-Baltimore-Arlington*, the UAM market potential was investigated considering a multi-modal transportation network in which UAM provides single legs of a commuting trip [83,84], respectively. Further, ref. [81] analyzes a sub-urban air mobility vertiport network setup in *Miami (U.S.)* based on work-home trip data-sets. A data driven optimization framework for defining and solving the Mixed-Integer-Programming based network problem was used while targeting to minimize the vertiport network setup costs. Lastly, ref. [85] established a six-piece vertiport network in *Islamabad (Pakistan)* focusing on vertiport site selections next to frequently used commute routes and places where traffic congestion is faced.

In order to reflect different time saving requirements and to develop resulting vertiport performance constraints, ref. [83] proposes to cluster commuting travellers into long distance commuters and short distance commuters. For long distance commuters a time saving of 25%, and at least 50% for short distance commuters is required due to their different value of time in order to switch to the UAM mode. The demand and vertiport distribution problem is formulated as an uncapacitated facility location problem which uses k-means algorithm for clustering.

This k-means approach was also used by [86], who investigated a vertiport network of 10, 40 and 100 vertiports in the metropolitan area of *Seoul (Korea)*. Areas like Han River Park, highway intersections, rooftops of parking lots and existing helipads on skyscraper rooftops have been utilized for vertiports. In order to evaluate how well the data is clustered, the silhouette technique is performed. Final vertiport locations are selected by re-positioning them to the appropriate sites near the centroid of the cluster to comply with geographical conditions. This caused frequent challenges due to most of the clusters are being residential areas.

Another "clustering approach" was defined by [81] who introduces the concept of a "catchment area" (3 miles (4.8 km) radius) where vertiport locations are paired up. The resulting time saving based on different numbers of work-home blocks and vertiport pairs is analyzed for the operating area of *Miami (U.S.)*. The larger the catchment area the bigger is the number of potential vertiport locations and routing options which then requires less vertiport pairs to satisfy the demand. On the contrary, larger catchment areas impose longer egress and access legs for the customer.

Since a change of transport modes is inevitable when considering a multi-modal transportation network, increasing overall time savings always asks for optimized transfer times between subsequent modes of transport.

Transfer times of 5, 10 and 15 min and varying numbers of vertiports (1 to 30) are considered for the *San Francisco Bay area (U.S.)* by [83]. The direct haversine between the origin and destination of each trip is computed and compared to the travel time on ground based on different ground traffic congestion levels extracted from the *Mobiliti* simulation by [87] and *Google Maps'* API. Focusing on short distance commuters, even if high transfer times of 15 min and high ground congestion are assumed, 45% of the short distance commuters in the *San Francisco Bay Area (U.S.)* will benefit from switching to UAM. However, it requires a rather large network of 30 vertiports in the east and 24 in the west. This benefited commuting share drops significantly to 3% if uncongested traffic and 10 min transfer time is assumed. A smaller network of 29 vertiports in the east and seven in the west is required instead. By contrast, no benefit is created if transfer times of 15 min and uncongested traffic are assumed. Additional time-saving and efficiency analyses about choosing UAM instead of ground taxis were conducted e.g., for New York City (U.S.) and Hamburg (Germany), and parameters affecting UAM mode choice were analyzed for the city of Munich (Germany) by [88–90], respectively.

Potential vertiports in the U.S. cities *Salt Lake City-Provo-Orem*, *Dallas-Fort Worth* and *Washington-Baltimore-Arlington* were examined by [84] and resulted into potential vertiport network sizes of 38, 407 and 207 vertiports, respectively. Census data and tracts are used to approximate the vertiport location in the centroid of census block groups. Those networks generated by different heuristic methods such as elimination heuristic, maximal edge-weighted subgraph heuristic, greedy heuristic, greedy heuristic with updates are compared. 1200 different cases are explored differing in input variables such as location, network type, battery range, number of vertiports and vehicle speed. Overall, the two greedy algorithms with update steps concluded as best-performing algorithms and produced solution networks with 91% of the optimal value. When selecting optimal vertiports the interdependence of vehicle attributes, potential locations, and desired network structure was considered.

Rather uniquely in this set of vertiport-network-publications, a noise analysis around the UAM route is performed on the basis of the day-evening average sound levels for the vertiport network in *Seoul (Korea)* [86]. To measure the percentage of the population affected by noise, a curve fitting function of the Shultz curve is used. By dividing the area of Gangseo-gu into hexagonal tiles, according to [86], noise will affect roughly 400,000 people in the 41.6 km^2 area. Due to the lack of eVTOL noise data, noise maps are created by using an aviation environmental design tool and by assuming noise characteristics of a five-seat helicopter. A noise priority scenario defined as a flight along the least populated area was compared to a business scenario describing a flight following the shortest distance; the number of affected people decreased by 76.9% for the noise priority scenario.

3.2. Vertiport Network in Support of Airports

Establishing a vertiport network in the vicinity of airports and operating as first or last leg of a multi-modal trip to or from an airport may be convenient for the passenger and lucrative in terms of time-saving.

Placing a single vertiport of the network directly next to an airport requires the identification of constraints which might be locally different but since a lot of aerodromes are following (inter)national standards, they may be transferred and adjusted quickly. Based on the exemplary operating environment of *Cologne Bonn (Germany)* airport, ref. [91] developed a rating system considering passenger accessibility, obstacle clearance, noise impact on adjacent buildings, expandability, applicability and strategic availability in order to evaluate the potential of each identified vacant area adjacent to the airport. This included parking garages, parking fields and rooftops of an existing bus terminal and of a future hotel. Based on that rating system, ref. [91] prioritized the rooftop level of an adjacent parking garage which provided the best passenger accessibility and may enable an almost unhindered UAM operation. During this process, several requirements deemed crucial for successful integration including vertiport connection to existing transportation modes and the proximity to terminal buildings.

Similar but a more detailed analysis was conducted by [92] who used a 2019 LAX passenger survey as primary data set to determine the optimal vertiport location and network size based on the passengers' selected top ten origin destinations in the area of *Los Angeles (U.S.)*. Restricted airspace boundaries prohibiting overflying or restricting the placement of vertiports are taken into account. A mode choice model with varying assumptions for the in-vehicle travel time, additional shuttle time and the out-of-vehicle time was created to capture a traveler's mode-choice to and from the airport. The demand-driven vertiport placement methodology by [93] was used. As a result, a mixed logic model with different parameters such as travel time, travel cost and the value of time is created. Together with the Fuzzy C-means clustering method which places a certain number of clusters in a specific area, ref. [92] concluded with an optimally placed vertiport set of three network sizes: 50, 75 and 100 vertiports. Those vertiports located adjacent to LAX attract zero demand due to the short travel distance or airspace restriction, whereas the vertiport in LA downtown expected the highest demand.

Of contrast, for the 25 vertiport network in *Dallas Fort-Worth*, the vertiport adjacent to Dallas Fort-Worth airport shares 28% of the total UAM operation and resulted into the most demanded node [94]. Taken into account peak and off-peak demand distribution, an average vehicle load factor of 67 and by using a M/M/1 queuing model together with a target waiting time of four minutes, a 76% utilization factor for a FATO is proposed in order to be able to absorb operational deviations. A FATO count per peak, off-peak and average hours was calculated and concluded with a required number of 27.5 FATOs for the vertiport located at the airport in order to serve peak hours. Operating multiple pads will require sufficient separation on the ground (over 200 ft (61 m)) based on helicopter operations) and separate arrival and departure paths with individual obstacle-free protection surfaces.

The vertiport network in *Dallas Fort-Worth* assumed a 5% shift of long distance transportation, but still intra-city, into the air while considering early operations of UAM [94]. Ref. [92] derived a potential 3.6% market share of UAM operating mainly as an airport shuttle and providing trips from and to LAX. To achieve this, a vertiport network size of 75 vertiports is required. For comparison, ref. [95] predicted a 0.5% mode share for airport shuttle and air taxi operations in the whole U.S.

Since operating in airport environment often leads to operating in controlled airspace with multiple other airspace users, a safe separation has to be maintained throughout the entire operation. Ref. [96] investigates different route designs for VTOL aircraft operating as an airport shuttle in a non-segregated airspace inside the terminal radar approach control (TRACON) airspace of *Tampa (U.S.)*. By using a Rapidly Exploring Random Tree optimization algorithm, those trajectories with minimum design costs and sufficient distance to manned operations, obstacles and ground are being selected. A user-specified distance was set to 25 ft (7.6 m) which increased incrementally by 25 ft (7.6 m). Based on those selected routes, possible vertiport locations are determined. For the airport and TRACON airspace of Tampa (U.S.), three vertiport locations, two inside airport area and one outside, were found. The algorithm identified 100 ft (30.5 m) being the largest available distance for those two vertiports located inside which does not provide sufficient distance of terrain and manned aircraft. Therefore, "[...] no acceptable airspace volumes could be found that would be permanently available for VTOL trajectories under current operating conditions" [96] for the selected airport (layout) in *Tampa (U.S.)*.

Adding environmental constraints, uncertainties and passenger interaction to the operation of individual vertiports located inside a UAM vertiport network, different vertiport layout and performance capabilities might be required to serve "nominal" demand [97,98]. An airport shuttle network in the *Washington D.C. (U.S.)* area was analyzed by [98] in regard to changing performances of vehicle speed, boarding time, vertiport operations times and arrival demand. A full set of requirements including historic travel demand, location constraints, capacity of vertiports, number of vehicles and charging limitations are considered. Additionally, the vertiport network "shall emit Day Night Average Sound Level (DNL) less than or equal to 65 dB", "[...] shall limit vehicles arriving at vertiports from waiting more than 20 min for an available landing pad" and "[...] system shall provide passenger transportation with 95% flights being within 5 min of expected time" [98]. The deterministic simulation concluded with a five node vertiport network, two FATOs and two parking spaces each and 70 vehicles in total being able to serve the demand of high value travelers. Using normal distributions for vehicle speed, boarding time and vertiport operations time and a Poisson distribution depicting passenger arrivals, the required number of landing pads increased from two to three in order to achieve same orders of throughput. In contrast to [98], ref. [97] conducted a sensitivity analysis for several variables (e.g., arrival/departure service time at pad and stall) by applying a lognormal distribution in order to evaluate the impact on vertiport capacity and operational efficiency (for additional details see [99]).

3.3. Holistic UAM Network Approaches

Despite vertiport networks serving a specific purpose such as providing alternative means of transport for commuter traffic or specifically operating in airport environment as airport shuttles, several contributions focus on a holistic development of a vertiport network. The overall goal is to provide a structured and generic process on how a vertiport network can be developed based on e.g., socio-demographic, local travel/commuting and city planning characteristics. According to [100], many U.S. cities of UAM interest are following a "wheel-and-spoke" design with interstate highways radiating out from the city center and circumferential concentric beltways connecting the suburbs. Therefore, the generalized model of vertiport placement proposes a UAM traffic network aligned to existing highway traffic configurations which can be adjusted to *every American metropolitan area* by customizing the size of the hexagon. Following this approach of a generic city model consisting of a hexagonal vertiport placement pattern, a UAM system of system network was developed by [101] enabling the analysis of a UAM network of seven vertiports in *Houston (U.S.)* and five vertiports in *Dallas Fort-Worth (U.S.)*.

Based on socio-demographic characteristics and expected developments for the year 2030 (used tool: SILO for modelling a synthetic population), an existing agent-based traffic simulation model (used tool: MATSim for trip assignment, MITO for generating travel demand) is used and extended to determine UAM demand and potential modal share for the metropolitan area of *Munich (Germany)* [102]. Within this study, the vertiport was inserted as a black box being able to accept and release UAM traffic. Serving four different business cases (business, commuting, tourism, leisure), three level of vertiport archetypes are considered; a low density network (24 vertiports) covering large agglomerations, transportation hub and densely populated areas with large share of high income; a medium density network (74 vertiports) including main subway and suburban lines and employment centers; a high density network (130 vertiports) covering all relevant trips and target groups [103]. Moreover, number of vehicles, cruise speeds, processing times and ticket fairs are varied. Potential vertiport locations are determined in the course of several workshops with representatives of Munich Airport, city of Munich and Ingolstadt and the Upper Bavarian Chamber of Industry and Commerce. For the medium density network a total UAM mode share of 1% was predicted, whereas targeting for longer distances, the mode share prediction increased to 3 to 4% [102].

A collaborative simulation approach is proposed by [104], in order to analyze a UAM network inside the metropolitan area of *Hamburg (Germany)*. It follows the objective of defining low-fidelity analysis components such as demand, vertiport design, vertiport integration, routing, scheduling and setting them into relation in order to analyze interdependencies. The vertiport integration is based on published 3D building data, which is then used to select a vertiport location in the centroid of every quarter in Hamburg. This is being reconciled with the expected demand, airspace structure and resulting routes, and general restrictions like no-fly zones.

A 3D geographic information system map was derived from lidar data and used by [105] to determine the optimal vertiport location for the *Tampa Bay area (U.S.)*. Both, regulation constraints for eVTOL operations at vertiports and socio-demographic characteristics were additionally considered. The potential UAM demand is analyzed and the UAM mode share is evaluated based on allocation of user to vertiport, access- and egress-mode choices and the interaction between vertiports. Ref. [105] concludes, that UAM ride shares are small therefore congestion relief will be limited, but the passengers who choose UAM will experience substantial time savings. Inside the network design, trips fully conducted by UAM or ground transportation modes as well as multi-modal ride shares are feasible. The network optimization follows the objective to minimize generalized travel cost for all network users no matter what transport mode was chosen. It is seen, that with increasing number of vertiports the overall accessibility and UAM mode share increases. However, this is saturated choosing a vertiport network of 80 vertiports. The transfer time between ground based modes and UAM plays a decisive role, which leads into a drastic reduction in numbers of customers if the transfer time is increasing.

3.4. Other-Vertiport Networks Based on Parcel Delivery and STOL Operations

In the following section, other air mobility operations are described such as parcel delivery and passenger transport with STOL aircraft. Even if those use-cases differ from the core theme of this manuscript, resulting ground infrastructure requirements may be comparable. Ref. [106] investigated the use of eVTOL aircraft for same-day/fast parcel delivery in the *San Francisco Bay Area (U.S.)*. The placement of vertiports is optimized based on the maximum package demand served. Vertiports should be placed near to the customer subject to minimizing the number of vertiports. This objective is additionally challenged by high building costs and limited building locations. The foundation of the optimization is the estimation of same-day delivery demands which is assumed to be the highest in areas with larger population and higher income. For this use-case, the San Francisco Bay area is discretized. For each census tract a scaled income measure, a combination of population and average per capita income is defined representing the demand for eVTOL aircraft parcel delivery. The ground-travel time of a customer's origin to the pick-up location, based on *Google Maps* Directions API, was determined as crucial limiting factor impacting the amount of customers served by one vertiport. Additionally, airspace restriction are taken into account, prohibiting a vertiport placement in a census track with a centroid inside class B and C airspace. A vertiport network of one to eight vertiports with an additional ten minute last-mile driving threshold is assumed. As a near-term implementation result, a network of seven vertiports with a distribution center and six distributed vertiports was elaborated.

Another vertiport network serving a package delivery scenario was analyzed by [107] but for the area of *Toulouse (France)*. Four warehouses/vertiports and individual delivery points are considered in order to optimize traffic flow management based on the key performance areas fairness and equity. Two highly dynamic demand scenarios of 50 and 25 flights per hour per vertiport were assumed.

A variety of airpark designs for STOL operations are proposed in [82] in order to fit different locations: vacant land construction, barge construction, additive construction type and the re-use of pre-existing ground infrastructure. The size and location of ground infrastructure accommodating STOL operations depend on runway dimension, faced environment (e.g., obstacles), local atmospheric impact (e.g., on noise propagation) and weather conditions (ice, snow, wind) including magnitude and direction. An airpark fitting algorithm was used to provide a first estimate of the potential of vacant places (using a Quantum geographic information system software together with a Boolean filter) and to derive to a resulting airpark geo-density in the *Miami (U.S.)* metropolitan area.

3.5. Summary

It can be seen that competing approaches and solving algorithms are available to determine the optimal vertiport placement. During theoretical analysis, vertiports are either assumed to be constrained by capacity or not. Some are focused on specific business cases of UAM such as airport shuttle (cf. Section 3.2), commuter (cf. Section 3.1), delivery (cf. Section 3.4), STOL operations (cf. Section 3.4), others follow a generic and holistic approach (cf. Section 3.3). Network designs may also learn from use-cases outside of passenger-carrying UAM operations such as delivery and STOL operations. Vertiport locations are mainly derived from (commuting) demand heat maps, 3D geographic information, frequently used traffic routes or vacant areas based on e.g., lidar data. Most of the analyzed areas are cities or metropolitan areas located in the U.S. Other cities of interest are located in Germany, Korea, France and Pakistan. The vertiport network development starts with a determination of the overall demand clustered into areas of interest. It is then followed by a specific location analysis for each vertiport serving the selected area of interest. Therefore, the specific location and the environment in which the vertiport is implemented in is a crucial step for initially setting up a vertiport network. Throughout the sighted publications, the constraint of transfer times was determined as important factor, which contributes significantly to the decision if a future traveler is taking a UAM mode or not. Next to socio-economic and demography characteristics of a certain area like population centres, commute routes and income distribution, current airspace utilization, time savings, and considered ticket prices are important attributes influencing UAM market shares and therefore a vertiport network's shape and size. Unfortunately, no vertiport networks exist yet, however, future vertiport network plans have been announced recently: *Ferrovial Airport's* 20-piece vertiport network in Spain [108], 25-piece vertiport network in the United Kingdom [109] and its plus 10 vertiport network in Florida [110]. In addition, a four to six-piece *VoloPort* network in Singapore was announced by *Volocopter* [111].

4. Vertiport Design and Operations

"We have a unique opportunity in aviation history to develop technical standards from scratch which will ensure that vertiports are safe and can be adapted to a succession of new VTOL aircraft types that we expect to be developed in the future." [22]

To conduct VTOL operations servicing UAM, not only infrastructure and procedures on the ground need to be elaborated, also procedures covering the airside operation in a strategic and tactical manner are required. Operational constraints affecting on-demand mobility may vary depending on where UAM should operate and topics such as ground infrastructure availability, scalability of air traffic control, emerging aircraft noise and community acceptance needs to be taken into account (see [112,113]).

Even though hundreds of VTOL aircraft designs are currently under development [45], only a handful flying prototypes are available and even fewer reached the process of certification. In terms of vertiports, the pool of available vertiport operators/manufacturers is even less. There are a few key players including *Skyports*, *Ferrovial* and *urban-Air Port*, contributing significantly to this development. But, the current development stage does not provide sufficient foundation to derive thorough conclusions regarding vertiport operations and designs especially under realistic environmental conditions. This will change rapidly once the first generation of VTOL aircraft and vertiports are available.

The following sections will provide a summary of vertiport design visions initially driven by architecture companies participating in *UBER* Elevate's UAM infrastructure challenge as well as by current infrastructure developers. Additionally, different approaches and concepts for vertiport airside air and airside ground operations will be discussed. This chapter will be concluded by first estimations of vertiport infrastructure costs.

4.1. Vertiport Design

After *UBER* Elevate's public UAM infrastructure challenge in 2016, many vertiport design proposals were developed and started circulating the web (e.g., [114–116]). One of the objective was to integrate all kinds of ride sharing in order to offer the customer a transfer to other individual and public transportation modes. Environmental integration as well as a neighbourhood's and customer's well-being, e.g., in terms of shopping, entertainment, relaxing areas, sound-barriers and sustainability, were also taken into account by the submitted design proposals. The vertiport was envisioned as a new public space for local residents rather than only providing UAM transportation services [117].

4.1.1. Visions

Following current vertiport design developments, proposals range from a ground-based single FATO (e.g., [118,119], left illustration of Figures 10 and 11), over one-story vertiports with multiple FATOs and stands (e.g., [117,120], middle illustration of Figure 10, and right illustration of Figure 11), to multiple/dozens of FATOs and stands distributed along multiple stories (e.g., [121,122] and right illustration of Figure 10). All serving different demand scales and operating environments.

Figure 10. Design visions ©MVRDV, Project "Airbus UAM" [118].

Figure 11. Design visions ©DLR, Project: "HorizonUAM" [123].

The "world's smallest airport" is provided by *urban-Air Port* [119] who partnered up with Hyundai Motor Group in order to provide an innovative, rapidly deployable, multi-functional and ultra-compact (fits in one container) infrastructure for manned and unmanned vehicles. The structure is cone shaped with a flat top part on which the FATO is located and which can be lowered to ground level. Additional access and egress is provided via staircases. The urban-air port provides charging, refuelling, as well as aircraft command and control suiting all kinds of UAM operations such as air taxi services, autonomous logistic services and disaster emergency management. Deployments on water (Marine One), on rooftops (Air One) and on ground (Terra One) are foreseen. The first fully operational Air One was unveiled in Conventry (UK) in April 2022 [124].

Multiple vertiport designs such as [117,120] consider the vertiport as extension of the public transportation network by re-using the roof of an already existing building or car park and turning it into an airside operating area with a passenger terminal. "Key to the designers' intent was creating a consistent, stress-free process that allows users to truly experience the joy of human flight. [...] Passengers' process of entering the building, rising to the waiting area, and boarding the aircraft is streamlined—and intentionally unlike a typical airport setup" [117]. By proposing the usage of a check-in app and biometric scanners integrated in the elevator, ref. [117] addresses the topic of safety and security. Ref. [117] vertiport design features an operating deck and a public area underneath which are connected by a terminal area in the centre. From there, the passenger follows a marked path towards the waiting VTOL aircraft. A designated sound barrier installed on the rim of the upper deck protecting the vicinity from noise and wind, caused by arriving and departing eVTOL aircraft, was incorporated into the design proposal.

If throughput needs to be increased drastically, modular and stackable vertiport concepts developed by [121,122] provide possible design options. [121]'s *The Hive 150*, a three-story high modular building including drop off, ride sharing, retail and public areas mainly on the ground level, provides two upper decks dedicated to air traffic operations. Each operating deck provides access to a terminal located in the center and offers several FATOs and the usage of aircraft parking stands connected by taxiways. On the top of the building, emergency FATOs are located offering an easy and quick access to the exit. A total of 168 take-offs and landings per hour (Deck 1: 108 landings/take-off, Deck 2: 60 landings/take-off) are envisioned. The *Hive* was developed in order to meet scalability constraints which enables different vertiport versions to accommodate different throughput levels. *UBER Hive 1000* may provide up to 1104 take-off and landings per hour while actively operating four operating decks.

Another stackable modular approach was designed by [122] consisting of 96 stands, six FATOs for landing and six FATOs for take-off, but here, all elements being connected to each other. A throughput of 1000 arrivals and departures each per hour is predicted. Instead of using lower levels for retail and entertainment purposes, they are used as vehicle parking stands. After landing, the vehicle will roll onto an elevator-pad which levels down and, similar to a car elevator, cycles through the parking position section until it finds its destination where the pad leaves the elevator and slides into the spot for disembarking and boarding. During the vehicle's turnaround time on the elevator-pad, it is charged automatically without any human in the loop. After boarding, the vehicle slides back on its designated elevator-pad into the elevator system and continues its way up to the area where it is leaving the vertiport. This way, different vertiport levels are servicing different destinations.

Next to architecture firms and infrastructure companies, eVTOL aircraft startups like *Lilium* and *Volocopter* are developing infrastructure requirements and design visions for vertiports. *Lilium*, a German eVTOL aircraft manufacturer, proposes a modular, adaptable and scalable vertiport concept tailored to their ducted electric vectored thrust aircraft design [125]. The vertiport needs to provide three key attributes: take-off and landing area, parking stands and a terminal. Ref. [125] proposes three vertiport configurations (courtyard, back-to-back, linear) based on the setup of stands at the terminal building. This setup can be scaled up to match the predicted/required throughput resulting into "micro", "small", "medium" and "standard" vertiport designs. All designs provide at least one FATO and two parking stands.

Different vertiport designs, based on size and location are also considered by *Volocopter*, another German eVTOL aircraft manufacturer naming them *VoloPort*. With the publication of the second whitepaper on the topic "Roadmap to scalable urban air mobility", ref. [28] highlights the first *VoloPort* demo case exhibited in Singapore in 2019 and introduces the development of a *VoloPort* in the area of Paris (France) for the 2024 Summer Olympic Games.

4.1.2. Sizing Approaches and Tools

Next to pure design visions, architecture firms, infrastructure companies, eVTOL aircraft startups and researchers are currently developing requirement catalogues and generic processes in order to provide a structured and automated way of designing a vertiport while still serving specific demand and implementation needs.

A very generic and systematic single vertiport design process was proposed by [33]. A six-step approach, including the systematic investigation of the topics *requirements, functions, architecture, validation/implementation, testing* and *usage/application*. Location criteria including building and infrastructure parameters, wind current, statics and building physics, space requirements, integration of charging infrastructure, noise protection, obstacles limitation surfaces, safety regulations, simultaneous VTOL operations and vertiport layout, have to be considered during the vertiport design process.

In order to support architecture groups in the trade-off between available vertiport surface area and attainable vehicle throughput, a vertiport design tool (behind paywall) was developed by [126]. The backbone of this analysis is defined by a stochastic Monte Carlo simulation calculating the vehicle throughput of three different vertiport design configurations: a multi-function single pad, a hybrid vertiport design consisting of a single landing pad and twin/trio staging areas, a solo/twin linear single function pads including a separate landing and take-off area and multiple parking spaces in single or double-row. Different design approaches result in varying noise contours depending on approach and departure flight paths and procedures. The more flight paths are available, the more distributed noise contours result into less impact to one specific residential area. For the multi-function single pad design, ref. [126] indicates an expected noise exposure at the center of the FATO of over 80 decibel (see [126]'s Figure 7). In addition, ref. [126] considers stakeholder interactions and tensions such as between community and property owners, between UAM transportation system and the user and three types of hazards eVTOL aircraft collision, charging and single pad operations. All constraints contribute to a certain vertiport operation followed by a specific design proposal. According to [126], the vertiport footprint has to increase by 420 m^2 in order to accommodate an additional vehicle per hour.

In a branch-and-bound fashion, the optimal gate to pad ratio for four topologies (single, satellite, linear, pier) is determined and the topology with the highest throughput capacity is selected by [127] based on mixed-integer programming. In this way, the optimal spatial layout of the vertiport airfield can be determined for any given area. In a follow-up work the vertiport "performance" indicator of "passenger throughput per hour and area" was defined in order to quantify the operational efficiency of any given vertiport airfield layout [128]. Through this indicator 10 prominent eVTOL aircraft (e.g., eHang, Lilium, Joby) are compared based on their operational "performance". Depending on the eVTOL aircraft design, one hourly passenger throughput needs 22–67 m^2 of airfield space, with the *CityAirbus* being the most favorable and *VoloCity* the least favorable performer [128]. In comparison, a small vertiport for 10 vehicles and a daily passenger throughput of 5400 was estimated to require an area of 4160 m^2, followed by a large vertiport for 50 vehicles and passenger throughput of 130,000 a day, resulting in over 20,000 m^2 footprint [102]. In contrast to VTOL operations, electric STOL operations might provide advantages in vehicle performance but are expected to require runway lengths between 100–300 ft (30–91 m) depending on the aircraft's technology level, desired cruise speed and battery performance [129].

Together with aviation industry-leading partners and architects, a *VoloPort* handbook was published to support vertiport design by guiding through design, constructions, material use, infrastructure adaptability and facility operations [130]. Operational needs are also discussed compliant with eVTOL designs, performance and ground handling needs like charging, maintenance and fire protection. This handbook is only available for Volocopter partners building UAM infrastructures.

4.2. Airside Ground Considerations and Operations

The vertiport airfield, or airside ground part of the vertiport, is a highly constrained element within the vertiport due to the limited inner-city space. High throughput demands are placed on this constrained space, which creates the need to optimize vertiport layouts under consideration of various boundary conditions towards maximum throughput capacity. Additionally, two processes are expected to be added to the airside ground operations, which are not or barely present on today's heliports: ground taxiing and charging of electric vehicles.

4.2.1. Airfield Layout and Capacity

The capacity of a vertiport is an important factor in the UAM system and depends on the type, number and dimensions of airfield elements (e.g., TLOFs, gates). Ref. [94] defines a vertiport as "taken to be one or more vertipads in close proximity that function as an integrated arrival/departure node within the UAM system". This statement reveals one of the major complexities, namely operating multiple take-off and landing pads simultaneously, who are in close spatial proximity. Ref. [131] did ground-breaking work in this area in 2019, suggesting three types of simultaneous pad operations: independent, dependent, partially dependent. Further airfield elements, next to pads, that are considered across the board are gates, parking stands, taxiways and the passenger terminal. Most sources derive their assumptions from the FAA heliport design guidelines [57] and some give a detailed treatment of airfield element dimensions [7,127,131,132].

Most publications determine the capacity of a vertiport analytically [91,132,133]. Ref. [131] on the other hand uses an integer-programming-based network flow approach. Ref. [127] developed an integer-programming-based branch-and-bound approach, which determines the number of pads and gates, the best suited topology and the anticipated throughput based on the shape and size of a given area. In the paper a range of generic scenarios is tabulated to determine the possible throughput on a given area or find the necessary area for a desired throughput.

Other publications use discrete-event-based [7,92] or agent-based [134] simulation approaches. In another work done by [135], the vertiport capacity is determined based on the different vertiport layouts, varying behavior of passengers and vehicles, imbalances in the vehicle fleet and magnitude and shape of the passenger demand profile with special focus on demand peaks.

The most common topologies proposed for vertiports are satellite, linear and pier topologies. Refs. [32,127,132] all give a detailed description of the different characteristics. Further topologies that are put forth are the remote apron topology [131], resembling today's commercial airports, the single topology [127], resembling today's helistops and a linear uni-directional flow topology (LIEDT [7], linear process configuration [20]) targeting for a high-throughput potential. Early contributions of [131] on the ratio between gates and pads have found the ratio to strongly depend on the turnaround time at the gates, which in turn depends on passenger boarding and vehicle charging. Ratios that are being put forth range from 2 to 8 gates per pad [104,132] and are therefore a novelty compared to today's heliports operations, which concerns itself almost exclusively with pad operations. Most publications place all elements on a two dimensional plane. Ref. [132] in turn suggests a level below the airfield, which is connected through staircases allowing the passengers to enter the airfield. Ref. [7] uses the same idea of a second level, but suggests elevators transporting the vehicle under deck for boarding and turnaround, freeing up space on the airfield.

There is a wide range of vertiport capacities being suggested from less than 10 to over 1000 operations per hour. A case study at Cologne airport determined an average of 9.6 movements per hour [91]. Another study focusing on business models in the Washington D.C. area considers 2–7 movements per half hour [98]. UAM network studies in San Francisco [97] and Los Angeles [92] found a maximum of 325 and 250 passengers, respectively, being serviced per day on the busiest vertiport. These studies showed that a vertiport network tends to have one vertiport with very high demand, a few semi-high-demand vertiports and a lot of low-demand vertiports. This was also depicted by [94] study for Dallas-Fort Worth. Ref. [84] also described this phenomenon differentiating between large vertiports and small vertistops while borrowing the hub-and-spoke concept from conventional aviation. Ref. [133] largest vertiport can handle up to 76 operations per 15 min and the use case study of [127] in northern Germany sees 60 to 780 passengers being processed per hour. The highest number found comes from [94] with 1400 passengers during the peak hour in Dallas-Fort Worth. Considering current operations, this number is in contrast to the Silverstone heliport, which becomes the "busiest heliport on earth" for a short moment each year during the Formula 1 British Grand Prix, with around 4200 helicopter operations in one day (average of around 260 helicopter operations per hour for a 16-h operational window) [136].

4.2.2. Ground Movement and Taxiing

A novel operational element on vertiports will be ground movement or taxiing of vehicles to free up landing and departure pads. The basic operation of a helicopter does not take ground movement into account to the extent we are familiar with fixed-wing commercial airliners. Following FAA's Helicopter Handbook [8], "taxiing" is conducted in three different ways: The first option is to "hover taxi", conducted above the surface and in ground effect at air speeds less than 20 knots. To reduce the ground effect, the height can vary up to 25 ft (7.6 m) AGL. The second option is to "air taxi", also above the surface but at greater heights (not above 100 ft (30.5 m) AGL) and at higher speeds (more than 20 knots). The third option is to "surface/ground taxi" describing taxiing on ground and a movement under the helicopter's own power.

When targeting high-density UAM operations, several vertiport designs consider a complex taxi-route system (e.g., [7,125,137]). It is assumed that the operating VTOL aircraft must somehow be able to taxi, which is an expected novelty compared to present helicopter operations. Different implementation approaches are already proposed including the use of e.g., conveyors [138] or autonomously towing platforms/carts [139]. Refs. [7,131,132] differentiate between vehicle taxiing under its own power (hover, ground taxiing) or being conveyed (ground taxiing). Yet, while different modes of taxiing are described, the speed is not differentiated: [132] gives an estimated 4 ft/s, ref. [131] assumes a median of 15 s taxiing time between pad and gate and [7] considers 2.6 m/s to meet the assumptions by [131]. Ref. [127] considers how taxiways and gates have different dimensions according to helicopter design guidelines depending in the mode of taxiing, which in turn affects to throughput capacity of a certain area. Ref. [7] further elaborates on the idea of towing vehicles on the ground and through elevators into levels below the airfield to safely process passenger handling and vehicle charging.

For the purpose of this review three types of taxiing will be differentiated: *hover, passive* and *active*. The authors are aware that these categories provide slightly different meaning in the context of helicopter operations. Yet, due to the expected novelty of vertiports operations and VTOL aircraft, new categories might be necessary. (1) "Hover taxiing" has been described above and combines all types of taxiing, where the *main engines* are in use. It might be possible to physically touch the ground while doing so, if the configuration has wheels/landing gears. In this exception, the used definition diverts from helicopter operations. In most cases though, hover taxiing is expected to be conducted without surface contact. The benefit of this way of taxiing is the low complexity and no need for external devices on the ground. The downsides are safety concerns and the energy intensity, in particular for tilt-wing or tilt-propeller configurations. (2) "Passive ground taxiing" sums up all the ways of moving an eVTOL aircraft on the ground with *all engines and motors shut down*. Conveyor belts or elevators have been mentioned before, but also towing bots and moving platforms are conceivable. This mode resembles the pushing of conventional aircrafts away from the gate onto the main taxiways, before they power up their main engines. (3) "Active ground taxiing" will be suggested as a third way, where the taxiing power comes from the vehicle, but from *motors other than the main engines*. One approach could be electric motors attached to the wheels of the eVTOL aircraft, which are powered by the on-board battery and let the vehicle taxi on the ground. Even though it is not common in conventional aviation, this approach has been investigated in the past and named alongside other modes of taxiing [140]. This novel taxiing approach might be of particular interest to vertiport operations.

A parameter value specification based on expert interviews has been conducted to determine the different taxiing speeds and related processes such as starting/stopping of engines or mounting/de-mounting devices for passive taxiing [141]. 17 Experts from the industry, research and active piloting were consulted with an average experience of over 10 years. Through statistical analysis the taxiing speeds were determined as follows: hover taxiing at 3.25 m/s, passive taxiing at 2.63 m/s and active taxiing at 2.15 m/s.

4.2.3. Turnaround at Gate: Boarding and Charging

Next to the operations on the pad, turnaround at the gate is the second most sensitive process on the vertiport airfield [141] and can encompass actions like passenger boarding and de-boarding, vehicle battery charging or swapping, pre- and post-flight checks and even minor MRO activities [126]. Ref. [131] found out that the turnaround time has a big impact on the ratio of gates to pads, which is one of the design drivers as discussed above. Several studies found the passenger processing time, which is directly linked to the vehicle turnaround time, to be one of the most relevant factors determining the market share UAM can achieve [102,105,142]. Parameter value specification for charging speed, swapping time, boarding, etc. are presented in a systematic fashion by [141].

The turnaround time assumed in scientific literature varies, but can be distinguished in short and long turnaround times. Short turnaround times take the perspective of a touch-and-go vertistop design, where only passenger boarding and de-boarding occurs at the gate as the minimal necessary operation. Turnaround times that are mentioned are 0.5–10 min [131], 2–10 min [105], 5 min [7] or 8 min [132]. Some of these studies leave the question open, whether charging/fueling might happen during this time, but full charging/fueling of vehicles is unlikely. Boarding of VTOL aircraft has not been studied in depth, but conventional aircraft boarding simulations could provide a starting point for initial assumptions [143,144]. Long turnaround times, in contrast, take the perspective of a well equipped vertiport or even vertihub design with 30 min [91] or more. Next to the charging of the vehicle, which will be discussed in the next paragraph, minor MRO activities might be conducted. Next to a few preliminary considerations [145,146] the question of eVTOL aircraft maintenance is not possible to be addressed in detail, yet, due to the missing experience of eVTOL aircraft operations.

One major question for turnaround length is the choice of primary energy source and its handling. While most current VTOL aircraft designs assume fully electric propulsion systems, a study conducted by [101] found LNG based designs to be more promising due to higher availability of LNG and lower occupancy times of vertiport infrastructure. Fully-electric designs, hybrid-electric fuel-cell based designs and direct combustion of LNG were considered. These variants are also conceivable with hydrogen instead of LNG. When choosing electric designs, the next question is direct charging of the vehicles or swapping of pre-charged battery packs. On the one hand, battery swapping might have potentials to mitigate peak loads on the electric grid and shorten turnaround times. On the other hand, charging is more easily implemented and the difficulties of defining battery pack standards in particular for mixed eVTOL aircraft fleets are unknown. Some studies considered the novel idea of battery swapping [98,147] and vehicle manufacturers such as *Volocopter* consider this approach for their vehicle design [148,149]. Further inspirations might be drawn from battery swapping in automotive applications [150,151]. Yet, during the time of writing, direct battery charging appears to be the preferred concept, possibly due to its lower complexity and wider application in related transportation modes.

4.3. Airspace Considerations and Airside Air Operations

Transitioning from vertiport airside ground considerations and operations to UAM airspace considerations and vertiport airside operations, it is important to define the structure of a UAM flight in order to decide on its operational framework. Following the classification of [79], a UAM flight is divided into six phases namely *pre-flight, departure, en route, approach, landing* and *post-flight*. A UAM flight starts with the pre-flight phase accommodating all actions related to flight planning and preparation including e.g., vehicle pre-flight checks, charging and boarding. It ends with the post-flight phase addressing all concluding actions after the particular flight is closed such as deboarding, vehicle servicing activities and log book updates.

Additional terms like *strategic* and *tactical* are used frequently between and inside different flight phases in order to address different time horizons and to refer to a certain scope of possible services available (e.g., in terms of U-space services) and actions choosable. For thorough description of both terms, please refer to [152,153]. Moreover, the term *pre-tactical* was defined to bridge the gap between strategic and tactical phases (e.g., used by [51,79]).

Providing on-demand UAM services require precise planning tasks on short time horizons under changing requirements. A quick and efficient exchange of relevant information between all involved stakeholders will be crucial. Since real UAM and vertiport operations are not existent yet, we do not have any planning approaches nor procedures in place. An impression on how it is currently conducted for commercial fixed-wing aviation is depicted in Figure 12. For commercial fixed wing operations, air traffic flow and capacity management tasks are conducted during four phases [154]. Passing each phase, uncertainties get more certain, adjustments can be made collaboratively by considering up-to-date information and the flight schedule created in the strategic phase gets more accurate. An optimized and automated conflict detection and resolution service will be of vital importance.

VTOL operations might follow a similar step-wise planning approach but addressing much shorter and highly-variable lead- and transition times.

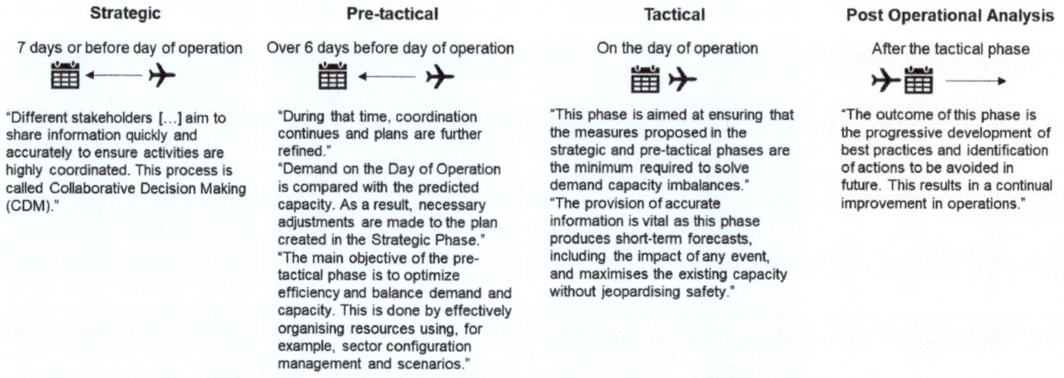

Strategic	**Pre-tactical**	**Tactical**	**Post Operational Analysis**
7 days or before day of operation	Over 6 days before day of operation	On the day of operation	After the tactical phase
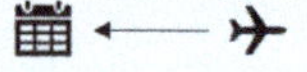	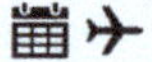		
"Different stakeholders [...] aim to share information quickly and accurately to ensure activities are highly coordinated. This process is called Collaborative Decision Making (CDM)."	"During that time, coordination continues and plans are further refined." "Demand on the Day of Operation is compared with the predicted capacity. As a result, necessary adjustments are made to the plan created in the Strategic Phase." "The main objective of the pre-tactical phase is to optimize efficiency and balance demand and capacity. This is done by effectively organising resources using, for example, sector configuration management and scenarios."	"This phase is aimed at ensuring that the measures proposed in the strategic and pre-tactical phases are the minimum required to solve demand capacity imbalances." "The provision of accurate information is vital as this phase produces short-term forecasts, including the impact of any event, and maximises the existing capacity without jeopardising safety."	"The outcome of this phase is the progressive development of best practices and identification of actions to be avoided in future. This results in a continual improvement in operations."

Figure 12. Air traffic flow and capacity management phases for commercial aviation according to [154]; all quotations by [154]; own depiction.

Especially during initial operations, UAM is facing very limited resources in terms of endurance capabilities and ground infrastructure availability. This will require a thorough analysis of demand and capacity balancing strategies on both strategic and tactical levels, deciding among others on the magnitude of possible UAM operations in the chosen operating environment (e.g., [88]). Furthermore, with rising UAM demand and increasing complexity of vertiport topologies (multiple FATOs, stands, taxiways, etc.), a highly automated flow and resource management will be necessary.

According to [155], flow management processes are seen as crucial operational services in order to provide future day-to-day UAM operations next to flight planning and authorization, dynamic airspace management and conformance monitoring. Vertiport capacity is declared to be initially the greatest limitation to the vertiport flow management service followed by airspace capacity when considering higher traffic densities.

A performance-based evaluation of a vertiport's airside traffic flow was conducted by [156]. For that purpose, a UAM tailored vertidrome airside level of service *VALoS* concept was developed in order to identify how well a specific vertiport setup can process a particular demand distribution based on a distinct vertiport layout, airside operational concept and emerging airside traffic flow. The multi-dimensional VALoS framework is build upon a set of stakeholder requirements, including but not limited to the VTOL aircraft operator, the vertidrome operator and the passenger. Based on those individual stakeholder constraints which are defining if an operation is acceptable or not, and a distinct definition of how a "flow" is measured, the processed airside traffic can be evaluated.

Furthermore, local airspace designs, current roles and responsibilities inside different airspace classes, as well as other airspace users need to be considered in order to establish a safe operation in- and outbound of vertiports. How current airspace classes will be modified or extended to fit UAM is not clear yet. In that regard, different airspace designs and management strategies such as density-based airspace management [157], full mix/layers/zones/four-dimensional tubes [158] (updates expected under [159]), ATM/U-space shared airspace *AUSA* [160] have been proposed and are currently under development. UAM airspace, whether it is going to be segregated or not, needs to be integrated safely and harmonized with already existing standards and airspace users. UAM airspace integration concepts and considerations for the U.S. airspace are currently developed addressing not only goals and objectives but also barriers and potential hazards [161].

Since eVTOL aircraft have significant short endurance characteristics, a detailed and highly precise scheduling and sequencing approach will be crucial. Scheduling and sequencing techniques can be conducted before departure but also during the flight. It may be assumed, the better an eVTOL aircraft flight is planned before take-off and strategic conflict detection and resolution strategies are applied, the less major tactical conflict resolution actions are required on a daily basis. Short UAM flight times of less than one hour could be favorable, nevertheless, all uncertainties can never be eliminated completely. Interaction with humans, appearing weather, CNS and technical degradation causing contingency or emergency situations are only predictable to a certain extent. Therefore, suitable strategic and tactical techniques and contingency measures like schedules and slots, buffers, aerial and ground delaying procedures, holding patterns and diversion to alternate vertiports need to be tested in order to investigate the potential of intercepting occurring deviations. Risk mitigation and maintaining the required safety standards are crucial.

Establishing a new ATM system coping with the peculiarities of on-demand, high density traffic in obstacle rich environment, CNS systems are technological key enablers. Ref. [85] identifies the need for fast and accurate communication between traffic controller and UAM vehicle, vehicle-to-vehicle, vertiport-to-vehicle and vertiport-to-vertiport. Additional needs are defined like self-position and situational awareness in the context of navigation and surveillance, vehicle tracking, position and identification updates. The overall CNS system must provide integrity, robustness, security and high geo-spatial accuracy.

Concluding, airspace and procedure design as well as information exchange are two substantial services in order to prepare the operating environment for upcoming UAM traffic [155].

In the following sub-sections, strategic and tactical measures as well as specialized approaches for operating UAM with respect to vertiports in airport environment are discussed.

4.3.1. Strategic Measures

In order to support strategic measures, several UAM mission and flight planning systems such as [162,163] and scheduling and sequencing approaches [107,133,164] have been developed .

A UAM mission planer algorithm considering capacity un-/limited origin and destination vertiports, flight trajectories, number of available vehicle, and constraints imposed by previously planned flights was developed by [162] and exercised for the Northern California region. After an available vehicle was matched to a request, a suitable take-off and landing time at the origin and destination vertiport will be determined. Subsequently, a conflict-free 4D trajectory connecting origin and destination vertiport will be calculated. The automated design and selection of the shortest strategically deconflicted 4D trajectory matching each UAM flight request is also provided by [163]'s low-altitude air traffic management system inside the developed automated flight planning system *AFPS*.

Strategic conflicts may occur, e.g., due to loss of separation or the crossing of no-fly zones. Several resolution actions may be applied such as departure delay, change of arrival/departure speed and direction, change in cruise speed and re-routing (for more resolution actions see [162]). Delay can be therefore generated on ground and in the air. Based on [162], a change in vertical speed during climb and descent appeared ineffective, whereas, using en route conflict resolution achieved 94% effectiveness. Departure delay was mainly used for resolving conflicts near the vertiport or in the first stages of take-off.

For a vertiport network in Dallas Fort-Worth (U.S.), ref. [164] concluded, when horizontal spatial separation values are reduced (0.3 nm to 0.1 nm) less conflicts and delay (-7.3%) were detected both on the ground and in the air. Instead, decreasing temporal separation (60 s to 45 s) resulted in even less conflicts and total delay (-28.4%) on the ground and in the air. Once the scheduling horizon was reduced (50 min to 8 min), total delay decreased and shifted its appearance from ground to mainly airborne delay since more conflicts have to be resolved post-departure. Considering a scheduling horizon greater than the actual flight time, most of the conflicts are resolved pre-departure generating ground delay.

Strategic conflicts may also occur due to multiple fleet operators utilizing same resources such as airspace and vertiport capacity. [163] introduces the *Unit Benefit Ratio* as a metric to measure the benefit of each operator instead of each flight due to possible market share differences. Under the aspects of system costs and operator equity, and based on formerly developed vertiport locations in Tampa Bay (U.S.) by [105], ref. [163] studied the applicability of a low-altitude traffic management system. Research on traffic flow management measures based on fairness and equity was also conducted by [107] for UAM delivery operations in *Toulouse (France)*.

The tension between multiple fleet operators may even increase if different business cases are operating simultaneously following different planning horizons such as expected for on-demand delivery, on-demand and scheduled air taxi services.

According to [107], on-demand delivery and UAM traffic may reduce efficiency and fairness of strategic UTM processes. Therefore, ref. [107] introduces three fairness metrics *reversals, overtaking, time-order deviation*. Furthermore a rolling horizon optimization framework is considered in order to include low (on-demand) and high lead time flights (scheduled) into the traffic flow. Therefore, a traffic flow management optimization problem is solved for each rolling horizon of the length of a certain time period allowing different ways of inserting or delaying demand pop-ups. The proposed approach is tested for the area of Toulouse (France) by exemplarily describing a drone package delivery scenario. If high number of pop-up demands are occurring on short horizons, inserting those pop-up demands should be preferred. Instead, if pop-ups are occurring less frequently under a short horizon, the option of inserting as well as delaying them are acceptable. It needs to be highlighted that the option of re-routing already airborne vehicles was not taken into account.

Following the most "natural" scheduling process and queuing approach, FCFS, [133] developed a theoretical model to evaluate the capacity of different vertiport configurations considering changing number of FATOs, parking spaces and occupancy times. A FCFS approach increases in inefficiency if numbers of resources increase. At least 80% throughput to capacity ratio can be captured by the FCFS model for most vertiport configurations in the 102 vertiport-network in Dallas Fort-Worth (U.S.).

4.3.2. Tactical Measures

Following the operational requirements made by EASA's *SC-VTOL-01*, VTOL aircraft certified in category enhanced and operating in European airspace, need to provide continuous safe flight and landing capabilities [21]. This means, once taken-off from the origin vertiport, a continuous flight to the destination vertiport or to an alternate vertiport must be possible after CFP. This will require additional extensive tactical contingency planning and information exchange.

Dividing flight path planning and trajectory computation into an online and offline phase, ref. [165] proposes a decision-based contingency approach calculating a tree of trajectories leading to the destination vertiport including branches leading to alternate vertiports. A Dubins path planner is used to ensure continuous transition between normal and contingency trajectories. Additional adjustments are made in order to enable diversion to other flight levels and local holding patterns for temporal de-confliction if velocity reduction is not sufficient anymore and would force the UAS into a hover state.

As soon as trajectory changes are executed during the active flight phase, separation violations and potentially occurring in-flight conflicts have to be evaluated and resolved prior. To do so, high situational awareness, precise and reliable tracking data and real-time traffic information is needed. This also means that airspace and safety conformance monitoring services need to be available ensuring safe conditions during all phases of the active flight. Since UAM operations are not yet conducted on a daily basis, the UAM and U-space/UTM community might consider emerging ideas proposed for traditional aviation such as [166–171].

Emerging in-flight separation conflicts of 40,000 simulated UAM flights in the area of Dallas Forth-Worth are being analyzed by [94]. During a three-hour time window, a departure scheduler ensured that emerging flights are not interfering with each other and causing immediate loss of separation due to their request time. A lateral separation bandwidth between 200 ft (61 m) to one nautical mile (1.85 km) and a cruise altitude ranging from 1000 ft (300 m) to 5000 ft (1500 m) was considered. The higher the separation value the higher the number and duration of conflicts. Flights with many occurring conflicts show, that many of those conflicts occur during the flight is approaching or leaving a vertiport and while interacting with flights towards and from vertiports located nearby.

Compared to [94] who focused on a departure scheduler and in-flight separation conflicts, the subsequent scientific contributions [172–177] are predominantly focusing on scheduling and sequencing the arrival stream towards a vertiport. Since in-flight changes may result into less-optimal flight paths (longer, additional maneuvers, varying wind conditions), critical delay can be accumulated. Assuming that UAM traffic is targeting a required time of arrival and is constraint by highly limited endurance capabilities, the arrival management may create a critical bottleneck [175]. For eVTOL aircraft, delay can be absorbed most energy efficiently if corrections procedures are conducted during the last leg of the cruise phase prior hovering directly above the vertiport [172]. Adding into operation various (e)VTOL aircraft designs such as tandem-tiltwing [172] and multicopter designs [173] may even increase the complexity of harmonizing the approach traffic flow.

Due to the fact, that winged aircraft have different cruise speeds than wingless eVTOL aircraft, ref. [174] proposes an airspace design in which both aircraft designs are operating but are separated into different traffic flows until they are merged at a metering fix. A sequencing and scheduling algorithm was developed in order to achieve the maximum on-demand arrival throughput of a mixed eVTOL aircraft fleet with different fleet mix ratios at a vertiport with only one FATO.

Building upon [173]'s energy-efficient trajectory optimization tool, a distinct vertiport terminal airspace structure and ConOps was developed in order to harmonize approaching UAM traffic [175]. The vertiport is assumed to be surrounded by a terminal airspace structured in concentric circles in which the innermost ring of the vertiport is controlled and designated for VTOL approach operations. The outmost ring defines the approach threshold at 3900 m (12,795 ft) distance from the vertiport at an altitude of 500 m (1640 ft) at which the arrival sequence is initiated. Each operation can adjust individually its descent angle to meet the requested time of arrival and to absorb delay (up to 3 min) if necessary without hovering or vectoring. Ref. [175]'s numerical experiment considered up to 40 arriving eVTOL aircraft per hour processed in a FCFS manner. It provides an optimal required time of arrival within a distinct planning horizon and selects arrival routes in order to minimize the total delay of all aircraft within a shared terminal airspace. This airspace concept was applied to a vertiport-hub with two FATOs located in the center of a hexagonal vertiport network [176]. A rolling-horizon scheduling algorithm was developed to support the tactical vertiport arrival management. It is highlighted that future work should be complemented by a departure scheduler and a conflict detection service in order to support planning and scheduling processes already in the strategic phase of a UAM flight and to ensure overall efficiency and safety.

Additional separation and collision avoidance services during the tactical arrival sequencing process were added by [177]. Each eVTOL aircraft is responsible for maintaining sufficient separation. Departing vehicles are assumed to operate either through distinct departure gates to separate both aircraft flows, or may operate below the altitude of the approach rings or may depart in hover mode through the center of the rings before transitioning into forward flight. Challenges are identified in the handover from the vertiport terminal area controller (responsible for flow through vertiport airspace structure) to the VTOL controller (responsible for sequencing the final approach). Proposals are made to change the first-in-first-out principle into a priority-based concept focusing on the remaining energy level and to dynamically add rings. It needs to be highlighted that the option of re-routing already airborne vehicles was not taken into account.

Ref. [178] identified "lacks" like the absence of an optimal airspace design for ATM and the neglect of a PAV capability of hovering while analyzing the approach of [175,177].

Additionally, ref. [178] highlights the concern "for safety in the surrounding urban areas due to unnecessary flights around the vertiport". Therefore [178] proposes not only dimensions of holding rings but also distinct holding points where PAV can hover in order to reduce unnecessary flights around the vertiport. Two different sequencing concepts for inward movement are developed: Sequence-Based Approach (SBA) and Branch Queuing Approach (BQA). For the SBA approach the PAV moves from the decision point into the inner circle based on the landing sequence and waits at the hover point. The SBA approach is more flexible and follows a clear landing sequence. In contrary, more conflicts are possible that require higher situational awareness and interventions by tactical de-confliction measures. For the BQA approach, only if a free holding point occurs which belongs to the starting point, the PAV is allowed to move to the inner circle which makes the landing sequence become inoperative. This will cause less conflicts and therefore less tactical de-confliction actions may be required. It creates a safer operating environment but neglects the landing sequence and therefore describes a more rigid and less flexible approach. For specific ring configuration and dimensions please refer to [178].

Furthermore, a third sequencing approach was analyzed by [179] by adding moving circles to the SBA approach (SBAM). After analyzing and comparing on-time performance and loss of separation, resulted into a non-favorable approach compared to SBAM and BQA of [178].

Following the prominent idea of a concentric airspace management structure, ref. [180] elaborated an adaptive control system to set up a multi-ring route ConOps including transition junctions inside the so called UAM multi-vertiport system terminal area and developed a corresponding scheduling model. The multi-ring concept includes approach, departure, emergency rings, junction points, approach and departure routes and waiting areas distributed at different heights and radius around a set of vertiports. Transition junctions are classified in different categories causing different levels of complexity and sets of transit conjunction control rules.

Expanding the focus from a departure and arrival scheduler at one vertiport towards a traffic management inside and between vertiport networks, ref. [181] proposes a decentralized, hierarchical approach to define ATM for UAM which allows the ATM concept to be scalable based on traffic densities and which can be used in a tactical and on-demand manner. Vertihubs, a conglomerate of individual vertiports and their corresponding local airspace "sector", are bundled into one control authority in which one vertihub is responsible for all operating vehicles in that local airspace as well as vehicle flows in and out of its sector. Thus, each vertiport is responsible itself for all vehicles taking-off and landing at their vertiport. Therefore, a UAM network can consist of multiple vertihub airspaces with differing capacity and changing responsibility which may result into several handovers between different vertihub controllers for specific UAM trips. A first application of the UAM ATM concept was conducted on the basis of large-volume UAM air traffic data addressing 1000 vertiports in the San Francisco area.

4.3.3. Measures in Airport Environment

Throughout the world, UAM is either envisioned to operate in a non-segregated airspace together with existing traffic (*U-space* in Europe) or is held separate by mandating UAM to operate within a corridor next to existing traffic (see *UTM* in the U.S.). The concepts of segregated and non-segregated may change over time when different maturity levels of UAM are approached. Specifically, the integration of UAM flights into controlled airspace and the consideration of vertiports located adjacent to airports may create additional challenges.

In this regard, ref. [182] "considers ATC as a critical barrier for the scaling of UAM operations (as opposed to terminal capacity or surface operations) [...]". Looking back in history, in 1960 both airports in Chicago (Midway International and O'Hare International) already processed an average of 135 helicopter flights per day [183].

In 1999, on one single day during the Formula 1 British Grand Prix, the temporary adjacent heliport recorded 4200 VFR aircraft movements [136]. It required the service of 24 air traffic controllers and the utilization of six ATC frequencies! In comparison, for general aviation airports, ref. [182] assumes that a single controller may be capable of managing 100 VFR helicopter operations per hour.

Official VFR routes and ATC protocols are used in order to manage theoretically UAM traffic to and from a vertiport adjacent to Koeln Bonn Airport (Germany) [91]. While the eVTOL aircraft is following the VFR route towards the destination vertiport, ATC needs to provide clearance to the aircraft to confirm final approach at a pre-defined way point. A similar clearance approach was proposed by [85] six-piece vertiport network in *Islamabad (Pakistan)*. For the vertiport adjacent to Koeln Bonn Airport, the UAM traffic should be able to operate in any cardinal direction which means, that no specific direction for approach and departure routes is defined prior. If the VFR approach is followed, the ATC would be able to create flexible flight routes, also distributed at a wider area where noise is able to expand within the controlled airspace. The separation between UAM to UAM and UAM to fixed-wing operations would be feasible, other than using special corridors designated only for UAM traffic. VFR routes and the corresponding compulsory reporting points forces the vehicle to comply with the safety minimum altitudes. Every UAM flight will be coordinated, managed and surveilled by an ATCO who is, in this case, now in charge of both the UAM and the conventional air traffic. This may increase fast in workload deteriorating a vertiport's airside to the predominant bottleneck.

What attributes are mainly contributing to the integration and scalability of UAM operations was investigated for the U.S. by [182]. The analysis addressed how existing arrival (SCIA, MAPt, PinS) and departure procedures can be used or adjusted to accommodate UAM traffic under either VFR or IFR. Next to separation minima and controller workload, ref. [182] also takes into account CNS capabilities (automatic dependent surveillance-broadcast (ADS-B), radio frequencies, traffic alert and collision avoidance system (TCAS) and performance based navigation) that may affect the density limit of concurrent operating UAM vehicles at airports. Five integration approaches are defined in which the UAM traffic is either mixed with conventional flights on a shared runway, closely or widely spaced from each other, operating independently or intersecting with conventional flights. After applying those operating schemes to Boston, San Francisco and Atlanta airport architectures, one departure (diverging departures) and four arrival procedures (converging arrivals, widely spaced VFR arrivals using an air taxiway, and widely spaced IFR arrivals following a PinS procedure) are concluded to be most suitable. From an ATC point of view, vertiports accommodating VFR or IFR UAM flight routes diverging by at least 15° from the conventional runway are not affected by wake vortices and therefore can be operated independently. Based on [182]'s insights, ref. [184] investigated different UAM implementation approaches at Hamburg Airport (Germany) and rated the achieved air taxi throughput while respecting the acceptable workload of an ATCO. A human in the loop study was conducted for the Dallas Fort-Worth Airport in order to elaborate the workload induced by integrating UAM flights in addition to existing commercial traffic [185].

Following standard procedures such as [182], ref. [20] adopts the point-in-space (PinS) approach, an existing standard for helicopter operation, to manage the inbound traffic inside the vertiport area. Here, "the PinS approach was taken as reference because it is used for existing helicopter operations, can be charted, and is rigid while allowing for some flexibility in arrival or departure procedure definition" [20]. The vertiport area is a dedicated airspace surrounding the vertiport and is located inside the vertiport operating area surrounding a single or multiple vertiports in which UAM traffic is assigned to UTM. Following the approach of a segregated airspace for UAM traffic, after the eVTOL exits the high-density UAM routes it starts descending into the vertiport operation area airspace at the initial approach fix. Afterwards, the vehicle proceeds its approach on a pre-defined pathway over the intermediate fix towards the final approach fix (FAF). The FAF is leading towards the decision point/PinS where it is decided if the aircraft proceeds the approach towards landing or if a missed approach will be conducted. Multiple FAF can converge towards a single PinS in order to develop a single stream towards the vertiport. Deciding to proceed with the final approach, the vehicle will enter the visual segment of the approach "where the vertiport has secured navigation and communication with the arriving aircraft" [20] which follows then a pre-defined landing procedure. Departure operations are not explicitly described.

4.4. Infrastructure Cost Estimation

Most of the building and operating costs of a vertiport are unclear as long as we do not know the demand and the VTOL aircraft's performance. Besides that, who is going to pay for it? A vertiport's cost heavily depends on what VTOL aircraft design and UAM "airline" needs to be accommodated, which VTOL aircraft fleet is being operated, what demand densities need to be served and where the vertiport is specifically located.

Considering all-electric and hybrid-electric propulsion systems, ref. [101] estimated energy operating costs as well as total cost per vertiport. Assuming a VTOL vehicle power level of 200 kW required in both propulsion systems, a vertiport network of five vertiports and 500 vehicles each, will require a total cost investment per vertiport of $72 million for operating only fully electric vehicles. This is assumed to decrease significantly to $2.25 million if purely refueling is needed. On the one hand, the amount of required chargers (160) will impose a significant burden on the city's electricity grid, but on the other hand, a fuel-based propulsion system will face non-revenue flights if refueling operations are centralized at a specific vertiport location. However, decoupling UAM transportation services from refueling operations would reduce a vertiport's footprint, creates faster turnarounds and therefore may increase potential throughput.

For a vertiport which offers only a multi-function single pad featuring the dimension of 39 by 69 m, the estimated costs are declared to be approx. $350,000 according to [126]. This increases to $750,000 and $950,000 if two or three additional parking areas are attached to the single FATO, respectively. The required footprint results into 72 by 99 m. Extending a vertiport to a linear design with one landing pad, one take-off pad and two disembarking, maintenance and embarking pads each, results into an expected vertiport cost of $1,600,000 and a footprint of 69 by 168. For the smallest configuration of *Lilium's* vertiports being ground-based with small terminal areas and a limited set of charging stands, an initial investment of €1–2 million is predicted [125]. Elevated vertiports with larger footprints and capabilities require investments between €7–15 millions depending on the resulting size and location [125].

4.5. Summary

Though many design proposals have been made and research papers have been published, there are no vertiports existing yet except of two single FATO designs such as the 2019's demo *VoloPort* in Singapore and Coventry's first *urban-Air Port*. However, the collection of vertiport designs displayed in Section 4.1 offer a wide range of ideas and approaches how to integrate UAM into urban and sub-urban environment and how to use already existent infrastructure. Keywords like *scalability, acceptance* and *sustainability* were raised frequently in this context. For those considered contributions, important topics influencing the vertiport design like energy grid capabilities, VTOL aircraft storage during non-operational hours, safety and security measures, contingency operations, check-in procedures, passenger flow and guidance from gate to the vehicle and operational weather dependencies are, if at all, described very briefly and not in detail. It is also unclear yet, on what basis a vertiport will be dimensioned; is it designed to accommodate peak hours, to fit the overall daily demand, or is the vertiport configuration dynamically adjustable to serve varying demand flows as proposed by [186]. Additional discrepancy is provided by the claimed footprint required for processing one vehicle per hour (cf. [102,126,128]). Vertiport throughput capacity has been studied both analytically as well es through simulation (cf. Section 4.2). There is a wide range of analyzed throughput addressing up to 1400 movements per hour. Various vertiport topologies, positioning pads, gates, and terminals, have been proposed such as satellite, linear and pier topologies. The ratio of gates to pads can vary from 2 to 8. It appears that vertiports will have strongly differing shapes and capacities depending on their location and demand profile they have to process. A novelty of vertiports compared to conventional heliports is the expected use of ground taxiing. Three types of taxiing are defined, namely hover taxiing, passive ground taxiing and active ground taxiing. Lastly, the turnaround at gates, which is driven by passenger de-/boarding and VTOL vehicle re-fueling will be of significant influence for the overall available capacity provided by the vertiport; the latter will depend on the primary energy source, which could be fully electric, hybrid-electric or LNG-powered. Transitioning from airside ground to airside air operations, high-density UAM operation itself is a challenging endeavor in terms of traffic management. But taking into account other airspace users such as commercial and general aviation, helicopter emergency and medical services will increase complexity immensely. This is even aggravated by first implementing piloted UAM operations and, over time, transitioning to automated and autonomous operations. The importance of harmonization between strategic and tactical measures of arrival and departure traffic is highlighted throughout Section 4.3. Different approaches how to structure a vertiport network airspace as well as a vertiport's local airspace and fair access to it was discussed. CNS and ATM capabilities are not only crucial for managing UAM traffic around vertiports, but also when merging UAM traffic with already existing airspace users and conventional traffic especially in airport environment. A need for a thorough strategic planning is discovered, but tactical measures cannot be neglected. The scientific publications discussed in Section 4.3 tend towards a FCFS scheduling and sequencing approach. However, it was clearly highlighted that certain parameters such as remaining endurance and agglomerated delay may impose critical constraints which may favor a priority-based sequencing concept. The transition from piloted to automated to autonomously operating UAM may impose additional implementation challenges especially in terms of traffic management, the distribution of roles and responsibilities, the way of communication and exchanging information while ensuring the highest standards for safety and cyber-/security. In Section 4.4, the prediction of vertiport costs was addressed, which seems to be not really part of scientific papers nor discussed frequently in the public. Neither are UAM and vertiport operations existent yet, nor does Europe has a mature high-volume urban air commuting market from which historic experience may provide reliable cost estimations. Current European research as well as UAM industry does not know what the real operation and traffic densities will look like. Existing aviation infrastructure like airports and heliports may be used initially. But, retrofitting and upgrading them to meet UAM needs and future standards, and integrating UAM traffic at those already existing traffic junctions may be limited and may result into even more additional investments.

5. Weather Impact on Vertiports

"Moreover, the weather enterprise needs champions in the aviation industry to embrace and promote weather as an integral component in the design, certification, and operation of aerial vehicles like eVTOLs or unmanned aerial systems (UAS)" [187]

Airborne operations performing in urban environment do not only face challenges due to a complex obstacle environment, but also due to so far unknown weather conditions arising in highly and densely built-up areas. Every operating environment in which UAM services should be offered, needs to be evaluated locally and regionally depending on the vertiport network size.

Other than for vertiports, STOL contributions are "more conscious" about weather influencing the placement and orientation of the take-off and landing strip. Based on an initial airpark placement which focused on identifying the largest vacant area [82], subsequent contributions like [188,189] use historical weather observation data together with a detailed obstacle analysis to determine the location and orientation of the runway within those areas of interest. For a single runway, its orientation needs to be defined so that the emerging crosswind vector does not exceed 10.5 kts (5.4 m/s) more than 95% of the time [188].

From a European regulatory perspective, EASA's *SC-VTOL-01* provides the requirement "[...] the applicant must demonstrate controllability in wind from zero to a wind limit appropriate for the aircraft type" [21]. In the subsequent *MOC-2 SC-VTOL*, performance data was considered under wind conditions defining "take-off until reaching VTOSS (see *MOC VTOL.2115*) and from below VREF (see *MOC VTOL.2130*) to landing (i.e., the ground referenced phase), at least 17 kts of relative steady wind should be considered" [38]. Additional high-level requirements regarding visibility during falling and blowing snow are displayed in [38]. Other than that, no further requirements are yet provided.

5.1. Meteorological Conditions in Different Operating Environment

Targeting a vertiport network operation 99% of the operating hours per year in the metropolitan area of Munich, future UAM vehicles have to withstand headwind of 20 m/s (39 kts) after the average hourly windspeed, measured at 66 weather stations in the area of interest between 2016–2018, was evaluated [102]. In order to compensate local bad weather conditions an blackouts in the charging infrastructure, a diversion reserve of 10 km (32,808 ft) is demanded.

Moving UAM operations to the U.S. and considering METAR data of 28 metropolitan areas, ref. [100] derived a headwind requirement of 10 kts (5.14 m/s) if at least 50% of the operational window should be covered. This requirement is followed due to the assumption that not all flights are fully facing headwind conditions and necessary reserves will account for uncertainties and additional deviation. Furthermore, if the eVTOL aircraft can withstand wind of 20 kts (10.3 m/s) and 35 kts (18 m/s) of gusts, the operation can be conducted in any of the 28 metropolitan areas a minimum of 95% of the time meeting wind constraints and 95% of the time in all but two cities meeting gust constraints.

The meteorological repercussion on UAM operations in various U.S. cities was further analyzed by [190], who determined the average number of weather-impacted hours for each area of interest. Considering an annual operation with a daily operational window of 7 a.m. to 6 p.m., seven years of METAR surface data (2010–2017) were examined together with supplemental data of pilot reports. In order to elaborate potentially impacted hours, a set of "impact scores" is elaborated rating the captured METAR observation from 1 (minimum impact) to 10 (significantly impactful). This includes among others temperature, rain, ceiling, visibility, wind, hazel and snow grains, but also appearances of dust storms, tornadoes and volcanic ash. An hourly average impact score of three was defined as a threshold between minimal and significant potential impact. Throughout the areas of interest, ref. [190] concluded that an average of 6.1 h per day during the winter, 7.3 h per day in the spring, 2.9 h per day in summer and 2.2 h per day in fall could be potentially affected by considerable impactful weather conditions.

All three examples show that different operating environments call for changing operating hours and vehicle requirements. [191] highlights regional and local variation of weather amongst others caused by geographic influences like latitude defining solar radiation and temperature, major water bodies being the source of moisture, mountains affecting range of altitude and air density and landcover gradients providing differential heating. Other influences are described as diurnal and seasonal cycles, weather systems (wind, clouds, precipitation) and the cityscape causing local scale wind and turbulence. Additional weather challenges need to be considered such as winds at and above ground level (turbulent eddies, extreme and rapid changes in wind speed and direction, microburst translation), ceilings and visibility (sub-grid micro climates) and temperature (heat island effect, effects on density altitudes) [192].

5.2. Meteorological Characteristics in Urban Environment

According to [193], the local climate in cities often differs from surrounding areas. The "urban heat island" effect is a feature of the urban climate which is amongst others characterized by differences in temperature of up to 10 Kelvins in large cities. Additional changes can also be seen in air humidity, radiation, wind, air quality and noise.

Prevailing weather characteristics may also change on very small scales inside city boundaries creating the phenomena of micro-weather. For this purpose, the investigation of wind channeling, turbulence from buildings and urban canyoning, and the development of smart city sensing, micro-grid networks/weather models as well as high computational resources and machine learning approaches are required [187,192]. One of the biggest challenges is that "it is recognised that the weather information for UAS operations may be different from the one provided by today's meteorological service providers [...]. UAS can fly near buildings and in areas where current aeronautical meteorological information is not always provided" [51].

According to [192], additional smart urban sensing can be achieved by optimally placed sensors. A contribution is expected in the development of urban climatology, the improvement of forecasts and the reduction of uncertainties, while targeting optimal UAM flight routes. Expected hurdles are communication bandwidth associated with high costs of expanding the network and possible congestion of current wireless networks due to the amount of data collectors required to achieve sufficient coverage. Processing and computational resources to sight and analyze collected data are needed. The "optimal placement" of weather sensors needs to be investigated thoroughly.

Equipping every VTOL aircraft with weather sensors and thereby increasing enormously the amount of real-time weather data could be a supplemental approach. This data could be then shared inside the UAM network e.g., through a U-space weather information service provider and can be used for weather analysis and forecasts. However, this also requires equipment investments and may probably lead to reduction of payload.

5.3. Weather Impact on Vertiport Elements and Procedures

Based on interviews with experienced helicopter pilots, ref. [132] concluded that eVTOL aircaft should not attempt departure nor arrival operations with a tailwind possibly causing the eVTOL aircraft to enter vortex ring state conditions and facing crosswind greater than 15 kts (7.7 m/s). In the context of UAM, "Vertiport operations are sensitive to wind conditions which may inhibit the use of one or more TLOFs for approach, departure or both" [132]. Thus, weather influences the maneuverability of the eVTOL aircraft and therefore may degrade the performance of the flight or specific flight phases.

How the performance of a vehicle is degraded during the final approach phase and what landing pad size is required to safely accommodate deviations from the nominal flight path was researched by [194]. The Drydon wind turbulence model is used to depict upcoming light, moderate and severe turbulence. Ref. [194] elaborates operational requirements for eVTOL aircraft and analyzes changes in approach angles and speeds leading towards a set of approach surfaces with minimum energy and time considerations. Landing accuracy under different weather constraints resulting into varying FATO sizes was analyzed statistically. An approach surface of a 5 degree approach angle and 40 ft/s (12.2 m/s) approach speed and "for general light turbulence conditions, 95% of the trajectories end up within a radius of 20–30 ft (6–9 m)" for a FATO [194].

Increasing automation will, most likely, increase accuracy, throughput and may lead to affordable UAM ticket prizes. In aviation, camera-based and visual recognition have been researched for decades especially to support and, at some point, to initiate and conduct fully automatic approach and landing operations.

For UAM operations, ref. [195] analyzes requirements and approaches how an enhanced vision system (EVS) can be used for landing procedures at vertiports. EVS is currently used for enhanced visual operations ensuring a safe flight under visual flight rules during night and adverse meteorological condition. According to [190], those conditions affect UAM operation in the U.S. for almost 16% of the operational time. Next to requirements of minimum converted meteorological visibility and the field of view, ref. [195] proposes to consider urban wind fields and wind gusts for EVS sensor requirements. A visual contact with the FATO has to be maintained continuously in order to operate safely but affecting possible take-off and landing directions. As a result, future UAM ground infrastructure and their FATOs need to make sure to be clearly distinguishable in the EVS imagery from surrounding buildings and infrastructure elements on the ground. Additional challenges can be imposed by the surrounding urban lightning and the limited amount and small size of the installed lightning systems at vertiports. With the implementation of fiducial markers and ad-hoc light patterns, a high pose estimation accuracy could be provided in the last 300 m of the nominal approach path.

5.4. Summary

All sighted sources addressing the impact of environmental constraints on UAM operations claim the need for real-time weather data collection and monitoring due to probably very sensitive UAM aircraft. Weather will not only constrain vertiport locations but may also affect directly operational procedures and flight directions towards and from vertiports. On a macro level, historic weather characteristics decide the selection of the operating environment and therefor which vertiport network and what VTOL aircraft performance is required. The specific vertiport design, its allocation of FATOs, approach and departure path orientation and operating concepts are influenced by the prevailing weather conditions and shape UAM on a micro level. Feasible operating hours of certain areas are derived from historic weather data which are then compared to assumed vehicle capabilities. Another approach is to examine historic weather data. Based on appearance and frequency of certain weather phenomena, VTOL aircraft requirements may be formulated in order to cover a certain proportion of the operational window. In both cases, weather considerations including wind, gusts, temperature etc. are not sufficiently addressed and researched yet in the context of UAM flights and vertiport operations. Micro-weather research and the development of fine scale urban weather models need to be pushed forward by current UAM development because *weather* will play a crucial role during the development of future UAM operational procedures and *U-space*/UTM services.

6. Conclusions

> *"Say goodbye to congested streets, traffic diversions, and frustrating journeys"* [196]

> vs.

> *"Ground infrastructure experts wrestle with vertiport challenges"* [197]

Urban Air Mobility needs vertiports to operate! This fact is unanimously acknowledged in the scientific community and industry, but at the same time, vertiports are not well understood and the research is scattered. This is the reason why we conducted a thorough literature review following the objective to summarize systematically the current state of the art and outline key areas where future research is needed. Due to the comprehensive collection of noteworthy UAM vertiport contributions, this manuscript provides the reader a structured setup, with each chapter concluded by a brief summary, which allows for selective reading.

Initial uncertainties in naming UAM ground infrastructure seem to be overcome since *vertiport* is now being predominantly used as the term of choice. After showing that vertiport is the most popular term for UAM ground infrastructure in Section 1.2, we continue to classify the field into eleven topics and analyze their prominence (see Section 1.3). In this manuscript, the scientific literature as well as industry and regulatory contributions such as existing vertiport and heliport design guidelines were reviewed extensively; All three bodies of publication are needed to frame the state of the art of UAM VTOL vertiports.

While searching for scientific publications in the database *Scopus* until the year 2021 (including), 49 scientific publications shared the overlap of "urban air mobility" on the one hand and "ground infrastructure" on the other hand which were used as a basis for this vertiport review manuscript. After analyzing all 49 scientific publications, it became apparent that airspace operations has been the strongest focus so far, followed by the general design of vertiports and its related considerations around throughput and capacity (see Section 4). Also the interaction between a UAM network and the choice of vertiport locations finds mention in the research as elaborated in Section 3. It was found that the majority of the vertiport network research considers U.S. UAM applications. Even German VTOL aircraft manufactures consider initial full-scale UAM applications outside Europe.

Vertiports are recognized as one of the critical elements of UAM by operating on limited spatial resources. Initial bottlenecks of a UAM network will be described by a vertiport's capacity and performance in the air and on ground. This will require thorough knowledge about the vertiport layout, dynamic behavior of airside air and ground operations and inter-dependencies of arrival, departure and passenger streams: who is responsible for coordinating arriving and departing VTOL aircraft traffic? How is a mixed VTOL aircraft fleet and multiple "UAM airlines" accommodated and managed fairly at a vertiport? What traffic densities can be processed and can UAM really reduce traffic congestion on ground?

Current vertiport designs, except of some early prototypes, are currently more describing a vision than providing a realistic and implementable proposal. And, although vertiport design and operations have been the predominant research focus, only few publications take into account non-nominal constraints and contingency incidences.

Continuing the review of current regulatory framework and design guidelines in Section 2, thorough content was virtually not existent until March 2022, when both FAA and EASA independently published a first engineering brief/prototype (respectively) covering only VFR vertiports. Discrepancies also arise when vertiport sequencing and scheduling procedures are discussed. On the one hand complex holding patterns and hover points are proposed for arriving VTOL aircraft traffic, but on the other hand UAM operations are considered using eVTOL aircraft currently providing very limited endurance characteristics. Therefore, further research is necessary to identify and quantify operating uncertainties and to evaluate the role and the limitation of strategic and tactical measures. The various UAM/vertiport design approaches are highlighted by contrasting similarities and differences of U.S., European and international standards (see Section 2.4). One crucial provider of uncertainty is described by the chosen operating environment and the prevailing weather conditions. Weather will be *the* factor constraining UAM and vertiport operational hours, consequently affecting throughput, ticket price and costumer segment. High efforts will be needed to understand urban weather behavior and phenomena in order to provide a safe but also efficient UAM operation. This review wants to highlight the importance of environmental constraints such as *weather* for future UAM and vertiport operations, since current vertiport research, except for a few publications described in Section 5, do not yet specifically focus on it.

The most underrepresented topic in the body of scientific research, but also in regulatory guidelines and vertiport design proposals is *noise* as well as *security*. None of the sighted contributions provide a distinct analysis of how noise is distributed at a vertiport considering e.g., different vertiport layouts, locations, arrival and departure paths/surfaces and VTOL aircraft designs. The same applies for the topic *security* which is mentioned rarely, and if so, only when passenger security checks are addressed. But, *security* means so much more especially when aviation eventually transitions towards a multi-connected, digitized and automated operating system. Implementing vertiports in densely populated environment will require thorough analyses in terms of noise propagation, safety and cyber-/security in order to create a business case finally being accepted by society.

Vertiport approaches and contributions considering different time horizons, maturity levels and traffic densities are currently available which need to be harmonized in order to allow for a structured development of UAM and to finally transition from vision to reality. A European UAM road-map is necessary in order to understand the (regulatory) complexity of UAM, the role of a vertiport and to derive realistic assumptions on societal implications. This literature review gathered a considerable amount of publications to depict the state of the art of UAM VTOL vertiports. The majority of them are of theoretical nature. At some point in the future of research, realistic operational constraints and requirements have to be considered which are going to require a lot of more research, testing, failing and lessons learned until we really reach the implementation of on-demand UAM.

This review manuscript will aid the harmonization process as it summarizes all major ongoing efforts and highlights both similarities and differences. We further hope that fellow researchers will find our work helpful to position their own work well into the context of vertiport and UAM research.

Author Contributions: Conceptualization, K.S., L.P.; methodology, K.S., L.P.; formal analysis, L.P.; investigation, K.S., L.P.; data curation, K.S., L.P.; writing—original draft preparation, K.S. (Sections 1.1, 2.1, 2.4, 3, 4.1.1, 4.1.2, 4.3, 4.4, 5 and 6), L.P. (Sections 1.2, 1.3, 2.2–2.4 and 4.2); writing—review and editing, K.S., L.P.; visualization, K.S., L.P.; funding acquisition, K.S. All authors have read and agreed to the published version of the manuscript.

Funding: This research received no external funding.

Acknowledgments: The authors thank Johannes Ernst, Franz Knabe, Reinhard Schmitz, Anna Straubinger and Antoine Habersetzer for their contribution during the internal review process, as well as MVRDV for providing us their vertiport visualizations. The authors would like to give special thanks to Tabitha Stephani for supporting this work with HorizonUAM renderings. This work was conducted under the cooperation agreement between DLR and Bauhaus Luftfahrt e.V. and contributes to the DLR project HorizonUAM.

Conflicts of Interest: The authors declare no conflict of interest.

Abbreviations

The following abbreviations are used in this manuscript:

AC	advisory circular
AGL	above ground level
API	application programming interface
ATC	air traffic control
ATM	air traffic management
CNS	communication, navigation and surveillance
ConOps	concept of operations
CFP	critical failure for performance
DOI	digital object identifier
EASA	European Union Aviation Safety Agency
eVTOL	electric vertical take-off and landing
EUROCAE	European Organization for Civil Aviation Equipment
FAA	U.S. Federal Aviation Administration
FATO	final approach and take-off area
FCFS	first come-first served
ICAO	International Civil Aviation Organization
IFR	instrument flight rules
LNG	liquefied natural gas
METAR	Meteorological Aerodrome Report
MRO	maintenance, repair and overhaul
NASA	National Aeronautical and Space Administration
PAV	personal aerial vehicle
PinS	point-in-space
SC	special condition
STOL	short take-off and landing

TLOF   touchdown and lift-off area
UAM   urban air mobility
UAS   unmanned aerial system
UTM   unmanned aircraft system traffic management
VFR   visual flight rules
VTOL   vertical take-off and landing

Appendix A

In Figure A1a, the top authors by number of publications in the field of vertiports are listed. Peng Wei, the number one, is an associate professor at the George Washington University in Washington, D.C. He published many papers with is co-author Priyank Pradeep. Another prominent institute is the Georgia Institute of Technology in Atlanta, Georgia: Cedric Justin, Dimitry Mavris and Brian German are associated with it. So far, it appears that the field is dominated by few strong players. In Figure A1b the top sources of publication are shown, which are both conference proceedings and journal issues. *Transportation Research Part C, CEAS Aeronautical* and *Aerospace Information Systems* are journals; the remaining major sources are conference proceedings. Minor sources are journals or proceedings with only one paper on vertiports. The 12 minor sources are the following with one source unknown:

- IEEE Transactions on Intelligent Transportation Systems
- International Journal of Aeronautical and Space Sciences
- MDPI Sustainability
- IEEE Metrology for Aerospace
- MDPI Applied Sciences
- Elsevier Engineering
- International Conference on Engineering Design
- Aerospace Science and Technology
- Transportation Research Record
- IEEE Transactions on Control of Network Systems
- MDPI Aerospace
- IEEE Chinese Guidance, Navigation and Control Conference

The top ten individual publications in the field of vertiports according to number of citations in *Scopus* are listed in Table A1. The reference day for the number of citations was 31 December 2021.

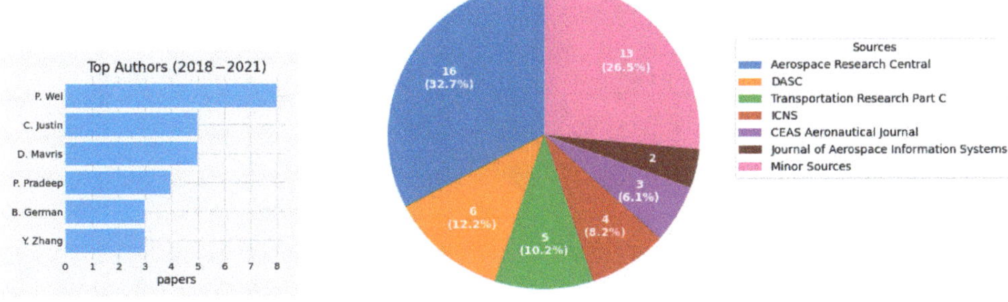

(**a**) Top publishing authors.   (**b**) Top publishing conferences and journals.

Figure A1. Data analytics in the field of vertiports.

Table A1. Top 10 papers according to citations in *Scopus* (as of 31 December 2021) addressing vertiports in the context of UAM.

DOI	Year	Authors	Title	Citations in Scopus
10.1109/DASC.2018.8569645	2018	I. C. Kleinbekman, M. A. Mitici, P. Wei	Evtol Arrival Sequencing And Scheduling For On-Demand Urban Air Mobility	30
10.2514/6.2018-3677	2018	L. W. Kohlman, M. D. Patterson	System-Level Urban Air Mobility Transportation Modeling And Determination Of Energy-Related Constraints	27
10.2514/6.2019-0526	2019	P. D. Vascik, R. J. Hansman	Development Of Vertiport Capacity Envelopes And Analysis Of Their Sensitivity To Topological And Operational Factors	25
10.2514/6.2018-2008	2018	P. Pradeep, P. Wei	Energy Efficient Arrival With Rta Constraint For Urban Evtol Operations	20
10.2514/6.2018-2006	2018	B. J. German, M. J. Daskilewicz, T. K. Hamilton, M. M. Warren	Cargo Delivery By Passenger Evtol Aircraft: A Case Study In The San Francisco Bay Area	19
10.1007/s13272-020-00468-5	2020	K. O. Ploetner, C. Al, C. Antoniou, F. Frank, M. Fu, S. Kabel, C. Llorca, R. Moeckel, A. T. Moreno, A. Pukhova, R. Rothfeld, M. Shamiyeh, A. Straubinger, H. Wagner, Q. Zhang	Long-Term Application Potential Of Urban Air Mobility Complementing Public Transport: An Upper Bavaria Example	15
10.2514/6.2018-3054	2018	J. N. Robinson, M. D. Sokollek, C. Y. Justin, D. N. Mavris	Development Of A Methodology For Parametric Analysis Of Stol Airpark Geo-Density	12
10.1109/GNCC42960.2018.9018748	2018	P. Pradeep, P. Wei	Energy Optimal Speed Profile For Arrival Of Tandem Tilt-Wing Evtol Aircraft With Rta Constraint	12
10.2514/1.I010710	2019	P. Pradeep, P. Wei	Energy-Efficient Arrival With Rta Constraint For Multirotor Evtol In Urban Air Mobility	12
10.2514/6.2021-1189	2021	R. C. Busan, P. C. Murphy, D. B. Hatke, B. M. Simmons	Wind Tunnel Testing Techniques For A Tandem Tilt-Wing, Distributed Electric Propulsion Vtol Aircraft	9

References

1. Johnston, T.; Riedel, R.; Sahdev, S. To Take Off, Flying Vehicles First Need Places to Land. Available online: https://www.mckinsey.com/industries/automotive-and-assembly/our-insights/to-take-off-flying-vehicles-first-need-places-to-land# (accessed on 18 November 2020).
2. European Union Aviation Safety Agency (EASA). *Study on Societal Acceptance of Urban Air Mobility in Europe*; Technical Report; EASA: Cologne, Germany, 2021.
3. Anandarajan, M.; Hill, C.; Nolan, T. *Practical Text Analytics: Maximizing the Value of Text Data*; Springer International Publishing: Berlin/Heidelberg, Germany, 2019.
4. Garrow, L.A.; German, B.J.; Leonard, C.E. Urban air mobility: A comprehensive review and comparative analysis with autonomous and electric ground transportation for informing future research. *Transp. Res. Part C Emerg. Technol.* **2021**, *132*, 103377. [CrossRef]
5. Straubinger, A.; Rothfeld, R.; Shamiyeh, M.; Büchter, K.D.; Kaiser, J.; Plötner, K.O. An overview of current research and developments in urban air mobility—Setting the scene for UAM introduction. *J. Air Transp. Manag.* **2020**, *87*, 101852. [CrossRef]

6. Hill, B.P.; DeCarme, D.; Metcalfe, M.; Griffin, C.; Wiggins, S.; Metts, C.; Bastedo, B.; Patterson, M.D.; Mendonca, N.L. UAM Vision Concept of Operations (ConOps) UAM Maturity Level (UML) 4-Version 1.0. Available online: https://ntrs.nasa.gov/api/citations/20205011091/downloads/UAM%20Vision%20Concept%20of%20Operations%20UML-4%20v1.0.pdf (accessed on 28 March 2021).

7. Schweiger, K.; Knabe, F.; Korn, B. An exemplary definition of a vertidrome's airside concept of operations. *Aerosp. Sci. Technol.* **2022**, *125*, 107144. [CrossRef]

8. Federal Aviation Administration (FAA). Helicopter Instructor's Handbook (FAA-H-8083-4). Available online: https://www.faa.gov/regulations_policies/handbooks_manuals/aviation/media/FAA-H-8083-4.pdf (accessed on 2 February 2021).

9. International Civil Aviation Organization (ICAO). Annex 14-Aerodromes-Volume I-Aerodrome Design and Operations. In *International Standards and Recommended Practices*, 4th ed.; ICAO: Montreal, QC, Canada, 2004; Volume I.

10. European Union Aviation Safety Agency (EASA). Certification Specification and Guidance Material for Aerodromes Design Issue 4. Available online: https://www.easa.europa.eu/sites/default/files/dfu/Annex%20to%20EDD%202017-021-R%20-%20CS-ADR-DSN%20Issue%204_0.pdf (accessed on 24 November 2020).

11. European Union Aviation Safety Agency (EASA). Aerodromes & Ground Handling. Available online: https://www.easa.europa.eu/light/topics/aerodromes-ground-handling (accessed on 24 November 2020).

12. European Union Aviation Safety Agency (EASA). Certification Specifications and Guidance Material for the Design of Surface-Level VFR Heliports Located at Aerodromes that Fall under the Scope of Regulation (EU) 2018/1139 (CS-HPT-DSN) Issue 1. Available online: https://www.easa.europa.eu/sites/default/files/dfu/Annex%20to%20ED%20Decision%202019-012-R%20CS-HPT-DSN.pdf (accessed on 7 May 2020).

13. Clay, B.; Baumgaertner, P.; Thompson, P.; Meyer, S.; Reber, R.; Berry, D. Civil Tiltrotor Missions and Applications: A Research Study. Contractor Report NASA-CR-177452, 1987. Available online: https://ntrs.nasa.gov/api/citations/19910004111/downloads/19910004111.pdf (accessed on 11 July 2022).

14. Thompson, P.; Neir, R.; Reber, R.; Scholes, R.; Alexander, H.; Sweet, D.; Berry, D. Civil Tiltrotor Missions and Applications Phase II: The Commercial Passenger Market. Contractor Report NASA-CR-177576, 1991. Available online: https://ntrs.nasa.gov/api/citations/19910016812/downloads/19910016812.pdf (accessed on 11 July 2022).

15. Civil Tiltrotor Development Advisory Committee (CTRDAC). *Civil Tiltrotor Development Advisory Committee-Report to Congress*; Defense Technical Information Center: Fort Belvoir, VA, USA, 1995; Volume 1.

16. FAA RE&D Committee Vertical Flight Subcommittee. *Tiltrotor and Advanced Rotorcraft Technology in the National Airspace System (TARTNAS)*; Technical Report; Federal Aviation Administration: Washington, DC, USA, 2001.

17. Federal Aviation Administration (FAA). Advisory Circular 150/5390-3—Vertiport Design (Cancelled). Available online: https://www.faa.gov/documentLibrary/media/advisory_circular/150-5390-3/150_5390_3.PDF (accessed on 5 May 2020).

18. Federal Aviation Administration (FAA). Memorandum subject to Engineering Brief No. 105, Vertiport Design. Available online: https://www.faa.gov/airports/engineering/engineering_briefs/drafts/media/eb-105-vertiport-design-industry-draft.pdf (accessed on 8 March 2022).

19. Federal Aviation Administration (FAA). Urban Air Mobility (UAM) Concept of Operations v1.0. Available online: https://nari.arc.nasa.gov/sites/default/files/attachments/UAM_ConOps_v1.0.pdf (accessed on 6 August 2020).

20. Northeast UAS Airspace Integration Research Alliance (NUAIR). High-Density Automated Vertiport Concept of Operations. Contractor or Grantee Report 20210016168. 2021. Available online: https://ntrs.nasa.gov/api/citations/20210016168/downloads/20210016168_MJohnson_VertiportAtmtnConOpsRprt_final_corrected.pdf (accessed on 11 July 2022).

21. European Union Aviation Safety Agency (EASA). Special Condition Vertical Take-Off and Landing (VTOL) Aircraft. Available online: https://www.easa.europa.eu/sites/default/files/dfu/SC-VTOL-01.pdf (accessed on 20 September 2021).

22. European Union Aviation Safety Agency (EASA). Vertiports Prototype Technical Specifications for the Design of VFR Vertiports for Operation with Manned VTOL-Capable Aircraft Certified in the Enhanced Category (PTS-VPT-DSN). Available online: https://www.easa.europa.eu/downloads/136259/en (accessed on 27 March 2022).

23. EUROCAE. EUROCAE Open Consultation ED-299. Available online: https://www.eurocae.net/news/posts/2021/december/eurocae-open-consultation-ed-299/ (accessed on 23 February 2022).

24. EUROCAE. *ED-293 Concept of Operations for VTOL Aircraft Volume 2: Commercial Passenger Air Taxi Transport*; EUROCAE: Saint-Denis, France, 2021.

25. International Organization for Standardization (ISO). ISO/AWI 5491 Vertiports—Infrastructure and Equipment for Vertical Take-Off and Landing (VTOL) of Electrically Powered Cargo Unmanned Aircraft System (UAS). Available online: https://www.iso.org/standard/81313.html (accessed on 20 September 2021).

26. ASTM International. ASTM WK59317 New Specification for Vertiport Design. Available online: https://www.astm.org/workitem-wk59317 (accessed on 8 June 2022).

27. Uber Elevate. Fast-Forwarding to a Future of On-Demand Urban Air Transportation. Available online: https://evtol.news/__media/PDFs/UberElevateWhitePaperOct2016.pdf (accessed on 20 September 2021).

28. Volocopter GmbH. The Roadmap to Scalable Urban Air Mobility White Paper 2.0. Available online: https://www.volocopter.com/content/uploads/Volocopter-WhitePaper-2-0.pdf (accessed on 13 March 2022).

29. Airbus; Boeing. A New Digital Era of Aviation: The Path Forward for Airspace and Traffic Management. Available online: https://storage.googleapis.com/blueprint/Airbus (accessed on 11 July 2022).

30.	Skyports; Wisk. *Concept of Operations: Autonomous UAM Aircraft Operations and Vertiport Integration*; Skyports: London, UK; Wisk: Mountain View, CA, USA, 2022. Available online: https://wisk.aero/wp-content/uploads/2022/04/2022-04-12-Wisk-Skyports-ConOps-Autonomous-eVTOL-Operations-FINAL.pdf (accessed on 11 July 2022).

31.	Lineberger, R.; Hussain, A.; Metcalfe, M.; Rutgers, V. Infrastructure Barriers to the Elevated Future of Mobility. Available online: https://www2.deloitte.com/content/dam/insights/us/articles/5103_Infrastructure-barriers-to-elevated-FOM/DI_Infrastructure-barriers-to-elevated-FOM.pdf (accessed on 9 June 2022).

32.	Vascik, P.D. Systems Analysis of Urban Air Mobility Operational Scaling. Ph.D. Thesis, Massachusetts Institute of Technology Cambridge, MA, USA, 2019.

33.	Salehi, V.; Wang, S. Application of Munich Agile Concepts for Mbse as a Holistic and Systematic Design of Urban Air Mobility in Case of Design of Vertiports and Vertistops. *Proc. Des. Soc.* **2021**, *1*, 497–510. [CrossRef]

34.	Skyports Limited. Landing Infrastructure. Available online: https://skyports.net/landing-infrastructure/ (accessed on 7 March 2022).

35.	Wei, L.; Justin, C.Y.; Mavris, D.N. Optimal Placement of Airparks for STOL Urban and Suburban Air Mobility. In Proceedings of the AIAA Scitech 2020 Forum, Orlando, FL, USA, 6–10 January 2020. [CrossRef]

36.	Plested, H. Vertiport Regulations and Standards: Creating the Rules Framework for UAM Infrastructure. Available online: https://skyports.net/2020/07/vertiport-regulations-and-standards-creating-the-rules-framework-for-uam-infrastructure/ (accessed on 7 February 2021).

37.	European Union Aviation Safety Agency (EASA). Proposed Means of Compliance with the Special Condition VTOL. Available online: https://www.easa.europa.eu/document-library/product-certification-consultations/special-condition-vtol (accessed on 22 October 2020).

38.	European Union Aviation Safety Agency (EASA). Second Publication of Proposed Means of Compliance with the Special Condition VTOL. Available online: https://www.easa.europa.eu/downloads/128938/en (accessed on 8 December 2021).

39.	Commission Delegated Regulation (EU) 2019/945 of 12 March 2019 on Unmanned Aircraft Systems and on Third-Country Operators of Unmanned Aircraft Systems (OJ L 152, 11.6.2019, p. 1). Available online: https://eur-lex.europa.eu/legal-content/EN/TXT/PDF/?uri=CELEX:02019R0945-20200809&from=EN (accessed on 23 February 2022).

40.	Commission Implementing Regulation (EU) 2019/947 of 24 May 2019 on the Rules and Procedures for the Operation of Unmanned Aircraft (OJ L 152, 11.6.2019, p. 45). Available online: https://eur-lex.europa.eu/legal-content/EN/TXT/PDF/?uri=CELEX:02019R0947-20210805&from=EN (accessed on 23 February 2022).

41.	European Union Aviation Safety Agency (EASA). Certified Category-Civil Drones. Available online: https://www.easa.europa.eu/domains/civil-drones/drones-regulatory-framework-background/certified-category-civil-drones (accessed on 11 March 2022).

42.	European Aviation Safety Agency (EASA). Terms of Reference for Rulemaking Task RMT.0230 Introduction of a Regulatory Framework for the Operation of Unmanned Aircraft Systems and for Urban Air Mobility in the European Union Aviation System, Issue 3. Available online: https://www.easa.europa.eu/downloads/126656/en (accessed on 22 February 2022).

43.	Alamouri, A.; Lampert, A.; Gerke, M. An Exploratory Investigation of UAS Regulations in Europe and the Impact on Effective Use and Economic Potential. *Drones* **2021**, *5*, 63. [CrossRef]

44.	European Union Aviation Safety Agency (EASA). Open Category-Civil Drones. Available online: https://www.easa.europa.eu/domains/civil-drones/drones-regulatory-framework-background/open-category-civil-drones (accessed on 24 May 2022).

45.	The Vertical Flight Society. eVTOL Aircraft Directory. Available online: https://evtol.news/aircraft (accessed on 30 May 2022).

46.	European Union Aviation Safety Agency (EASA). VTOL Trajectories and Vertiports-Rotorcraft & VTOL Symposium 2021. Available online: https://www.youtube.com/watch?v=e_fsxgWIENI (accessed on 21 February 2022).

47.	European Union Aviation Safety Agency (EASA). Annexes to the draft Commission Regulation on 'Air Operations-OPS'. Available online: https://www.easa.europa.eu/sites/default/files/dfu/Annexes%20to%20Regulation.pdf (accessed on 9 December 2021).

48.	CORUS. U-Space Concept of Operations. Available online: https://ext.eurocontrol.int/ftp/?t=714bd3ca21914c619387f1811a6b2f24 (accessed on 22 October 2020).

49.	Barrado, C.; Boyero, M.; Brucculeri, L.; Ferrara, G.; Hately, A.; Hullah, P.; Martin-Marrero, D.; Pastor, E.; Rushton, A.P.; Volkert, A. U-Space Concept of Operations: A Key Enabler for Opening Airspace to Emerging Low-Altitude Operations. *Aerospace* **2020**, *7*, 24. [CrossRef]

50.	Commission Implementing Regulation (EU) 2021/664 of 22.4.2021 on a Regulatory Framework for the U-Space (OJ L 139, 23.4.2021, pp. 161–183). Available online: https://eur-lex.europa.eu/legal-content/EN/TXT/PDF/?uri=CELEX:32021R0664&from=EN (accessed on 23 February 2022).

51.	European Union Aviation Safety Agency (EASA). Notice of Proposed Amendment 2021-14 in accordance with Articles 6(3), 7 and 8 ('Standard Procedure': Public Consultation) of MB Decision No 18-2015 Development of Acceptable Means of Compliance and Guidance Material to Support the U-Space Regulation. Available online: https://www.easa.europa.eu/downloads/134303/en (accessed on 24 February 2022).

52.	European Commission. Concept Of Operations For European U-Space Services—Extension For Urban Air Mobility. Available online: https://cordis.europa.eu/project/id/101017682 (accessed on 27 May 2022).

53.	European Commission. Tactical Instrumental Deconfliction and in flight Resolution. Available online: https://cordis.europa.eu/project/id/101017677 (accessed on 27 May 2022).

54. European Commission. Demand and Capacity Optimisation in U-Space. Available online: https://cordis.europa.eu/project/id/893864 (accessed on 27 May 2022).
55. European Commission. PJ34-W3 AURA "ATM U-SPACE INTERFACE". Available online: https://cordis.europa.eu/project/id/101017521/de (accessed on 27 May 2022).
56. Central Intelligence Agency. The World Factbook: Heliports. Available online: https://www.cia.gov/the-world-factbook/field/heliports/ (accessed on 7 April 2021).
57. Federal Aviation Administration (FAA). 150/5390-2C-Heliport Design. Available online: https://www.faa.gov/airports/resources/advisory_circulars/index.cfm/go/document.current/documentnumber/150_5390-2 (accessed on 13 June 2022).
58. Federal Aviation Administration (FAA). Draft 150/5390-2D-Heliport Design. Available online: https://www.faa.gov/regulations_policies/advisory_circulars/index.cfm/go/document.information/documentID/1038739 (accessed on 13 June 2022).
59. National Air Transportation Association (NATA). Urban Air Mobility: Considerations for Vertiport Operation. Available online: https://www.nata.aero/assets/Site_18/files/GIA/NATA%20UAM%20White%20Paper%20-%20FINAL%20cb.pdf (accessed on 13 June 2022).
60. Zoldi, D.M.K. Vertiport Infrastructure: New Tech, Old Regulations. Available online: https://insideunmannedsystems.com/vertiport-infrastructure-new-tech-old-regulations/ (accessed on 24 February 2022).
61. HeliOffshore. Approach Path Management Guidelines. Available online: www.helioffshore.org (accessed on 8 July 2021).
62. Goodrich, K.H.; Theodore, C.R. Description of the NASA Urban Air Mobility Maturity Level (UML) Scale. In Proceedings of the AIAA Scitech 2021 Forum, Reston, VA, USA, 11–21 January 2021; [CrossRef]
63. Massachusetts Institute of Technology (MIT). Concepts Studies for Future Intracity Air Transportation Systems. Available online: https://dspace.mit.edu/handle/1721.1/68000 (accessed on 13 June 2022).
64. Peisen, D.J.; Ferguson, S.W. Vertiport Design Characteristics for Advanced Rotorcraft Technology. *SAE Trans.* **1996**, *105*, 1313–1319.
65. Peisen, D.J. Analysis of Vertiport Studies Funded by the Airport Improvement Program (AIP). Available online: https://apps.dtic.mil/sti/pdfs/ADA283249.pdf (accessed on 13 June 2022).
66. Giligan, M.; Grizzle, J.D.; Cox, V.H. Integration of Unmanned Aircraft Systems into the National Airspace System: Concept of Operations v2.0. Available online: https://www.suasnews.com/wp-content/uploads/2012/10/FAA-UAS-Conops-Version-2-0-1.pdf (accessed on 13 June 2022).
67. Kopardekar, P.H. Unmanned Aerial System (UAS) Traffic Management (UTM): Enabling Low-Altitude Airspace and UAS Operations. Available online: https://ntrs.nasa.gov/api/citations/20140013436/downloads/20140013436.pdf (accessed on 13 June 2022).
68. Kopardekar, P.H.; Rios, J.; Prevot, T.; Johnson, M.A.; Jung, J.; Robinson III, J.E. UAS Traffic Management (UTM) Concept of Operations to Safely Enable Low Altitude Flight Operations. In Proceedings of the 16th AIAA Aviation Technology, Integration, and Operations Conference, Washington, DC, USA, 13–17 June 2016; p. 201. [CrossRef]
69. Kopardekar, P.H.; Bradford, S. UAS Traffic Management (UTM): Research Transition Team (RTT) Plan. Available online: https://www.faa.gov/uas/research_development/traffic_management/media/FAA_NASA_UAS_Traffic_Management_Research_Plan.pdf (accessed on 13 June 2022).
70. Federal Aviation Administration (FAA). Low Altitude Authorization and Notification Capability (LAANC) Concept of Operations. Available online: https://www.faa.gov/uas/programs_partnerships/data_exchange/laanc_for_industry/media/laanc_concept_of_operations.pdf (accessed on 13 June 2022).
71. Chan, W.N.; Barmore, B.; Kibler, J.; Lee, P.U.; O'Connor, N.; Palopo, K.; Thipphavong, D.P.; Zelinski, S. Overview of NASA's ATM-X Project. In Proceedings of the 18th AIAA Aviation Technology, Integration, and Operations Conference 2018, Atlanta, GA, USA, 25–29 June 2018; Curran Associates Inc.: Red Hook, NY, USA, 2018. [CrossRef]
72. Federal Aviation Administration (FAA); U.S. Department of Transportation (DoT). Unmanned Aircraft System (UAS) Traffic Management (UTM): Concept of Operations v1.0. Available online: https://www.faa.gov/uas/research_development/traffic_management/media/UTM_ConOps_v2.pdf (accessed on 13 June 2022).
73. International Civil Aviation Organization (ICAO). *Heliport Manual (Doc 9261)*, 5th ed. Available online: https://store.icao.int/en/heliport-manual-doc-9261 (accessed on 13 June 2022).
74. Balakrishnan, K.; Polastre, J.; Mooberry, J.; Golding, R.; Sachs, P. Blueprint For The Sky: The Roadmap for the Safe Integration of Autonomous Aircraft. Available online: https://www.airbusutm.com/uam-resources-airbus-blueprint (accessed on 13 June 2022).
75. Zhang, J. UOMS in China. Available online: www.eu-china-app.org (accessed on 13 June 2022).
76. Civil Aviation Administration of China (CAAC). Map of UTM Implementation: China. Available online: http://gutma.org/maps/index.php?title=China (accessed on 13 June 2022).
77. Ushijima, H. UTM Project in Japan. Available online: https://gutma.org/montreal-2017/wp-content/uploads/sites/2/2017/07/UTM-Project-in-Japan_METI.pdf (accessed on 13 June 2022).
78. Japan Aerospace Exploration Agency (JAXA). Development of UAS Traffic Management System (UTM) in Progress. Available online: https://global.jaxa.jp/activity/pr/jaxas/no079/08.html (accessed on 13 June 2022).

79. Airservices Australia; Embraer Business Innovation Center. Urban Air Traffic Management Concept of Operations Version 1. Available online: https://daflwcl3bnxyt.cloudfront.net/m/3dc1907d3388ff52/original/PPJ016561-UATM-Concept-of-Operations-Design_D11-FINAL.pdf (accessed on 24 February 2022).
80. International Transport Forum (ITF); Organisation for Economic Co-Operation and Development (OECD). Ready for Take-Off? Integrating Drones into the Transport System. Available online: https://www.itf-oecd.org/integrating-drones-transport-system (accessed on 13 June 2022).
81. Venkatesh, N.; Payan, A.P.; Justin, C.Y.; Kee, E.; Mavris, D. Optimal Siting of Sub-Urban Air Mobility (sUAM) Ground Architectures using Network Flow Formulation. In Proceedings of the AIAA AVIATION 2020 FORUM, Virtual Event, 15–19 June 2020. [CrossRef]
82. Robinson, J.N.; Sokollek, M.D.R.; Justin, C.Y.; Mavris, D.N. Development of a Methodology for Parametric Analysis of STOL Airpark Geo-Density. In Proceedings of the 2018 Aviation Technology, Integration, and Operations Conference, Atlanta, GA, USA, 25–29 June 2018. [CrossRef]
83. Bulusu, V.; Onat, E.B.; Sengupta, R.; Yedavalli, P.; Macfarlane, J. A Traffic Demand Analysis Method for Urban Air Mobility. *IEEE Trans. Intell. Transp. Syst.* **2021**, *22*, 6039–6047. [CrossRef]
84. Willey, L.C.; Salmon, J.L. A method for urban air mobility network design using hub location and subgraph isomorphism. *Transp. Res. Part C Emerg. Technol.* **2021**, *125*, 102997. [CrossRef]
85. Gillani, R.; Jahan, S.; Majid, I. A Proposed Communication, Navigation & Surveillance System Architecture to Support Urban Air Traffic Management. In Proceedings of the 2021 IEEE/AIAA 40th Digital Avionics Systems Conference (DASC), San Antonio, TX, USA, 3–7 October 2021; pp. 1–7. [CrossRef]
86. Jeong, J.; So, M.; Hwang, H.Y. Selection of Vertiports Using K-Means Algorithm and Noise Analyses for Urban Air Mobility (UAM) in the Seoul Metropolitan Area. *Appl. Sci.* **2021**, *11*, 5729. [CrossRef]
87. Chan, C.; Wang, B.; Bachan, J.; Macfarlane, J. Mobiliti: Scalable Transportation Simulation Using High-Performance Parallel Computing. In Proceedings of the 2018 21st International Conference on Intelligent Transportation Systems (ITSC), Maui, HI, USA, 4–7 November 2018; pp. 634–641. [CrossRef]
88. Alvarez, L.E.; Jones, J.C.; Bryan, A.; Weinert, A.J. Demand and Capacity Modeling for Advanced Air Mobility. In Proceedings of the AIAA AVIATION 2021 FORUM, American Institute of Aeronautics and Astronautics, Virtual Event, 2–6 August 2021. [CrossRef]
89. Naser, F.; Peinecke, N.; Schuchardt, B.I. Air Taxis vs. Taxicabs: A Simulation Study on the Efficiency of UAM. In Proceedings of the AIAA AVIATION 2021 FORUM, American Institute of Aeronautics and Astronautics, Virtual Event, 2–6 August 2021. [CrossRef]
90. Fu, M.; Rothfeld, R.; Antoniou, C. Exploring Preferences for Transportation Modes in an Urban Air Mobility Environment: Munich Case Study. *Transp. Res. Rec. J. Transp. Res. Board* **2019**, *2673*, 427–442. [CrossRef]
91. Feldhoff, E.; Soares Roque, G. Determining infrastructure requirements for an air taxi service at Cologne Bonn Airport. *CEAS Aeronaut. J.* **2021**, *12*, 821–833. [CrossRef] [PubMed]
92. Rimjha, M.; Hotle, S.; Trani, A.; Hinze, N.; Smith, J.C. Urban Air Mobility Demand Estimation for Airport Access: A Los Angeles International Airport Case Study. In Proceedings of the 2021 Integrated Communications Navigation and Surveillance Conference (ICNS), Dulles, VA, USA, 19–23 April 2021; pp. 1–15. [CrossRef]
93. Rimjha, M.; Li, M.; Hinze, N.; Tarafdar, S.; Hotle, S.; Swingle, H.; Trani, A. *Demand Forecast Model Development and Scenarios Generation for Urban Air Mobility Concepts*; Contractor or Grantee Report, 20205005881; Virginia Tech Air Transportation Systems Laboratory: Blacksburg, VA, USA, 2020.
94. Goodrich, K.H.; Barmore, B. Exploratory Analysis of the Airspace Throughput and Sensitivies of an Urban Air Mobility System. In Proceedings of the 2018 Aviation Technology, Integration, and Operations Conference, Atlanta, GA, USA, 25–29 June 2018. [CrossRef]
95. Goyal, R.; Reiche, C.; Fernando, C.; Cohen, A. Advanced Air Mobility: Demand Analysis and Market Potential of the Airport Shuttle and Air Taxi Markets. *Sustainability* **2021**, *13*, 7421. [CrossRef]
96. Frej Vitalle, R.; Zhang, Y.; Normann, B.; Shen, N. A Model for the Integration of UAM operations in and near Terminal Areas. In Proceedings of the AIAA AVIATION 2020 FORUM, Virtual Event, 15–19 June 2020. [CrossRef]
97. Rimjha, M.; Trani, A. Urban Air Mobility: Factors Affecting Vertiport Capacity. In Proceedings of the 2021 Integrated Communications Navigation and Surveillance Conference (ICNS), Dulles, VA, USA, 19–23 April 2021; pp. 1–14. [CrossRef]
98. Taylor, M.; Flenniken, L.; Nembhard, J.; Barreal, A. Design of a Rapid, Reliable Urban Mobility System for the DC Region. In Proceedings of the 2020 Integrated Communications Navigation and Surveillance Conference (ICNS), Virtual Conference, 8–10 September 2020. [CrossRef]
99. Rimjha, M.; Hotle, S.; Trani, A.; Hinze, N. Commuter demand estimation and feasibility assessment for Urban Air Mobility in Northern California. *Transp. Res. Part A Policy Pract.* **2021**, *148*, 506–524. [CrossRef]
100. Patterson, M.D.; Antcliff, K.R.; Kohlman, L.W. A Proposed Approach to Studying Urban Air Mobility Missions Including an Initial Exploration of Mission Requirements. In Proceedings of the 74th Annual American Helicopter Society International Forum and Technology Display 2018 (FORUM 74), Phoenix, AZ, USA, 14–17 May 2018; Curran Associates Inc.: Red Hook, NY, USA, 2018.

101. Kohlman, L.W.; Patterson, M.D. System-Level Urban Air Mobility Transportation Modeling and Determination of Energy-Related Constraints. In Proceedings of the 2018 Aviation Technology, Integration, and Operations Conference, Atlanta, GA, USA, 25–29 June 2018. [CrossRef]
102. Ploetner, K.O.; Al Haddad, C.; Antoniou, C.; Frank, F.; Fu, M.; Kabel, S.; Llorca, C.; Moeckel, R.; Moreno, A.T.; Pukhova, A.; et al. Long-term application potential of urban air mobility complementing public transport: An upper Bavaria example. *CEAS Aeronaut. J.* **2020**, *11*, 991–1007. [CrossRef]
103. Ploetner, K.O.; Al Haddad, C.; Antoniou, C.; Frank, F.; Fu, M.; Kabel, S.; Llorca, C.; Moeckel, R.; Moreno Chou, T.; Pukhova, A.; et al. Erforschung des Langfristigen Anwendungspotenzials von Urban Air Mobility als Ergänzung zum öffentlichen Personennahverkehr am Beispiel Oberbayern (OBUAM). Available online: https://www.bauhaus-luftfahrt.net/fileadmin/user_upload/OBUAM_Final_Project_Review_external_kom.pdf (accessed on 16 April 2020).
104. Niklaß, M.; Dzikus, N.; Swaid, M.; Berling, J.; Lührs, B.; Lau, A.; Terekhov, I.; Gollnick, V. A Collaborative Approach for an Integrated Modeling of Urban Air Transportation Systems. *Aerospace* **2020**, *7*, 50. [CrossRef]
105. Wu, Z.; Zhang, Y. Integrated Network Design and Demand Forecast for On-Demand Urban Air Mobility. *Engineering* **2021**, *7*, 473–487. [CrossRef]
106. German, B.; Daskilewicz, M.; Hamilton, T.K.; Warren, M.M. Cargo Delivery in by Passenger eVTOL Aircraft: A Case Study in the San Francisco Bay Area. In Proceedings of the 2018 AIAA Aerospace Sciences Meeting, Kissimmee, FL, USA, 8–12 January 2018. [CrossRef]
107. Chin, C.; Gopalakrishnan, K.; Balakrishnan, H.; Egorov, M.; Evans, A. Efficient and fair traffic flow management for on-demand air mobility. *CEAS Aeronaut. J.* **2021**, *13*, 359–369. [CrossRef]
108. Ferrovial Launches a Project to Develop more than 20 Sustainable Vertiports in Spain Ferrovial. Available online: https://newsroom.ferrovial.com/en/news/ferrovial-launches-a-project-to-develop-more-than-20-sustainable-vertiports-in-spain/ (accessed on 13 March 2022).
109. Ferrovial Airports will Deploy a Network of Vertiports in the United Kingdom Ferrovial. Available online: https://newsroom.ferrovial.com/en/news/ferrovial-airports-will-deploy-a-network-of-vertiports-in-the-united-kingdom/ (accessed on 13 March 2022).
110. Ferrovial and Lilium to develop US Vertiport Network-Lilium. Available online: https://lilium.com/newsroom-detail/ferrovial-and-lilium-develop-us-vertiport-network (accessed on 13 March 2022).
111. Volocopter GmbH. Volocopter Expects to Generate SGD 4.18 billion for Singapore by 2030. Available online: https://www.volocopter.com/newsroom/vc-generates-sgd-4bn/ (accessed on 13 March 2022).
112. Vascik, P.D.; Hansman, R.J. Evaluation of Key Operational Constraints Affecting On-Demand Mobility for Aviation in the Los Angeles Basin: Ground Infrastructure, Air Traffic Control and Noise. In Proceedings of the 17th AIAA Aviation Technology, Integration, and Operations Conference, Denver, CO, USA, 5–9 June 2017; American Institute of Aeronautics and Astronautics: Denver, CO, USA, 2017. [CrossRef]
113. Vascik, P.D.; Hansman, R.J.; Dunn, N.S. Analysis of Urban Air Mobility Operational Constraints. *J. Air Transp.* **2018**, *26*, 133–146. [CrossRef]
114. Benkö, L. Mobilität der Zukunft Lufttaxi Uber Air Skyports. Available online: https://www.ubm-development.com/magazin/uber-air-skyport/ (accessed on 30 August 2020).
115. Foster + Partners unveils Uber Air Skyport for Santa Clara. Available online: https://www.dezeen.com/2019/06/19/uber-air-skyport-foster-partners-santa-clara/ (accessed on 22 February 2021).
116. SHoP and Gensler reveal designs for Uber Air Skyports. Available online: https://www.dezeen.com/2019/06/12/uber-air-skyports-shop-architects-gensler/ (accessed on 22 February 2021).
117. BOKA Powell. Uber Air 2023 Skyport Mobility Hub Concepts. Available online: http://www.bokapowell.com/project/uber-air-2023-skyport-mobility-hub-concepts/ (accessed on 22 February 2021).
118. MVRDV. MVRDV-Airbus UAM. Available online: https://www.mvrdv.nl/projects/421/airbus-uam (accessed on 30 May 2022).
119. Urban-Air Port. Available online: https://www.urbanairport.com (accessed on 13 March 2022).
120. Gannett Fleming. Uber SKYPORT by Gannett Fleming. Available online: https://www.youtube.com/watch?v=WxBmpCwngVI (accessed on 22 February 2021).
121. The Hive: Uber Elevate Design Competition. Available online: https://www.beckgroup.com/projects/the-hive-uber-elevate-design-competition/ (accessed on 22 February 2021).
122. Chilton, P. Uber Sky Tower. Available online: https://www.pickardchilton.com/work/uber-sky-tower (accessed on 28 January 2022).
123. Institute of Flight Guidance-HorizonUAM. Available online: https://www.dlr.de/fl/en/desktopdefault.aspx/tabid-1149/1737_read-69326/ (accessed on 10 June 2022).
124. World-First Hub for Flying Taxis, Air-One, Opens in Coventry, UK, Heralding a New Age of Zero-Emission Transport. Available online: https://www.urbanairport.com/uap-blog/world-first-hub-for-flying-taxis-air-one-opens-in-coventry-uk (accessed on 30 May 2022).
125. Lilium GmbH. Designing a Scalable Vertiport. Available online: https://lilium.com/newsroom-detail/designing-a-scalable-vertiport (accessed on 16 August 2021).

126. Taylor, M.; Saldanli, A.; Park, A. Design of a Vertiport Design Tool. In Proceedings of the 2020 Integrated Communications Navigation and Surveillance Conference (ICNS), Virtual Conference, 8–10 September 2020; pp. 2A2-1–2A2-12. [CrossRef]

127. Preis, L. Quick Sizing, Throughput Estimating and Layout Planning for VTOL Aerodromes—A Methodology for Vertiport Design. In Proceedings of the AIAA Aviation 2021 Forum, Virtual Event, 2–6 August 2021. [CrossRef]

128. Preis, L.; Hack Vazquez, M. Vertiport Throughput Capacity under Constraints caused by Vehicle Design, Regulations and Operations. In Proceedings of the Delft International Conference on Urban Air-Mobility (DICUAM), Delft, The Netherlands, 22–24 March 2022.

129. Courtin, C.; Burton, M.J.; Yu, A.; Butler, P.; Vascik, P.D.; Hansman, R.J. Feasibility Study of Short Takeoff and Landing Urban Air Mobility Vehicles using Geometric Programming. In Proceedings of the 2018 Aviation Technology, Integration, and Operations Conference, Atlanta, GA, USA, 25–29 June 2018. [CrossRef]

130. Volocopter GmbH. Volocopter VoloPort: The Efficient & Ready-Made Vertiport Network Solution for Urban eVTOL Operations. Available online: https://www.volocopter.com/newsroom/voloport-efficient-vertiport/ (accessed on 13 March 2022).

131. Vascik, P.D.; Hansman, J.R. Development of Vertiport Capacity Envelopes and Analysis of Their Sensitivity to Topological and Operational Factors. In Proceedings of the AIAA Scitech 2019 Forum, San Diego, CA, USA, 7–11 January 2019; p. 409. [CrossRef]

132. Zelinski, S. Operational Analysis of Vertiport Surface Topology. In Proceedings of the 2020 AIAA/IEEE 39th Digital Avionics Systems Conference (DASC) Proceedings, Virtual Conference, 11–16 October 2020; p. 10. [CrossRef]

133. Guerreiro, N.M.; Hagen, G.E.; Maddalon, J.M.; Butler, R.W. Capacity and Throughput of Urban Air Mobility Vertiports with a First-Come, First-Served Vertiport Scheduling Algorithm. In Proceedings of the AIAA AVIATION 2020 FORUM, Virtual Event, 15–19 June 2020. [CrossRef]

134. Preis, L.; Amirzada, A.; Hornung, M. Ground Operation on Vertiports - Introduction of an Agent-Based Simulation Framework. In Proceedings of the AIAA SciTech 2021 Forum, Virtual Event, 11–21 January 2021. [CrossRef]

135. Preis, L.; Hornung, M. Identification of Driving Processes for Vertiport Operations Using Agent-Based Simulation. In Proceedings of the AIAA SciTech 2022 Forum, San Diego, CA, USA, 3–7 January 2022. [CrossRef]

136. Drwiega, A. A Day at the (Motor) Races Beats the Recession. Available online: https://www.rotorandwing.com/2012/05/29/a-day-at-the-motor-races-beats-the-recession/ (accessed on 7 March 2022).

137. Swanson, D. Designing 'Vertiports' To Cater For The Evtol Revolution. Available online: https://www.linkedin.com/pulse/designing-vertiports-cater-evtol-revolution-darrell-swanson (accessed on 2 February 2021).

138. Volocopter GmbH. Infrastructure to Integrate and Scale Air Taxi Services in Cities. Available online: https://volocopter-statics.azureedge.net/content/uploads/2018_04_05_Volo_Hub_Overview_without_roof1-scaled.jpg (accessed on 26 August 2021).

139. Volocopter GmbH. First Air Taxi Volo-Port to be Built by End of 2019. Available online: https://volocopter-statics.azureedge.net/content/uploads/190522_VOL_View05_Interior021-scaled.jpg (accessed on 24 February 2022).

140. Guo, R.; Zhang, Y.; Wang, Q. Comparison of emerging ground propulsion systems for electrified aircraft taxi operations. *Transp. Res. Part C Emerg. Technol.* **2014**, *44*, 98–109. [CrossRef]

141. Preis, L.; Hornung, M. Vertiport Operations Modeling, Agent-Based Simulation and Parameter Value Specification. *Electronics* **2022**, *11*, 1071. [CrossRef]

142. Rothfeld, R.; Fu, M.; Balac, M.; Antoniou, C. Potential Urban Air Mobility Travel Time Savings: An Exploratory Analysis of Munich, Paris, and San Francisco. *Sustainability* **2021**, *13*, 2217. [CrossRef]

143. Engelmann, M.; Hornung, M. Boarding Process Assessment of the AVACON Research Baseline Aircraft. In Proceedings of the Deutscher Luft- und Raumfahrtkongress DLRK. Deutsche Gesellschaft fur Luft- und Raumfahrt—Lilienthal-Oberth e.V. Available online: https://publikationen.dglr.de/?tx_dglrpublications_pi1%5bdocument_id%5d=490049 (accessed on 24 February 2022).

144. Schultz, M.; Reitmann, S. Prediction of passenger boarding progress using neural network approach. In Proceedings of the 8th International Conference on Research in Air Transportation (ICRAT), Barcelona, Spain, 26–29 June 2018.

145. Sieb, P.; Michelmann, J.; Flöter, F.; Kai, W. Towards Minimum Expenditure MRO Concepts for UAM through Vehicle Design and Operational Modelling. In Proceedings of the Deutscher Luft- und Raumfahrtkongress (DLRK), Bremen, Germany, 31 August–2 September 2021.

146. Naru, R.; German, B. Maintenance Considerations for Electric Aircraft and Feedback from Aircraft Maintenance Technicians. In Proceedings of the 18th AIAA Aviation Technology, Integration, and Operations Conference 2018, Atlanta, GA, USA, 25–29 June 2018; Curran Associates Inc.: Red Hook, NY, USA, 2018. [CrossRef]

147. Justin, C.Y.; Payan, A.P.; Briceno, S.I.; German, B.J.; Mavris, D.N. Power optimized battery swap and recharge strategies for electric aircraft operations. *Transp. Res. Part C Emerg. Technol.* **2020**, *115*, 102605. [CrossRef]

148. Volocopter. VoloCity: Design specifications, Calculated Approximations not yet Tested in Flight. Available online: https://www.volocopter.com/solutions/volocity/ (accessed on 22 July 2021).

149. Volocopter. Pioneering The Urban Air Taxi Revolution. Available online: www.volocopter.com/content/uploads/Volocopter-WhitePaper-1-01.pdf (accessed on 15 March 2022).

150. NIO. Nio launcht Akkutausch-Station der zweiten Generation. Available online: https://www.electrive.net/2021/04/15/nio-launcht-akkutausch-station-der-zweiten-generation/ (accessed on 13 June 2022).

151. Geely. Gone in 90 Seconds: Geely's Solution to Vehicle Charging. Available online: http://zgh.com/media-center/story/gone-in-90-seconds/?lang=en (accessed on 6 July 2021).

152. Paul, C.S. Definitions of Tactical and Strategic: An Informal Study. Technical Memorandum NASA/TM-2004-213024. 2004. https://ntrs.nasa.gov/api/citations/20040191538/downloads/20040191538.pdf (accessed on 11 July 2022).

153. Windhorst, R.D.; Lauderdale, T.A.; Sadovsky, A.V.; Phillips, J.; Chu, Y.C. Strategic and Tactical Functions in an Autonomous Air Traffic Management System. In Proceedings of the AIAA AVIATION 2021 FORUM, Virtual Event, 2–6 August 2021. [CrossRef]

154. Why Air traffic Flow & Capacity Management. Available online: https://www.icao.int/MID/Documents/2019/ACAO-ICAO%20ATFM%20Workshop/1.4.3-%20ACAO%20last%20V7presentation%20-%20Copy-converti-1.pdf (accessed on 22 December 2021).

155. Urban Air Mobility Concept of Operations for the London Environment. Available online: https://eveairmobility.com/wp-content/uploads/2022/03/UK_Air_Mobility_Consortium_CONOPS.pdf (accessed on 4 March 2022).

156. Schweiger, K.; Knabe, F.; Korn, B. Urban Air Mobility: Vertidrome Airside Level of Service Concept. In Proceedings of the AIAA AVIATION 2021 FORUM, Virtual Event, 2–6 August 2021. [CrossRef]

157. Geister, D.; Korn, B. Density based Management Concept for Urban Air Traffic. In Proceedings of the 2018 IEEE/AIAA 37th Digital Avionics Systems Conference (DASC), London, UK, 23–27 September 2018; pp. 1–9. [CrossRef]

158. Sunil, E.; Hoekstra, J.; Ellerbroek, J.; Vidosavljevic, A.; Arntzen, M.; Aalmoes, R. Metropolis WP5 Results of Simulations and Data Analysis. Available online: https://homepage.tudelft.nl/7p97s/Metropolis/ (accessed on 13 March 2022).

159. Metropolis 2. Deliverables. Available online: https://metropolis2.eu/deliverables/ (accessed on 31 May 2022).

160. Blamey, J.; Sánches-Escalonilla, P.; Chornique Sanchez, J.; Hampson, C.; Hervías Vallejo, P.; Martínez López, M.; Janisch, D.; Löhr, F.; Stridsman, L. AURA Solution 2 Workshop. 2021. Available online: https://www.pj34aura.com/sites/aura/files/documents/aura_solution_2_workshop_slides.pdf (accessed on 11 July 2022).

161. Thipphavong, D.P.; Apaza, R.; Barmore, B.; Battiste, V.; Burian, B.; Dao, Q.; Feary, M.; Go, S.; Goodrich, K.H.; Homola, J.; et al. Urban Air Mobility Airspace Integration Concepts and Considerations. In Proceedings of the 2018 Aviation Technology, Integration, and Operations Conference, Atlanta, GA, USA, 25–29 June 2018; American Institute of Aeronautics and Astronautics: Atlanta, GA, USA, 2018. [CrossRef]

162. Guerreiro, N.M.; Butler, R.W.; Maddalon, J.M.; Hagen, G.E. Mission Planner Algorithm for Urban Air Mobility—Initial Performance Characterization. In Proceedings of the AIAA Aviation 2019 Forum, Dallas, TX, USA, 17–21 June 2019. [CrossRef]

163. Tang, H.; Zhang, Y.; Mohmoodian, V.; Charkhgard, H. Automated flight planning of high-density urban air mobility. *Transp. Res. Part C Emerg. Technol.* **2021**, *131*, 103324. [CrossRef]

164. Bosson, C.; Lauderdale, T.A. Simulation Evaluations of an Autonomous Urban Air Mobility Network Management and Separation Service. In Proceedings of the 2018 Aviation Technology, Integration, and Operations Conference, Atlanta, GA, USA, 25–29 June 2018. [CrossRef]

165. Ortlieb, M.; Adolf, F.M.; Holzapfel, F. Computation of a Database of Trajectories and Primitives for Decision-Based Contingency Management of UAVs over Congested Areas. In Proceedings of the 2021 IEEE/AIAA 40th Digital Avionics Systems Conference (DASC), San Antonio, TX, USA, 3–7 October 2021; pp. 1–8. [CrossRef]

166. Gariel, M.; Srivastava, A.N.; Feron, E. Trajectory Clustering and an Application to Airspace Monitoring. *IEEE Trans. Intell. Transp. Syst.* **2011**, *12*, 1511–1524. [CrossRef]

167. Barratt, S.T.; Kochenderfer, M.J.; Boyd, S.P. Learning Probabilistic Trajectory Models of Aircraft in Terminal Airspace From Position Data. *IEEE Trans. Intell. Transp. Syst.* **2019**, *20*, 3536–3545. [CrossRef]

168. Li, L.; Hansman, R.J.; Palacios, R.; Welsch, R. Anomaly detection via a Gaussian Mixture Model for flight operation and safety monitoring. *Transp. Res. Part C Emerg. Technol.* **2016**, *64*, 45–57. [CrossRef]

169. Krozel, J. Intelligent Tracking of Aircraft in the National Airspace System. In Proceedings of the AIAA Guidance, Navigation, and Control Conference and Exhibit, Monterey, CA, USA, 5–8 August 2002; American Institute of Aeronautics and Astronautics: Monterey, CA, USA, 2002. [CrossRef]

170. Georgiou, H.; Pelekis, N.; Sideridis, S.; Scarlatti, D.; Theodoridis, Y. Semantic-aware aircraft trajectory prediction using flight plans. *Int. J. Data Sci. Anal.* **2020**, *9*, 215–228. [CrossRef]

171. Weinert, A.; Underhill, N.; Serres, C.; Guendel, R. Correlated Bayesian Model of Aircraft Encounters in the Terminal Area Given a Straight Takeoff or Landing. *Aerospace* **2022**, *9*, 58. [CrossRef]

172. Pradeep, P.; Wei, P. Energy Optimal Speed Profile for Arrival of Tandem Tilt-Wing eVTOL Aircraft with RTA Constraint. In Proceedings of the 2018 IEEE CSAA Guidance, Navigation and Control Conference (CGNCC), Xiamen, China, 10–12 August 2018; pp. 1–6. [CrossRef]

173. Pradeep, P.; Wei, P. Energy-Efficient Arrival with RTA Constraint for Multirotor eVTOL in Urban Air Mobility. *J. Aerosp. Inf. Syst.* **2019**, *16*, 263–277. [CrossRef]

174. Pradeep, P.; Wei, P. Heuristic Approach for Arrival Sequencing and Scheduling for eVTOL Aircraft in On-Demand Urban Air Mobility. In Proceedings of the 2018 IEEE/AIAA 37th Digital Avionics Systems Conference (DASC), London, UK, 23–27 September 2018; pp. 1–7. [CrossRef]

175. Kleinbekman, I.C.; Mitici, M.A.; Wei, P. eVTOL Arrival Sequencing and Scheduling for On-Demand Urban Air Mobility. In Proceedings of the 2018 IEEE/AIAA 37th Digital Avionics Systems Conference (DASC), London, UK, 23–27 September 2018; p. 7. [CrossRef]

176. Kleinbekman, I.; Mitici, M.A.; Wei, P. A Rolling-horizon eVTOL Arrival Scheduling for On-demand Urban Air Mobility. *J. Aerosp. Inf. Syst.* **2019**, *17*, 150–159. [CrossRef]

177. Bertram, J.; Wei, P. An Efficient Algorithm for Self-Organized Terminal Arrival in Urban Air Mobility. In Proceedings of the AIAA Scitech 2020 Forum, Orlando, FL, USA, 6–10 January 2020. [CrossRef]
178. Song, K.; Yeo, H.; Moon, J.H. Approach Control Concepts and Optimal Vertiport Airspace Design for Urban Air Mobility (UAM) Operation. *Int. J. Aeronaut. Space Sci.* **2021**, *22*, 982–994. [CrossRef]
179. Song, K.; Yeo, H. Development of optimal scheduling strategy and approach control model of multicopter VTOL aircraft for urban air mobility (UAM) operation. *Transp. Res. Part C Emerg. Technol.* **2021**, *128*, 103181. [CrossRef]
180. Shao, Q.; Shao, M.; Lu, Y. Terminal area control rules and eVTOL adaptive scheduling model for multi-vertiport system in urban air Mobility. *Transp. Res. Part C Emerg. Technol.* **2021**, *132*, 103385. [CrossRef]
181. Bharadwaj, S.; Carr, S.; Neogi, N.; Topcu, U. Decentralized Control Synthesis for Air Traffic Management in Urban Air Mobility. *IEEE Trans. Control. Netw. Syst.* **2021**, *8*, 598–608. [CrossRef]
182. Vascik, P.D.; John Hansman, R. Evaluating the Interoperability of Urban Air Mobility Systems and Airports. *Transp. Res. Rec. J. Transp. Res. Board* **2021**, *2675*, 1–14. [CrossRef]
183. Fitzek, R.A. Lessons Gained in Helicopter Air Traffic Control from Federal Aviation Agency Activities. *J. R. Aeronaut. Soc.* **1962**, *66*, 499–502. [CrossRef]
184. Ahrenhold, N. Entwurf und Evaluation einer luftseitigen Kapazitätsprognoserechnung für den Einfluss von Lufttaxis am Beispiel des Flughafens Hamburg. Master's Thesis, Technische Universität Carolo-Wilhelmina zu Braunschweig, Braunschweig, Germany, 2021.
185. Verma, S.; Keeler, J.; Edwards, T.E.; Dulchinos, V. Exploration of Near term Potential Routes and Procedures for Urban Air Mobility. In Proceedings of the AIAA Aviation 2019 Forum, Dallas, TX, USA, 17–21 June 2019. [CrossRef]
186. Petersen, J.D.; Alexander, R.J.; Swaintek, S.S. Dynamic vertiport configuration. U.S. Patent US20200226937A1, 16 July 2020.
187. Steiner, M. Urban Air Mobility: Opportunities for the Weather Community. *Bull. Am. Meteorol. Soc.* **2019**, *100*, 2131–2133. [CrossRef]
188. Justin, C.Y.; Mavris, D.N. Environment Impact on Feasibility of Sub-Urban Air Mobility using STOL Vehicles. In Proceedings of the AIAA Scitech 2019 Forum, San Diego, CA, USA, 7–11 January 2019. [CrossRef]
189. Somers, L.A.; Justin, C.Y.; Mavris, D.N. Wind and Obstacles Impact on Airpark Placement for STOL-based Sub-Urban Air Mobility. In Proceedings of the AIAA Aviation 2019 Forum, Dallas, TX, 17–21 June 2019. [CrossRef]
190. Reiche, C.; Brody, F.; McGillen, C.; Siegel, J.; Cohen, A. An Assessment of the Potential Weather Barriers of Urban Air Mobility (UAM). Final Report, 21 November 2018. Available online: https://escholarship.org/uc/item/2pc8b4wt (accessed on 24 February 2022).
191. Steiner, M. Weather Challenges for Advanced Aerial Mobility in Urban Environments. Available online: https://vtol.org/files/dmfile/20200422---matthias-steiner---ncar---weather-challenges---no-animations2.pdf (accessed on 14 March 2022).
192. Berchoff, D. Weather-Resilient AAM Operations in Urban Environments. Available online: https://vtol.org/files/dmfile/202004 22---don-berchoff---truweather---weather-solutions.pdf (accessed on 9 September 2021).
193. Deutscher Wetterdienst. Urban Heat Islands. Available online: https://www.dwd.de/EN/research/climateenvironment/climate_impact/urbanism/urban_heat_island/urbanheatisland.html (accessed on 3 February 2021).
194. Yilmaz, E.; Warren, M.; German, B. Energy and Landing Accuracy Considerations for Urban Air Mobility Vertiport Approach Surfaces. In Proceedings of the AIAA Aviation 2019 Forum, Dallas, TX, USA, 17–21 June 2019. [CrossRef]
195. Veneruso, P.; Opromolla, R.; Fasano, G.; Burgio, G.; Gentile, G.; Tiana, C. Extending Enhanced Visual Operations to Urban Air Mobility: Requirements and Approaches. In Proceedings of the 2021 IEEE/AIAA 40th Digital Avionics Systems Conference (DASC), San Antonio, TX, USA, 3–7 October 2021; pp. 1–9. [CrossRef]
196. Volocopter GmbH. Urban Air Mobility. Available online: https://www.volocopter.com/urban-air-mobility/ (accessed on 22 May 2022).
197. Alcock, C. Ground Infrastructure Experts Wrestle With Vertiport Challenges. Available online: https://www.futureflight.aero/news-article/2021-12-20/ground-infrastructure-experts-wrestle-vertiport-challenges (accessed on 27 March 2022).

Article

Systemic Performance Analysis on Zoning for Unmanned Aerial Vehicle-Based Service Delivery

Casper Bak Pedersen †, **Kasper Rosenkrands** †, **Inkyung Sung** * and **Peter Nielsen**

Operations Research Group, Department of Materials and Production, Aalborg University, 9220 Aalborg, Denmark; cbpe17@student.aau.dk (C.B.P.); krosen17@student.aau.dk (K.R.); peter@mp.aau.dk (P.N.)
* Correspondence: inkyung_sung@mp.aau.dk
† These authors contributed equally to this work.

Abstract: A zoning approach that divides an area of interest into multiple sub-areas can be a systemic and strategic solution to safely deploy a fleet of unmanned aerial vehicles (UAVs) for package delivery services. Following the zoning approach, a UAV can be assigned to one of the sub-areas, taking sole ownership and responsibility of the sub-area. As a result, the need for collision avoidance between units and the complexity of relevant operational activities can be minimized, ensuring both safe and reliable execution of the tasks. Given that the zoning approach involves the demand-server allocation decision, the service quality to customers can also be improved by performing the zoning properly. To illuminate the benefits of the zoning approach to UAV operations from a systemic perspective, this study applies clustering techniques to derive zoning solutions under different scenarios and examines the performance of the solutions using a simulation model. The simulation results demonstrate that the zoning approach can improve the safety of UAV operations, as well as the quality of service to demands.

Keywords: unmanned aerial vehicle; zoning; unmanned aircraft system traffic management; clustering; collision avoidance; drone package delivery

Citation: Pedersen, C.B.; Rosenkrands, K.; Sung, I.; Nielsen, P. Systemic Performance Analysis on Zoning for Unmanned Aerial Vehicle-Based Service Delivery. *Drones* **2022**, *6*, 157. https://doi.org/10.3390/drones6070157

Academic Editors: Ivana Semanjski, Antonio Pratelli, Massimiliano Pieraccini, Silvio Semanjski, Massimiliano Petri and Sidharta Gautama

Received: 9 June 2022
Accepted: 23 June 2022
Published: 26 June 2022

1. A Zoning Approach: A Systemic Solution for Successful Airspace Control

Unmanned aerial vehicles (UAVs), or drones, are a game changer in many business and public sectors because of their ability to exploit aerial dimensions and inexpensive operating cost. However, the applications of UAVs are often made with a single or a few UAVs in a relatively safe operation area. One of the reasons for such limited UAV applications is the difficulty in air traffic control and collision avoidance for UAVs [1] for large-scale deployments. Although there have been dramatic advances in technologies for collision avoidance (e.g., artificial intelligence for path finding [2]) and UAV flight system automation, the technologies are not at the desired level to resolve safety issues completely.

This difficulty becomes more significant when multiple UAVs are deployed in a relatively small area at the same time in a dynamic environment. When no cyclical stability on UAV tasks can be identified, this difficulty becomes even worse [3]. This situation can be found in UAV application scenarios with a delivery function, such as UAV-based logistics [4,5], humanitarian/emergency aid operations [6,7], or in cooperation with other unmanned systems [8,9]. These applications are where the demands for services are expected to exceed the supply levels that can be fulfilled with traditional means of transportation, and the task can be performed better with exploitation of the aerial dimension. Therefore, a large-scale UAV fleet deployment is desired to handle the increasing service demands and to maximize the service quality.

A key solution to realize large-scale UAV operations is an unmanned aircraft traffic management (UTM) system, which manages airspace for multiple UAV operations through real-time control. Following the increasing volume of UAV operations and the need for

UTM systems, federal regulatory agencies such as the European Aviation Safety Agency (EASA) and the Federal Aviation Administration (FAA) in the United States have conducted several projects and discussed the concepts of UTM and regulations of UAVs [10,11].

While airspace for UAVs is increasingly regulated, the feasibility and performance of UTM is still far behind the desired level, especially considering the envisioned future air traffic densities. Unlike the well-established and -functioning air traffic management (ATM) systems for manned aircraft, the following aspects of UAVs challenge the development of UTM [12]:

- Different types of UAVs with different performances;
- Limited capability to carry heavy or power-intensive equipment;
- Limited capability to automatically detect and avoid collisions;
- Increasing needs for UAV operations beyond visual line of sight (BVLOS) for greater economic values;
- High density of operations, likely flying very close to other units;
- High degree of vulnerability to operational environments (e.g., wind).

These key differences between manned aircraft and UAVs underline the complexity of the development and implementation of a UTM system. UTM implementation involves novel challenges, including decentralization of governing authority over large-scale UAV operations and interactions with pilots to share crucial flight and safety data [13]. When narrowing down the UTM for the safety aspects of UAVs, safety separation standards, collision risk prediction, and collision avoidance can be listed as the critical research topics, and there needs to be significant advances for full- and large-scale UAV operations [14].

Considering the complexity of the involved design and decision-making problems in the UTM, it is clear that a centralized air traffic control (ATC) with a full synchronization of large-scale UAV operations is extremely difficult to achieve. In the context of ATM, the ATC-related decision-making authority is distributed among flight crews, the air traffic service providers, and aeronautical operational control organizations in order to reduce reliance on the centralized ATC [15]. In the context of UTM, however, it is difficult to give UAVs the freedom for their path and speed selection in real time, because full automation of path finding and conflict resolution for large-scale UAV operations in a decentralized manner is highly complex [14,16,17].

As a systemic solution to the difficulties of UTM implementation, a zoning approach, where an operational area for UAVs is decomposed into a set of sub-areas and a maximum of one UAV can be deployed within a sub-area at a given time, can be considered [18]. By this approach, the flight paths of UAVs in different zones do not overlap, and it is likely that UAVs remain well clear with or without minimum control efforts. The flight planning and control problems involved in a UTM can also be made simple, as the interaction level between UAVs are minimized by the zoning approach.

As such, the workload for airspace control and collision avoidance can be dramatically reduced by the zoning approach. Importantly, this reduced workload can be further distributed to the sub-UTMs for each zone in favor of the zoning approach. Given that most of UAV control systems are still operated by humans, this zoning approach also brings advantages to UAV service providers, that is, to reduce the manpower for UAV deployment and corresponding operating costs. The expected benefits of the zoning approach are summarized as follows:

- Reduced burden to control/monitor UAVs;
- Reduced risk to safety;
- Minimum communication between UAVs;
- Increased chance of UAV flight automation;
- Scalability for a large-scale UAV deployment;
- Reduced complexity of relevant planning and control problems in a UTM.

On the other hand, one can question the negative impacts of the zoning approach on the service-to-customers level, as the means to serve demands is restricted by the approach.

Indeed, the set of demands or the area a UAV can cover is limited by a zoning solution in terms of volume, size, and location of the demands. Therefore, considering the zoning approach as a demand-server allocation problem, it is important to derive a zoning solution so that the resulting service level to customers is kept at an acceptable level, while fully exploiting the advantages of the zoning approach.

The aim of this study is to demonstrate the advantages of the zoning approach from the UAV safety and service quality perspectives. Specifically, we answer the following questions:

- How much separation between UAVs is expected to be achieved by the zoning approach?
- Will the zoning approach result in service quality degradation?
- What would be the determining factors for the zoning approach performance?

To answer these questions, we analyze the performance of the zoning approach from a systemic perspective. We first generate package-delivery-like scenarios, where multiple demand nodes in an area are served by multiple UAVs. We group the demand nodes using clustering techniques and derive a zoning solution accordingly. The performance of the proposed zoning approach is examined under multiple demand configurations using a simulation model, compared to other benchmark service strategies.

2. Related Work: Zoning in Literature

UAV application areas are numerous and include those domains where automation, low operation costs, and aerial dimension exploitation can bring values. Otto et al. [19] present a comprehensive review on UAV applications, classifying the UAV-involved tasks in the applications and relevant planning problems. Mukhamediev et al. [20] also present detailed UAV use cases in various industry applications, highlighting relevant data management and processing tasks.

The UAV applications where the concept of zoning is relevant can be further classified into two classes: area coverage and package delivery. The area coverage class addresses tasks for searching and monitoring an area, whereas the package delivery class addresses transportation-related activities.

By the nature of the zoning approach, one can easily find its link to the area coverage class, where efficient use of units by properly assigning responsible areas to the units is critical to achieve a service/mission goal (e.g., to minimize the time it takes to search an area). Fu et al. [21] propose a local Voronoi decomposition algorithm for exploration task allocation to multi-agents. In this approach, Voronoi regions of each agent are calculated based on the agents' positions, and the agents move only within their regions to avoid overlapping tasks. Miao et al. [22] apply a map decomposition approach to assign sub-maps to cleaning robots so that large-scale cleaning areas are effectively distributed and assigned to the robots. Xiao et al. [23] apply area segmentation for UAV coverage planning in a grid map. Once a take-off location is determined, an area of interest is specified as a set of grids and further divided into sub-areas so that the energy required to cover the sub-areas are balanced and collisions between UAVs can be avoided. For three-dimensional coverage path planning, where different coverage paths depending on altitudes of a region of interest are generated [24], multiple UAVs can be assigned to different altitudes for coverage operations.

The zoning approach is also found in the package delivery application class, the focus of the zoning approach in this study. UAV-based package delivery is a UAV application with increasing demands and demonstrated profits. To realize this service concept with minimum barriers to its implementation, various practical issues, such as preserving privacy, have been addressed [25].

The main focus of the zoning approach in this class is collision avoidance and ease of UAV traffic management, the critical challenges for package delivery by UAVs [26]. Amazon, who first introduced the concept of package delivery by UAVs, proposes an airspace model where civil airspace is segregated by altitudes based on vehicle capability [27]. In this model, a low-speed localized traffic area is used for the UAVs without sophisticated sense-and-avoid technology, whereas a high-speed transit area is used for well-equipped vehicles. The model also includes a no fly zone to buffer the UAV operations from current aviation operations. Feng and Yuan [28] apply space zoning to a low-altitude airspace that divides the space into upper, buffer, safe, and bottom zones according to space height restrictions in order to construct flight corridors for UAVs. Sung and Nielsen [18] propose a zoning approach that divides a service area with a single UAV station into zones. They allow at most a single UAV to fly within a zone and investigate the expected service level with these zoning practice using a simulation.

Note that the clustering approaches applied to UAV routing problems for package delivery services can also be seen as a special case of the zoning approach for tactical decision-making. In particular, clustering has been actively applied to truck-assisted UAV package delivery services, where a truck visits multiple spots following a path to support UAVs' operations (e.g., recharging and (un)loading payloads). Under this service delivery scheme, delivery points are grouped into a set of clusters, and a route for a truck to visit the clusters and routes for UAVs within the clusters are derived to maximize the service delivery performance [29–31].

As reviewed, the concept of zoning has been addressed from different perspectives in the literature. In general, the zoning approach is designed to optimize the UAV performance for a single service instance (tactical/operational) under the area coverage class, whereas the zoning approach is applied to design a UAV traffic management system for multiple service instances (strategic) under the package delivery class. The zoning approach for the package delivery class is further separated by the restriction level on a zone. Sung and Nielsen [18] apply the most restricted practice, which allows only a single UAV to fly within a zone at a given time, whereas the other studies separate airspace by altitudes and allow multiple UAV operations in a zone.

Our study is aligned with the work of Sung and Nielsen [18]. It is difficult to assume a full connectivity between UAVs during their operations and 100% reliable UAV control logic mainly due to the dynamics in the airspace and the absence of a decent UTM. Considering this fact, operating UAVs with zoning, as proposed by Sung and Nielsen [18], seems more appropriate for safety and service reliability reasons and can increase the feasibility of large-scale UAV deployment.

Based on the review, the contributions of our study can be noted as follows. We first describe a systemic and strategic solution for application of UAVs to package delivery scenarios, which allows UAV deployment at a large scale and reduces reliance on dramatic advances in UAV navigation and control technology. Next, in contrast to the work of Sung and Nielsen [18], we apply clustering techniques for zoning to address multiple UAV stations in a service area, which can naturally be seen in a large-scale service scenario. Last, we test the performance of the zoning approach under different demand distribution configurations to clearly illuminate the expected benefits of the proposed zoning approach.

3. The Zoning Problem and the Considered Solution Approach

3.1. Problem Description

Assume that multiple demand nodes are distributed in a two-dimensional service area. Demand node x has demand rate λ_x and is independent from the other demand nodes in the area. K number of UAVs are available for operation, and a UAV will serve a single demand (a demand instance at a demand node) per trip.

Given the setting, we seek a set of zones $\mathbf{Z}$ and base locations μ such that (1) all the demand nodes are divided into K number of zones $\mathbf{Z} = \{Z_1, \ldots Z_K\}$, (2) a flight path between the base of a zone and a demand node within the zone is guaranteed, and (3) the total service level for the demand nodes by the zoning solution is maximized. The zoning problem of this study can be formally described by

$$P_{zoning} = \arg\max_{\mathbf{Z},\,\mu} \sum_{k=1}^{K} \sum_{x \in Z_k} \mathcal{L}(x|\mu_k, \mathbf{Z}), \tag{1}$$

where $\mathcal{L}(x|\mu_k, \mathbf{Z})$ is the service level for demand node x given a base location μ_k and a set of zones $\mathbf{Z}$.

From the safety perspective, the most relevant measure for the service level would be the distance between operating UAVs in different zones. Considering this idea, a zoning solution that maximizes the separation between demand nodes in different zones would be desired. This concept is well-aligned with the k-means clustering algorithm, which partitions observations into k clusters such that observations are assigned to the clusters with the nearest mean. Based on this, the original zoning problem can be rewritten as

$$P^*_{zoning} = \underset{\mathbf{Z},\, \mu}{\arg\min} \sum_{k=1}^{K} \sum_{x \in Z_k} \lambda_x \cdot ||x - \mu_k||^2, \tag{2}$$

which is intended to minimize the weighted within-cluster variances (which is the same as maximizing between-cluster variances). It should be clear that this formulation is one of the options to approximate the original zoning problem (Equation (1)); thus, a different formulation can be designed based on the use case's specific constraints, for example, the minimum distance between demands and the center of a zone or no-fly zone.

3.2. Solution Approaches: Weighted k-Means Algorithm

The target zoning problem P^*_{zoning} is solved by a weighted K-means algorithm (WKMA), based on the Hartigan algorithm for k-means clustering [32]. Instead of using the mean and squared distance when updating centroids, we use the weighted mean and squared weighted distance in order to account for the arrival rate. The WKMA implementation is described in Algorithm 1. The computational complexity of the algorithm is $\gamma \cdot \mathcal{O}(K \cdot D)$, where γ is the number of iterations performed by the algorithm until the clustering solution is converged, and D is the number of demand nodes.

Algorithm 1: WKMA

Require: demand set, the number of centroids (k)
1: Randomly assign all demand nodes to a centroid
2: Calculate centroids as weighted mean of their assigned nodes
3: **while** *Not converged* **do**
4:     **for** demand node d in demand set **do**
5:         **for** centriod c in centroid set **do**
6:             Assign demand node d to centroid c
7:             Compute the sum of squared weighted distances from each node to its centroid
8:         **end for**
9:         Assign d to the centroid which resulted in the smallest sum
10:         Recalculate centroids as weighted mean over all nodes assigned
11:     **end for**
12: **end while**
13: **return** demand allocation to centroids

3.3. Benchmark UAV Deployment Strategies

As a benchmark solution to the zoning approach, a strategy termed closest, which assigns the closest available UAV to a demand without restrictions in its operation area, can be considered. Note that from our initial experiments, the closest strategy showed poor performance, because distant demands are often assigned to UAVs, hampering the efficient use of UAVs.

To avoid such a distant demand allocation to a UAV, a relaxed version of the closest strategy, termed closest with thresholds, is introduced. Given base locations, we first set a maximum UAV flight range such that all demand nodes can be covered by at least one UAV. For each demand node, any UAVs placed within the maximum flight range are then considered as a candidate server for the demand. Illustrations of the service areas derived by the three different UAV deployment strategies—the zoning approach, the closest, and the closest with thresholds—are given in Figure 1. In the figure, the base and demand node are represented as a red box and a black circle, respectively.

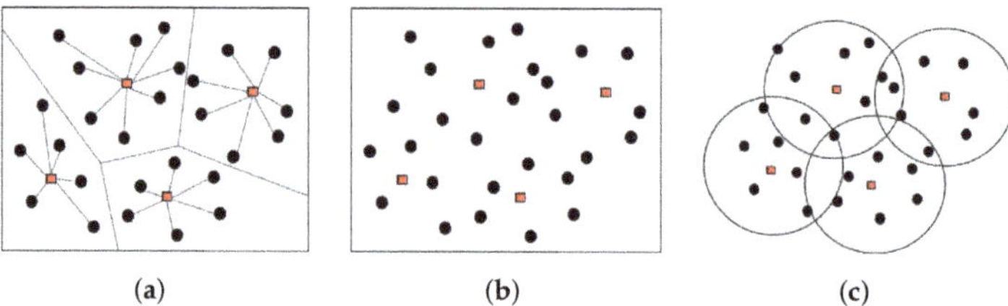

(a)   (b)   (c)

Figure 1. Different service area configurations by the different UAV deployment strategies: (**a**) zoning; (**b**) closest; (**c**) closest with thresholds.

4. Experimental Results

Following the aim of this study (i.e., to demonstrate the performance of the zoning approach), we simulate UAV trajectories for package delivery scenarios and measure how close to each other UAVs are supposed to fly based on the trajectories. We also evaluate the service quality level for customers based on the simulated trajectories to evaluate the service quality degradation by zoning approach.

4.1. Experimental Setting

We generate four problem classes, varying the settings of the following two design factors: a geographical distribution of demands and a representative package delivery scenario.

For the geographical distribution of demands, we first create demand set U, where 100 demand nodes are uniformly distributed over a square-shaped 2D area. To see the impact of densely located demand nodes on the performance of the zoning approach, we also create demand set C, where demands are primarily located in the center of a service area. This demand set is created by sampling 100 demand nodes from the york dataset available from the r-package `maxcovr`. Figure 2 shows the two different demand node distributions.

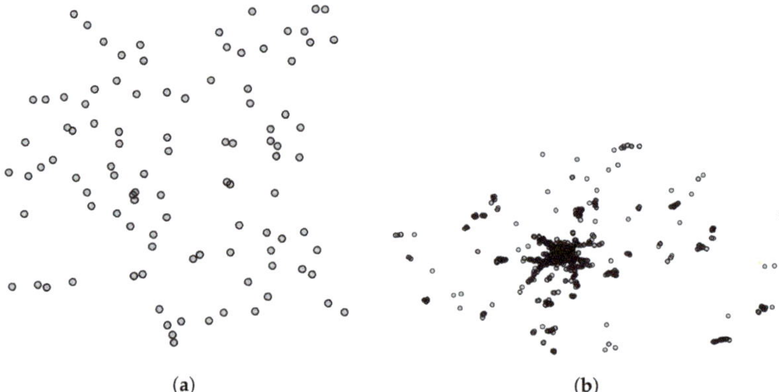

(a)   (b)

Figure 2. The distributions of demands: (**a**) demand distribution U; (**b**) demand distribution C.

Next, we apply two different service modes to test the zoning approach under different service scenarios. For scenario NQ (no queue), we assume a situation where a service request, which cannot be served immediately by a responsible UAV, will leave the system without receiving service. This scenario represents an emergency situation, where a timely response to a demand is critical (e.g., visual surveillance of a traffic accident). For scenario FCFS, we assume a situation where demands wait until they receive service by UAVs. In this scenario, demands in a queue are handled by the first come first serve (FCFS) policy. This scenario is implemented to represent a commercial package delivery scenario.

Following the setting, we generate the four different problem classes ($\{\mathtt{U}, \mathtt{C}\} \times \{\mathtt{NQ}, \mathtt{FCFS}\}$) and test the performance of the zoning approach in a simulation environment. Note that the objective function of the zoning problem, that is, to minimize the within-zone variances, is a proxy for the actual UAV safety level. Therefore, a simulation model is implemented to investigate the performance of the proposed zoning approach close to its actual performance. In principle, the zoning approach can be applied to a UAV-based service system with any type of UAV. Since we examine the systemic performance of the zoning approach, the detailed dynamics of UAVs (e.g., the minimum turning radius of a unit, energy consumption as a function of weather conditions, and collision avoidance logic during operation) are simplified in the simulation. The simulation model generates demands following the demand rates of demand nodes and assigns available UAVs by the applied UAV deployment strategy, updating the operational status (busy/idle) of the UAVs.

For each problem class, we replicate a simulation run 20 times. The length of the simulation is four hours, and the status of UAVs and demands are updated every second in a simulation run. In the simulation, the demand rate of a demand node (the number of demand requests per minute) is drawn from a uniform distribution with the bounds $[0.4, 0.6]$. The number of UAVs is set by ten to provide a sufficient service capacity for the demands.

The output of a simulation run is the UAV flight trajectories for all time steps (every second), including the positions of UAVs and their status, and the records regarding how demands are served. Based on the output, the separation level between UAVs is analyzed to see how much the safety can be improved by the zoning approach, which is the main topic of this study (Section 4.2). The service quality with regard to the demands is also analyzed to evaluate the performance of the zoning approach from the customers' perspective (Section 4.3).

4.2. UAV Safety Improvement by the Zoning Approach

Given the UAV flight trajectories obtained from simulation runs, we compute the distances between two operating UAVs at each time stamp to measure the proximity between UAVs. Given that a key performance measurement from the safety perspective would be the separation level between UAV flights, we compute the 5th percentile of the proximity between UAVs (i.e., the distance between two flying UAVs) to identify the minimum distance gap guaranteed for the majority of the UAV flights. Figure 3 shows the distribution of the 5th percentile proximity values for 20 simulation runs under the four different problem classes and the three different UAV deployment strategies (zoning vs. closest vs. closest with thresholds).

From the figure, it is observed that for all the problem classes, the zoning approach shows greater distance gaps than the other considered strategies. While the distance gap increases when the demand distribution $\mathtt{C}$ is applied (which seems inevitable due to the density of the demand nodes in the dataset), the zoning approach can still provide greater distance gaps between UAVs than the other strategies. Given that collision avoidance action is taken when a UAV detects an object within a certain range, a greater distance between UAVs clearly indicates that UAVs are likely to be cleared with less efforts for collision avoidance from both UAV operators and UTMs. The safety improvement by the zoning approach becomes clearer in Figure 4, where distributions of the closest distance between two operating UAVs from the experiments are presented.

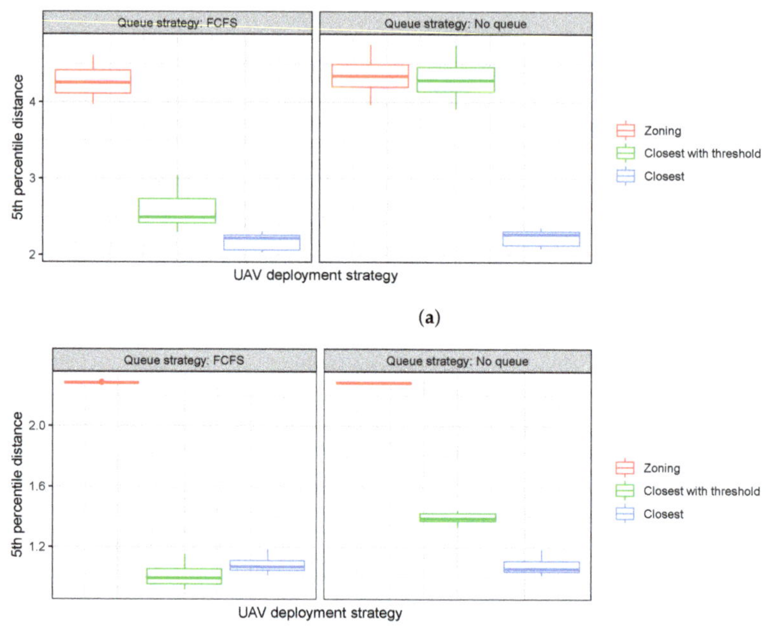

(a)

(b)

Figure 3. The 5th percentile of the distances between operating UAVs: (**a**) demand distribution U; (**b**) demand distribution C.

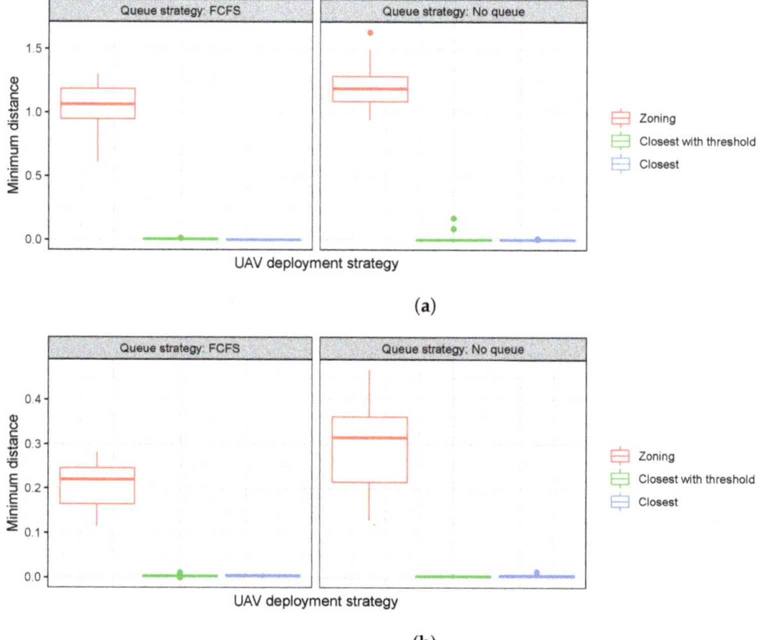

(a)

(b)

Figure 4. The worst case proximity between operating UAVs: (**a**) demand distribution U; (**b**) demand distribution C.

Figure 4 shows that the zoning approach always guarantees positive distance gaps between UAVs, which reduces the relevant collision avoidance efforts. On the other hand, the benchmark service strategies incur very close and even crossing UAV flights (i.e., zero distance gap). Recall that under the zoning approach, a flight path is generated within a zone, and thus it does not overlap with other paths in other zones. As collision avoidance is complex and might be infeasible when the distance between UAVs is tight, the safety net provided by the zoning approach is promising. In other words, the zoning approach can guarantee the safety of UAVs and the reliability of the UAV-based system, especially when large-scale UAV operations are considered.

4.3. Expected Service Quality to Demands by the Zoning Approach

The service quality for customers regarding UAV operations is another key factor that determines the success of a UAV-based service system. Given that immediate responses to demands determine service quality in many contexts, for emergency response in particular, we measure UAV response time to demands (i.e., the time interval between a service request receipt and a UAV arrival at the corresponding location). Figure 5 shows distributions of the mean response time over 20 simulations under the different problem classes and UAV deployment strategies.

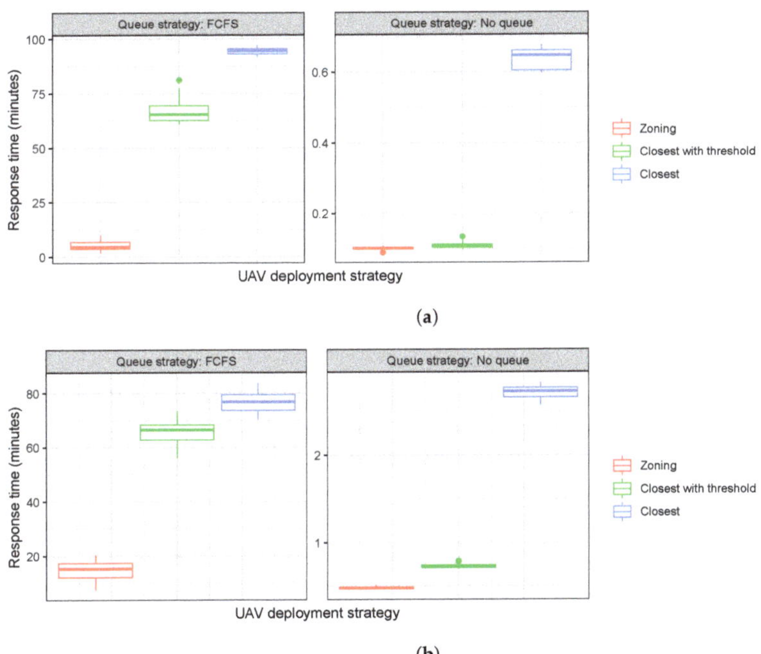

Figure 5. The mean response time to demands: (**a**) demand distribution U; (**b**) demand distribution C.

The results shows that the zoning approach outperforms the other considered service strategies in terms of the mean response time for customers. This is because the benchmark strategies allow UAVs to serve distant demands, whereas the zoning approach restricts such behavior by limiting the UAV operation area. Naturally, long-distance travel increases the individual UAV workload and the likelihood of being busy. Under the scenario FCFS, where there is a waiting queue, such a practice significantly increases the response time for customers (see the response time differences between the FCFS and NQ scenarios).

It should be noted that while the zoning approach can avoid such long-distance and inefficient travels, it is true that a demand generated when the responsible UAV is busy will not be served under the zoning approach, even if there could be other UAVs available nearby. When there is no waiting queue, such a practice might result in many demand nodes being abandoned.

To verify this issue, we compute the percentage of demands that left the system without receiving service, with regard to the total number of demands generated in a simulation run. Figure 6 shows the results for scenario NQ. As shown in the figure, this negative impact by the zoning approach seems marginal, and interestingly, the zoning approach even serves more demands than the other strategies. This is due to the efficient use of UAVs and corresponding low utilization levels of the units.

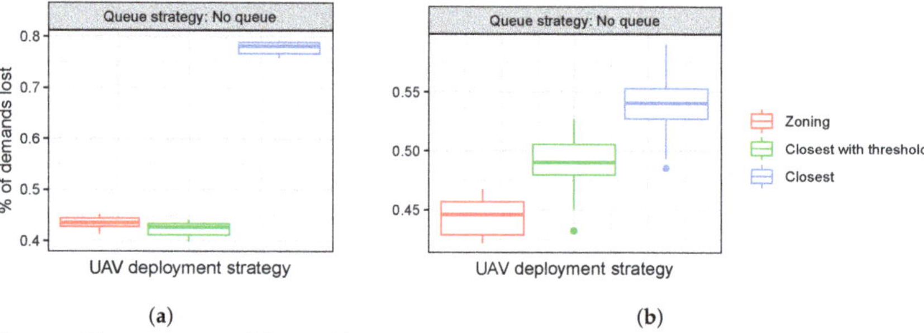

Figure 6. The percentage of demand loss under scenario NQ: (**a**) demand distribution U; (**b**) demand distribution C.

4.4. Summary

For all the problem classes, the zoning approach outperforms the other considered UAV deployment strategies in terms of the safety of UAVs and the service quality for demands. The zoning approach shows that it can guarantee a sufficient gap between operating UAVs and that it can serve more demands at a faster rate than the other benchmark strategies. All of these findings are critical to implement a UTM with distributed workloads for collision avoidance and to provide acceptable service quality to customers.

5. Discussion

5.1. The Trade-Off between the Safety of UAVs and the Service Quality for Demands

We observed that the distance gap between flying UAVs becomes decreased when most of the demand nodes in a service area are densely located in the center of the area, which is natural considering the distribution of the demand nodes. If the safety of UAVs should be placed before the service quality to customers, one can consider further restricting the zoning approach. For example, adding a constraint that restricts the allocation of neighborhood demand nodes to different zones or treating the neighborhood demand nodes as abstract nodes could be considered for the zoning approach. Figure 7 illustrates two zoning solutions obtained with and without such additional constraints for UAV safety.

It should be noted that the zoning approach can also be implemented to maximize the service quality for demands. For example, one may want a zoning solution that minimizes the total distance between demand nodes and their responsible bases. The workload balance between zones can also be considered to derive a zoning solution.

Following this idea, we obtain different zoning solutions by applying the objective functions—to minimize the weighted total distance between demand nodes and bases and to minimize the between-zone variance of the total weighted distance between demand nodes and bases. We apply a genetic algorithm (GA) to derive the solutions. By the nature of a meta-heuristic algorithm, the implemented GA can easily address different types of objective functions and additional constraints (e.g., the limited flight time of UAVs) of the zoning problem. Please refer to Appendix A for details of the GA implementation. Figure 8 shows the solutions obtained; differences to the solutions in Figure 7 are clearly observed.

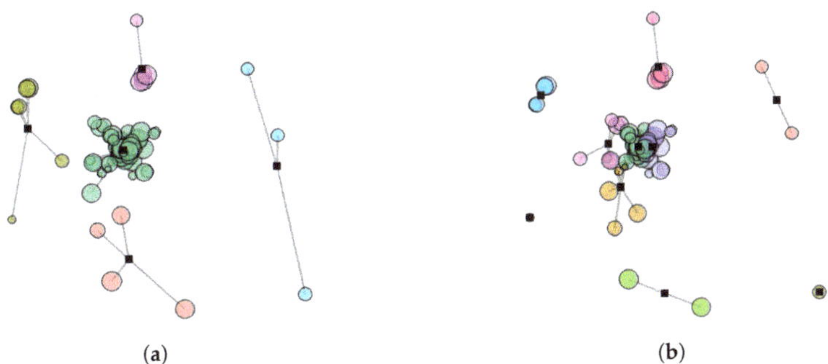

(a) (b)

Figure 7. Zoning solutions with and without additional safety constraints: (**a**) with safety constraints; (**b**) without safety constraints.

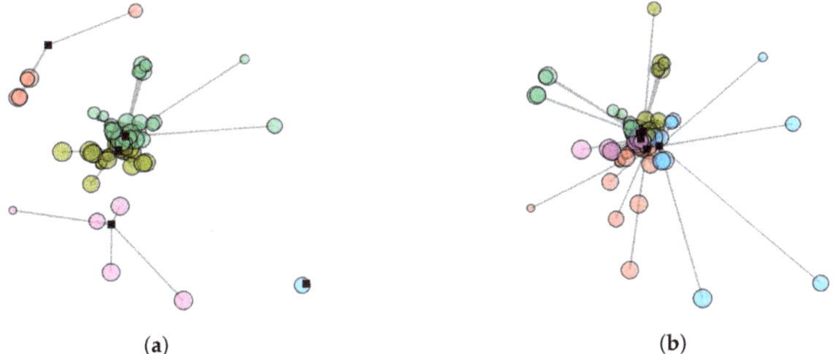

(a) (b)

Figure 8. Illustrations of zoning solutions with the service quality-oriented objective functions: (**a**) a solution to minimize the total distance between demands and bases; (**b**) a solution to evenly distribute demands to bases.

Importantly, the shift in the objective function of the zoning problem (P_{zoning}), which leads a different zoning solution, can bring a different degree of trade-off between UAV safety and service quality to demands. This phenomena is conceptually visualized in Figure 9.

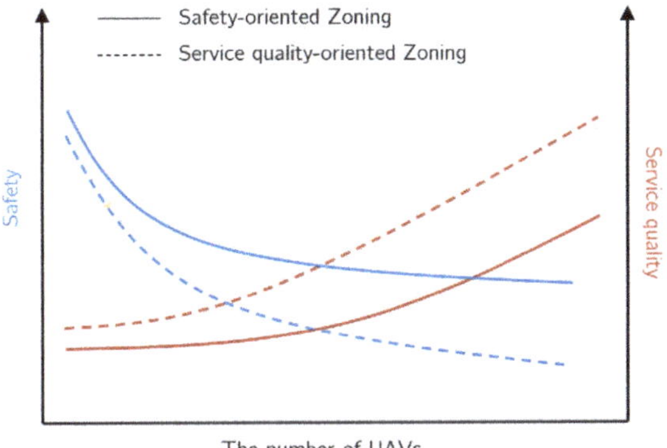

Figure 9. The performance trade-off between different zoning solutions.

In the figure, the performance of two different zoning solutions with different numbers of available UAVs is plotted. The solutions are distinguished by their line type (a solid line for a safety-oriented zoning solution and a dotted line for a service quality-oriented zoning solution), and their performances with respect to different perspectives are represented with different colors (blue for UAV safety and red for service quality for demands).

As illustrated in Figure 9, the UAV safety and the service quality for demands are indeed difficult maximize at the same time. The degree of the trade-off would also vary by the objective function applied to the zoning problem and the number of available UAVs. Therefore, it is important to properly formulate a target zoning problem and tune a solution algorithm for the zoning based on the priority, preference, and constraints of a target service system, so that the proposed approach can produce a valid and effective zoning solution. A whole process to successfully implement the zoning approach is illustrated in Figure 10.

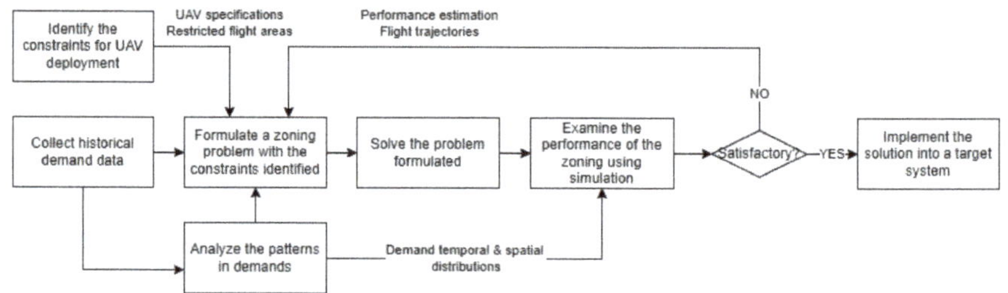

Figure 10. A process for zoning approach implementation.

5.2. Further Considerations for Zoning

The followings can be further considered for the practical implementation and improved performance of the zoning approach. First, the connectivity between zones is important to guarantee relatively easy transition between zones. If the bases of zones were placed outside of the zones, the connectivity would be guaranteed by a common and shared area acting as a flight corridor for crossing UAVs.

Next, the shape of zones can be considered for potential flight trajectories within a zone. For smooth and efficient movement within a zone, convexity of the zone would be desired. This can be guaranteed by known methods, such as convex decomposition of an area [33]. When there is a clear temporal pattern in demand arrivals, dynamic zoning that changes a zoning solution over time can be considered. In this case, keeping the relevant adjustments (e.g., frequency of updated zoning solution) at a minimum should be pursued.

Finally, aligned with the proposed and implemented solution for a UTM that separates airspace by altitudes, a three-dimensional zoning approach can be designed. This approach can provide flexibility in forming a zoning solution and scalability for increasing UAV operation volumes.

6. Concluding Remarks

Considering that the zoning approach is initially proposed to guarantee the safety of UAVs during operations, the relevant service quality degradation for customers is expected due to the restricted movement of UAVs by the zoning approach. Unlike this concern, however, the zoning approach shows a dominant performance in our experiments compared to the other current UAV deployment strategies in terms of both the safety of UAVs and the service quality for customers. This is accordance with Sung and Nielsen [18]'s observation, where they examine the zoning approach with a single UAV station.

We also observed the trade-off between the performance criteria (safety vs. service quality) in the experiments, a natural phenomena of UAV-based service systems. To meet a desired service level of a UAV-based system while addressing this trade-off, further assistance from the tactical part of the UAV operation (e.g., UAV trajectory planning optimization) and a well-tuned zoning solution algorithm based on the system's priorities are essential for successful implementation of the zoning approach for a UAV-based service system.

Finally, let us highlight that the proposed zoning approach is a systemic solution for a UAV-based service system, which includes a tactical and operational solution for UAV deployment. The proposed approach does not require significant investments for a UTM, nor dramatic advances in UAV navigation and control technologies. Therefore, based on the demonstrated performance of the zoning approach and its implementation simplicity for a UAV-based service system, we believe that the zoning approach is a breakthrough for currently limited UAV applications and their deployment at large scale.

Author Contributions: Conceptualization, C.B.P., K.R., I.S. and P.N.; methodology, C.B.P., K.R., I.S. and P.N.; software, C.B.P. and K.R.; validation, C.B.P., K.R. and I.S.; writing—original draft preparation, C.B.P., K.R. and I.S.; writing—review and editing, I.S. and P.N.; visualization, C.B.P., K.R. and I.S.; supervision, I.S. and P.N. All authors have read and agreed to the published version of the manuscript.

Funding: This research received no external funding.

Institutional Review Board Statement: Not applicable.

Informed Consent Statement: Not applicable.

Data Availability Statement: Publicly available datasets were analyzed in this study. This data can be found here: https://github.com/Rosenkrands/zav.

Conflicts of Interest: The authors declare no conflict of interest.

Appendix A. Genetic Algorithm Implementation

We develop the GA such that it determines centroids of clusters from a predetermined set of candidate centroids in a service area. By doing so, a clustering solution can be derived while excluding the area where UAVs cannot fly (buildings, mountains, no-fly zones, etc.). With the area discretization, the zoning problem is to choose K number of centroids among a set of candidate centroids and to assign demand nodes to the selected centroids. To address this zoning problem with a discrete solution space, we implement the GA following the general purpose implementation presented by Scrucca [34].

A zoning solution is represented as a bit string for all candidate centroids, forming a chromosome of the GA. A value of one represents that a corresponding centroid is used for clustering; otherwise, it is zero. For the initial population we use 100 randomly generated chromosomes. The fitness for each chromosome is computed by the corresponding objective value. To generate the next population, the selection operator selects chromosomes from the current population following the probabilities, assigned to each chromosome of the population, inversely proportional to their fitness value. The selected chromosomes are further updated by applying the crossover operator with an 80% probability. The mutation operator that flips a bit of a chromosome is also applied to chromosomes with a probability of 10% in order to escape from a local optima.

References

1. Shakhatreh, H.; Sawalmeh, A.H.; Al-Fuqaha, A.; Dou, Z.; Almaita, E.; Khalil, I.; Othman, N.S.; Khreishah, A.; Guizani, M. Unmanned aerial vehicles (UAVs): A survey on civil applications and key research challenges. *IEEE Access* **2019**, *7*, 48572–48634. [CrossRef]
2. Sung, I.; Choi, B.; Nielsen, P. On the training of a neural network for online path planning with offline path planning algorithms. *Int. J. Inf. Manag.* **2021**, *57*, 102142. [CrossRef]
3. Bocewicz, G.; Nielsen, P.; Banaszak, Z.; Thibbotuwawa, A. Routing and scheduling of unmanned aerial vehicles subject to cyclic production flow constraints. In Proceedings of the International Symposium on Distributed Computing and Artificial Intelligence, Toledo, Spain, 20–22 June 2018; Springer: Cham, Switzerland, 2018; pp. 75–86.
4. Murray, C.C.; Chu, A.G. The flying sidekick traveling salesman problem: Optimization of drone-assisted parcel delivery. *Transp. Res. Part C Emerg. Technol.* **2015**, *54*, 86–109. [CrossRef]
5. Sah, B.; Gupta, R.; Bani-Hani, D. Analysis of barriers to implement drone logistics. *Int. J. Logist. Res. Appl.* **2021**, *24*, 531–550. [CrossRef]
6. Kim, S.J.; Lim, G.J.; Cho, J.; Côté, M.J. Drone-aided healthcare services for patients with chronic diseases in rural areas. *J. Intell. Robot. Syst.* **2017**, *88*, 163–180.

7. Ghelichi, Z.; Gentili, M.; Mirchandani, P.B. Logistics for a fleet of drones for medical item delivery: A case study for Louisville, KY. *Comput. Oper. Res.* **2021**, *135*, 105443. [CrossRef]

8. Dang, Q.V.; Nielsen, I.E.; Bocewicz, G. A genetic algorithm-based heuristic for part-feeding mobile robot scheduling problem. In *Trends in Practical Applications of Agents and Multiagent Systems*; Springer: Berlin/Heidelberg, Germany, 2012; pp. 85–92.

9. Huang, H.; Savkin, A.V.; Huang, C. Scheduling of a parcel delivery system consisting of an aerial drone interacting with public transportation vehicles. *Sensors* **2020**, *20*, 2045. [CrossRef]

10. European Aviation Safety Agency. *Concept of Operations for Drones: A Risk Based Approach to Regulation of Unmanned Aircraft*; Technical Report; European Aviation Safety Agency: Cologne, Germany, 2015.

11. Federal Aviation Administration. *Unmanned Aircraft Systems (UAS) Traffic Management (UTM) Concept of Operation*; Technical Report; U.S. Department of Transportation, Federal Aviation Administration: Washington, DC, USA, 2020.

12. Radanovic, M.; Omeri, M.; Piera, M.A. Test analysis of a scalable UAV conflict management framework. *Proc. Inst. Mech. Eng. Part G J. Aerosp. Eng.* **2019**, *233*, 6076–6088. [CrossRef]

13. Jiang, T.; Geller, J.; Ni, D.; Collura, J. Unmanned Aircraft System traffic management: Concept of operation and system architecture. *Int. J. Transp. Sci. Technol.* **2016**, *5*, 123–135. [CrossRef]

14. Xiangmin, G.; Renli, L.; Hongxia, S.; Jun, C. A survey of safety separation management and collision avoidance approaches of civil UAS operating in integration national airspace system. *Chin. J. Aeronaut.* **2020**, *33*, 2851–2863.

15. Ballin, M.; Hoekstra, J.; Wing, D.; Lohr, G. NASA Langley and NLR research of distributed air/ground traffic management. In Proceedings of the AIAA's Aircraft Technology, Integration, and Operations (ATIO) 2002 Technical Forum, Los Angeles, CA, USA, 1–3 October 2002; p. 5826.

16. Dorling, K.; Heinrichs, J.; Messier, G.G.; Magierowski, S. Vehicle Routing Problems for Drone Delivery. *IEEE Trans. Syst. Man Cybern. Syst.* **2017**, *47*, 70–85. [CrossRef]

17. Peng, K.; Du, J.; Lu, F.; Sun, Q.; Dong, Y.; Zhou, P.; Hu, M. A hybrid genetic algorithm on routing and scheduling for vehicle-assisted multi-drone parcel delivery. *IEEE Access* **2019**, *7*, 49191–49200. [CrossRef]

18. Sung, I.; Nielsen, P. Zoning a Service Area of Unmanned Aerial Vehicles for Package Delivery Services. *J. Intell. Robot. Syst.* **2020**, *97*, 719–731. [CrossRef]

19. Otto, A.; Agatz, N.; Campbell, J.; Golden, B.; Pesch, E. Optimization approaches for civil applications of unmanned aerial vehicles (UAVs) or aerial drones: A survey. *Networks* **2018**, *72*, 411–458. [CrossRef]

20. Mukhamediev, R.I.; Symagulov, A.; Kuchin, Y.; Zaitseva, E.; Bekbotayeva, A.; Yakunin, K.; Assanov, I.; Levashenko, V.; Popova, Y.; Akzhalova, A.; et al. Review of Some Applications of Unmanned Aerial Vehicles Technology in the Resource-Rich Country. *Appl. Sci.* **2021**, *11*, 10171. [CrossRef]

21. Fu, J.G.M.; Bandyopadhyay, T.; Ang, M.H. Local Voronoi decomposition for multi-agent task allocation. In Proceedings of the 2009 IEEE International Conference on Robotics and Automation, Kobe, Japan, 12–17 May 2009; IEEE: Piscataway, NJ, USA, 2009; pp. 1935–1940.

22. Miao, X.; Lee, H.S.; Kang, B.Y. Multi-cleaning robots using cleaning distribution method based on map decomposition in large environments. *IEEE Access* **2020**, *8*, 97873–97889. [CrossRef]

23. Xiao, S.; Tan, X.; Wang, J. A simulated annealing algorithm and grid map-based UAV coverage path planning method for 3D reconstruction. *Electronics* **2021**, *10*, 853. [CrossRef]

24. Zhou, Q.; Lo, L.Y.; Jiang, B.; Chang, C.W.; Wen, C.Y.; Chen, C.K.; Zhou, W. Development of Fixed-Wing UAV 3D Coverage Paths for Urban Air Quality Profiling. *Sensors* **2022**, *22*, 3630. [CrossRef]

25. Ding, G.; Berke, A.; Gopalakrishnan, K.; Degue, K.H.; Balakrishnan, H.; Li, M.Z. Routing with Privacy for Drone Package Delivery Systems. *arXiv* **2022**, arXiv:2203.04406.

26. Hayat, S.; Yanmaz, E.; Muzaffar, R. Survey on unmanned aerial vehicle networks for civil applications: A communications viewpoint. *IEEE Commun. Surv. Tutorials* **2016**, *18*, 2624–2661.

27. Amazon Prime Air. *Revising the Airspace Model for the Safe Integration of Small Unmanned Aircraft Systems*; Technical Report; Amazon Prime Air: Seattle, WA, USA, 2015.

28. Feng, D.; Yuan, X. Automatic construction of aerial corridor for navigation of unmanned aircraft systems in class G airspace using LiDAR. In *Airborne Intelligence, Surveillance, Reconnaissance (ISR) Systems and Applications XIII*; International Society for Optics and Photonics, SPIE: Bellingham, WA, USA, 2016; Volume 9828, pp. 133–140. [CrossRef]

29. Chang, Y.S.; Lee, H.J. Optimal delivery routing with wider drone-delivery areas along a shorter truck-route. *Expert Syst. Appl.* **2018**, *104*, 307–317. [CrossRef]

30. Gu, Q.; Fan, T.; Pan, F.; Zhang, C. A vehicle-UAV operation scheme for instant delivery. *Comput. Ind. Eng.* **2020**, *149*, 106809.

31. Deng, X.; Guan, M.; Ma, Y.; Yang, X.; Xiang, T. Vehicle-Assisted UAV Delivery Scheme Considering Energy Consumption for Instant Delivery. *Sensors* **2022**, *22*, 2045. [CrossRef]

32. Hartigan, J. *Clustering Algorithms*; Wiley: Hoboken, NJ, USA, 1975.

33. Nielsen, L.D.; Sung, I.; Nielsen, P. Convex decomposition for a coverage path planning for autonomous vehicles: Interior extension of edges. *Sensors* **2019**, *19*, 4165. [CrossRef] [PubMed]

34. Scrucca, L. GA: A Package for Genetic Algorithms in R. *J. Stat. Softw.* **2013**, *53*, 1–37. [CrossRef]

MDPI

St. Alban-Anlage 66

4052 Basel

Switzerland

www.mdpi.com

Drones Editorial Office

E-mail: drones@mdpi.com

www.mdpi.com/journal/drones